*Pseudopotential
Theory of
Atoms and
Molecules*

Pseudopotential Theory of Atoms and Molecules

Levente Szasz
Fordham University

A WILEY-INTERSCIENCE PUBLICATION
JOHN WILEY & SONS
New York • Chichester • Brisbane • Toronto • Singapore

86000/1987

Library of Congress Cataloging in Publication Data:

Szasz, Levente, 1931–
   Pseudopotential theory of atoms and molecules.

   "A Wiley-Interscience publication."
   Bibliography: p. 302
   Includes index.
   1. Atomic theory.   2. Molecular theory.   3. Pseudopotential method.   I. Title.

QD461.S988916   1985   539'.1   85-5333
ISBN 0-471-82417-8

Printed in the United States of America

10 9 8 7 6 5 4 3 2 1

Preface

In this book the word *pseudopotential* refers exclusively to that quantum mechanical technique in which the Pauli exclusion principle is replaced by operators and potential functions jointly called pseudopotentials. In quantum physics the word is used to describe many different quantities; in this book it will be used exclusively in the sense defined above.

The purpose of the pseudopotential theory of atoms and molecules is to provide a "valence-only" theory for these systems. Thus, in this theory, the valence electrons of atoms and molecules are treated in accurate detail, while the role of the core electrons is to provide the potential fields in which the valence electrons move. This theory covers a very large number of atomic and molecular problems—problems connected with the valence electrons—although, of course, its applicability does not extend to all atomic and molecular problems. In addition to providing a "valence-only" theory, the formalism has been recently extended to an all-electron treatment of atoms. This development, presented in Chapter 5, raises the possibility that pseudopotential theory will someday cover not only "valence-only" problems but all-electron problems of atoms and molecules as well.

The goal of the book is to present, for the first time, a detailed and comprehensive treatment of pseudopotential theory. Its primary audience is quantum chemists and theoretical physicists interested in atoms and molecules. It is also potentially useful for quantum biologists and quantum pharmacologists whose work requires atomic and molecular structure calculations.

In addition to the presentation of pseudopotential theory, the book also contains a detailed discussion of the *ab initio* theory of atoms and molecules with two or more valence electrons. The presentation of such a theory, in which the electron correlation between the valence electrons is fully taken into account, is usually omitted from textbooks; it is missing, for example, from the otherwise brilliant set of treatises on atomic and molecular structure by J.C. Slater. Thus the discussion of such theories is intrinsically useful, besides being necessary for the development of pseudopotential methods.

Pseudopotential methods were first developed in the mid-thirties, and since science was not compartmentalized in those days to the extent it is today, the method was introduced simultaneously for atomic, molecular, and solid-state problems. Although the new ideas were always introduced by using atomic examples, the theory was most extensively applied first in solid-state physics. This does not mean that pseudopotential techniques are primarily for solid-state problems; it means only that the advantages of these techniques were recognized earlier by solid-state physicists. In the last twenty years the usefulness of the method has been fully recognized by atomic and molecular scientists; indeed, these years have seen so many applications of the theory that today we can speak of a pseudopotential theory of atoms and molecules.

The book is constructed in such a way that, starting from simple quantum mechanical principles, the detailed development of the theory is presented, leading to the presentation of the most up-to-date, "state of the art" techniques. The emphasis is everywhere on a detailed discussion of the theory; calculations are presented only as demonstrative examples. The reader is assumed to possess a fair knowledge of quantum mechanics.

Finally, a personal remark. The division between "pure" and "applied" science is a topic that is often discussed. A monograph of this kind can be used in two different ways. A scientist may use it to study the mathematical and conceptual aspects of the theory—that is, those parts of the book which deal with "pure" science. Another reader may be interested in the computational aspects—that is, in the applications of the theory to concrete atomic and molecular problems. Unfortunately, there is a tendency today to emphasize the latter aspects at the expense of the former; that is, to emphasize applied science. I hope that this book will not be viewed as a mere collection of computational recipes, although, it can be used as such. I would like to emphasize that, in writing the book, my main goal has been to provide the reader with a tool that can be used as the starting point of new, *bona fide* scientific research.

LEVENTE SZASZ

August 1985, New York

Contents

*Pseudopotential
Theory of
Atoms and
Molecules*

chapter 1

Introduction

1.1. THE IDEA OF PSEUDOPOTENTIALS; HELLMANN'S WORK

In many problems of atomic and molecular physics the electrons of the system can be divided into valence and core electrons. In many cases the important physical properties are determined by the valence electrons. A typical example is a random selection of atomic spectra, such as those shown on pp. 89, 106, and 124 of the National Bureau of Standards tables of Moore.[1] The tables on those pages show the optical spectra of the neutral atoms Na, Mg, and Al, and it is evident that they are 1, 2, and 3-electron spectra (i.e., they are generated by the 1, 2, and 3 valence electrons of these atoms which are 11-, 12-, and 13-electron systems).

In the quantum mechanical treatment of such systems it is natural to try to reduce the problem from an N-electron problem, where N is the total number of electrons, to an n-electron problem, where n is the number of valence electrons. Since in many cases $n \ll N$, such a reduction means a significant conceptual and mathematical simplification.

In a series of pioneering papers Hellmann attempted to develop a computational model in which the treatment of such atoms and molecules is reduced to the treatment of valence electrons.[2] Hellmann demonstrated, using the Thomas–Fermi model, that the Pauli exclusion principle for the valence electrons can be replaced by a nonclassical potential (*Abstossungspotential*) which is now called the *pseudopotential*. Since the requirement that the valence orbital be orthogonal to the core orbitals is equivalent to the Pauli exclusion principle, Hellmann's idea was to replace the orthogonality requirement by the pseudopotential. Besides the simplification mentioned above, which was inherent in the reduction of the problem to the valence electrons, the introduction of the pseudopotential represents another considerable mathematical simplification, since the orthogonality requirement between valence and core wave functions may lead to mathematical difficulties, especially in the case of large cores.

For the actual purpose of atomic and molecular calculations Hellmann suggested a simple analytic formula. Let V be the sum of electrostatic, exchange, and polarization potentials, representing the interaction between a valence electron and the core of an atom. Let V_p be the pseudopotential. The total potential V_H may be expressed

$$V_H = V + V_p = -\frac{z}{r} + A\frac{e^{-\kappa r}}{r}. \tag{1.1}$$

Here z is the ionic charge of the core; that is, if the nucleus contains Z positive charges and the core contains N electrons then $z = Z - N$. The constants A and κ are determined from the requirement that the potential V_H should reproduce the energy spectrum of the valence electron as accurately as possible.

The physical meaning of Eq. (1.1) can be understood from Figures 1.1 and 1.2. In Figure 1.1 we have plotted V_H for the valence electron of the Na atom. As we see, in contrast to the Hartree–Fock potential, which is an everywhere negative, attractive potential, the potential V_H has a positive

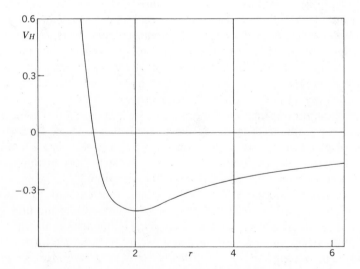

FIGURE 1.1. The Hellmann potential for the Na atom.

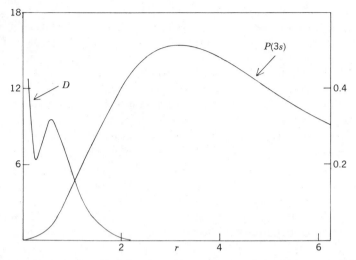

FIGURE 1.2. The pseudo-orbital for the valence electron of the Na atom and the radial density of the Na core.

potential barrier at about $r \approx 1$ a.u., a deep negative well between $r \approx 1$ and $r \approx 4$, and is Coulombic for larger r values.

Now let us consider the Schroedinger equation for the valence electron with the potential V_H. It will have form

$$\left\{ -\frac{1}{2}\Delta - \frac{z}{r} + A\frac{e^{-\kappa r}}{r} \right\} \psi = E\psi, \qquad (1.2)$$

where Δ is the Laplacian. Because of the presence of the pseudopotential, the solutions of Eq. (1.2) do not need to satisfy any orthogonality conditions relative to the core orbitals; Eq. (1.2) can be solved as if the core did not exist. In the case of the Na atom, the lowest solution of Eq. (1.2) will correspond to the (3s) state of the valence electron and the higher solutions to the excited states. Because of the absence of orthogonality requirements the lowest eigenfunction of Eq. (1.2) will be nodeless, in contrast to the HF orbital which has two nodes for the 3s state; the higher solutions will have as many nodes as are necessary to make them orthogonal to the lower, unfilled valence states, but there will be no nodes corresponding to the core states. Such eigenfunctions as the solutions of Eq. (1.2) are called pseudo-orbitals or pseudo-wave functions.

In Figure 1.2 we have the pseudo-orbital for the 3s state of the Na atom. In the diagram we have $P(3s)$ which is the radial part of the eigenfunction multiplied by r. In the inner part of the diagram, we also have the total radial density of the core of the neutral Na atom according to the HF calculations.[3] (Note the different scales for the two quantities: the scale for D is on the left while the scale for $P(3s)$ is on the right.) We see clearly the absence of nodes in the pseudo-orbital in the core region. Looking at the two diagrams together, we see that the pseudopotential "keeps the valence electron out of the core", that is, the position of the potential barrier is such that the maximum of the pseudo-orbital is well outside of the core. The barrier "prevents the valence electron from falling into the core" despite the lack of orthogonality with respect to the core orbitals.

The method may also be applied to an atom with more than one valence electrons. Let us consider the Mg atom. Hellmann assumed that the electron distribution of the Mg core is the same as the electron distribution of the Mg^+ core; that is, that the addition of the second valence electron does not change the electron distribution of the core ("frozen core" approximation). Then he used the empirical spectrum of the Mg^+ ion to adjust A and κ in Eq. (1.1). Having obtained the potential for the Mg^+ he assumed that the same potential can be used to represent the core-valence interaction in the neutral atom. The Hamiltonian for the two valence electrons of the neutral atom becomes

$$H_{12} = -\frac{1}{2}\Delta_1 - \frac{1}{2}\Delta_2 - \frac{z}{r_1} + A\frac{e^{-\kappa r_1}}{r_1} - \frac{z}{r_2} + A\frac{e^{-\kappa r_2}}{r_2} + \frac{1}{r_{12}}, \qquad (1.3)$$

and the Schroedinger equation will be

$$H_{12}\Psi(1, 2) = E_{12}\Psi(1, 2), \qquad (1.4)$$

where this equation can again be solved as if the core did not exist; that is, the eigenfunction $\Psi(1, 2)$ does not have to satisfy any orthogonality conditions with respect to the core orbitals. The original 12-electron problem of the Mg atom is reduced to a 2-electron problem of the two valence electrons.

The generalization to a diatomic molecule is made in a similar fashion. Let us consider Na_2. Hellmann assumed that the Na^+ cores do not participate in the binding process. In this case the binding of the molecule is entirely due to the two valence electrons. The parameters of the potential V_H can be adjusted by matching the spectrum of the valence electron of the neutral atom; that is, the same parameters can be used as in Eq. (1.2). The Hamiltonian of the two binding electrons becomes

$$H_{12} = \sum_{i=1}^{2}\left\{-\frac{1}{2}\Delta_i - \frac{z}{r_{ai}} + A\frac{e^{-\kappa r_{ai}}}{r_{ai}} - \frac{z}{r_{bi}} + A\frac{e^{-\kappa r_2}}{r_2}\right\} + \frac{1}{r_{12}}, \qquad (1.5)$$

where a and b represent the position of the two nuclei. The Schroedinger equation again has the form of Eq. (1.4) where the existence of the two cores can again be ignored in the calculation of the solutions. In the case of the Na_2 molecule the original 22-electron problem is effectively reduced to a 2-electron problem.

The discussion above presents the barest outline of Hellmann's work, in order to introduce the concept of pseudopotentials. A detailed presentation follows in Chapter 4. Summing up, Hellmann developed a model based on three ideas:

1. It was assumed that an N-electron problem can be reduced to an n-valence-electron problem, where $n < N$.

2. It was assumed (on the basis of the Thomas–Fermi model) that the Pauli exclusion principle can be replaced by a pseudopotential.

3. It was assumed that the Hamiltonian of a 2-valence-electron atom can be built up from the Hamiltonian of the corresponding 1-valence electron ion, and the Hamiltonian of a 2-binding-electron molecule from the Hamiltonians of the constituent neutral atoms.

1.2. AB INITIO VERSUS MODEL METHODS IN THE QUANTUM THEORY OF ATOMS AND MOLECULES

Having outlined the basic idea of the pseudopotential method in the preceding section, we shall attempt now to define its place in the quantum theory of atomic and molecular electronic structure. The purpose of atomic and molecular theory is to understand the structure and interactions of atoms and molecules by applying quantum mechanics to them. This is essentially, though not exclusively, a mathematical task, where the word *mathematical* implies analytic as well as computer techniques. It is not exclusively mathematical: physical and/or chemical intuition is often useful in the application of quantum mechanical techniques.

From the mathematical point of view we shall divide the quantum mechanical techniques into two groups: *ab initio* and *model* methods. The division will be, to a certain extent, arbitrary, since the definitions of these two groups cannot be formulated with high precision. We designate a method as ab initio if it is based on an exact nonrelativistic Hamiltonian operator; that is, on a Hamiltonian which is defined according to the basic rules of quantum mechanics. We also designate a method as ab initio if it is based on the exact Hamiltonian and on a wave function which is not an exact eigenfunction of this Hamiltonian but is a clearly defined approximation of the exact eigenfunction.

We shall define a quantum mechanical model as a computational procedure which does not belong to the category above. Usually a model will be defined either by a model Hamiltonian (model potential) or by a model energy expression. Examples for typical ab initio methods are the variational, Morse-type calculations for atoms, the Heitler–London (valence bond)-type calculations for the H_2 molecule, and, in general, the configuration interaction (CI)-type calculations for atoms and molecules. Typical models are the $X\alpha$ model and the Thomas–Fermi model. In both of these, the starting point is a model energy expression. One of the most important computational techniques, the Hartree–Fock (HF) approximation, occupies a halfway position between the two basic types. It is ab initio, since the HF wave functions are approximate solutions of an exact Hamiltonian (they can be used as a first step in a CI calculation); but it is also a model, since the HF wave functions are exact solutions of a model Hamiltonian defining the best one-electron orbitals for the system.

It is not our purpose here to go deeper into the discussion of these two basic types of computational methods. We want to make one observation, however. Models are generally simpler mathematically than the ab initio methods. Their relative simplicity is the reason for their introduction into quantum mechanics. Thus, while working with ab initio methods almost always means working with approximate wave functions and energies, exact solutions of model equations are frequently obtainable because of their relative simplicity.

We emphasize that the juxtaposition of ab initio verus model methods does not imply a value judgment. It is our opinion that both techniques are equally useful and necessary for atomic and molecular theory. It is important, however, that a computational method be defined as either ab initio or model with as much clarity as possible. (In both cases it should also possess a sound theoretical justification.)

Where does the pseudopotential method fit in? It will be demonstrated in this book that *the pseudopotential method can be developed either ab initio or as a model.* This fact must be noted. Pseudopotential methods are still viewed by some scientists as crude models, not on a par with ab initio techniques; but while this assumption is accurate in a few cases, it is by no means inherent in the pseudopotential concept. It is not generally recognized that there are ab initio pseudopotential methods (even for atoms with many valence electrons); or, more accurately, that pseudopotential models have ab initio justifications as their background.

There is one point which must be mentioned in connection with models. The concept of model does not a priori imply being semiempirical. Unfortunately, these two concepts are usually coupled. There are models which contain adjustable parameters that must be determined from empirical data. There are others, however, which do not require the use of empirical data; the density-dependent pseudopotentials, which are discussed in Chapter 3, are typical examples. Other examples include the effective Hamiltonians derived in Chapters 6, 7, and 8; these are models which do not contain, or do not necessarily contain, empirical data. Thus, in talking about models, it must be kept in mind that the class of model methods is a much broader category than the class of semiempirical procedures.

In this book the ab initio and model formulations will be presented side by side, with the ab initio methods serving as the justifications for the models. In each area of atomic and molecular theory we shall start with the conventional ab initio method. A method will be called conventional if the wave functions are orthogonalized to each other. Having formulated the conventional theory, we shall transform it exactly into a pseudopotential formalism; that is, into a formalism in which the orthogonality requirements are replaced by pseudopotentials. This will be called the exact, ab initio pseudopotential formulation. Next, this ab initio formulation will be replaced or approximated by a model. We shall call this step *modelization*. In those cases in which the model was developed prior to the development of ab initio theory, the relationship between the two will be established in such a way that the model will be justified on the basis of the ab initio theory; that is, we shall derive the equations of the model from the ab initio theory.

1.3.   THE GOALS OF PSEUDOPOTENTIAL THEORY; SCOPE OF THE BOOK

It is evident from the first section that the goal of atomic and molecular pseudopotential theory is to develop a "valence only" formalism for atoms and molecules. By this we mean a theory which is capable of handling the valence-electron properties with high accuracy while minimizing the role played by the core electrons in the formalism. While this is the main purpose, the theory is not restricted to the treatment of valence electrons. As we shall see in Chapter 5, the method can also be used to develop an all-electron model which has given good results for the bulk properties of light atoms.

The plan of this book is to present the pseudopotential theory of atoms and molecules at its current stage of development. As we see from the Table of Contents, the presentation will proceed from the simplest system to the more complex cases. In Chapters 2, 3, and 4, we develop the theory for atoms with a single valence electron. In Chapter 2 we present the exact, ab initio theory, while in Chapters 3 and 4 we discuss the model formulations. The discussion of atoms with more than one valence electron begins in Chapter 6 with the atoms which contain two valence electrons. In Chapter 7 we turn to atoms with an arbitrary number of valence electrons, while the general treatment of molecules follows in Chapter 8. Chapter 9 contains a specialized discussion of the excited states of atoms and molecules. In the middle of this progression, in Chapter 5, we have placed the all-electron pseudopotential model, which is based on the Hartree–Fock model and is thus essentially a one-electron approximation for which only Chapters 2–4 are needed. Those parts of the theory which require a heavier dose of mathematics are placed in the appendixes. The presentation everywhere follows the method described in the preceding section. It consists of treating the ab initio theory first and moving from that to the model formulation. Thus, the sequence will be: conventional ab initio theory → exact pseudopotential theory → model formulation of pseudopotential theory.

There is question of priorities that must be resolved in a work of this kind. The topic under discussion is the pseudopotential theory of atoms and molecules. The theory, of course, is a mathematical formalism. The final goal is to apply this theory to the calculation of atomic and molecular properties. Thus the presentation should include the description of the theory as well as the description of how it can be applied to the calculation of atomic and molecular properties. We have to decide the relative weight to give to the theoretical and computational aspects of the presentation.

Our plan is to treat the description of the theory—that is, the mathematical formalism—in detail. The applications of the formalism to atomic and molecular problems will not be presented in detail. The main reason for this omission is that the computational techniques required for such applications are not very different from those used in conventional methods. For example, a scientist familiar with the generalized-valence-bond method in general will be able to apply this method in the pseudopotential theory as soon as he learns how to set up the effective pseudopotential Hamiltonian for the system. We consider it to be our task to explain this last problem; it is not our goal to discuss known computational techniques.

Therefore, we shall discuss such applications only to the extent necessary to demonstrate the quality of results which can be expected. This book contains sections entitled "Test Calculations" and "Representative Calculations." These titles are accurate; the sections describe calculations which were done to test certain parts of the formalism, or were selected from a larger number of applications in order to demonstrate the quality of results that can be expected from the application of certain parts of the formalism.

A typical example for our treatment of calculations is the construction of Chapter 9. The goal of that chapter is to describe the pseudopotential theory of excited states and Rydberg states. The chapter also contains a fair amount of numerical results. These results serve to demonstrate that the theory can be used in practical calculations, and to show how accurate such calculations can be. Thus the presentation of these results serves to demonstrate the usefulness of the formalism; no attempt is made to collect the results of all pseudopotential calculations for Rydberg states.

The pseudopotential theory was first developed to be used in solid-state physics. By the 1930s and 1940s the statistical and model formulations were used to build up a theory of metals. Large-scale applications in solid-state physics began in the 1960s. Atomic and molecular pseudopotential theory began to be developed, independently from solid-state theory, around 1966. Naturally, there was a considerable overlap between the two fields during the period before 1966.

The goal of this book is to present the atomic and molecular pseudopotential theory; it is not our purpose to go into the development of the theory in solid-state physics. For the pre-1966 period, during which there was a considerable overlap between the two fields, we shall cover everything in detail—regardless of whether it was developed for solid-state physics or for atomic and molecular physics—that had a demonstratable effect on the development of atomic and molecular pseudopotential theory.

There are two areas of atomic and molecular physics that will not be discussed despite the fact that pseudopotential methods have been applied to them. These are the relativistic calculations and scattering theory. It is our opinion that the relativistic pseudopotential calculations are straightforward generalizations of the nonrelativistic techniques, and, as such, belong under the heading of applications. Their presentation is not necessary for the understanding of the basic theory, which is our main concern. The application of pseudopotential methods in scattering theory is not very extensive, and would fit more logically into a monograph on scattering theory.

Finally, having outlined the goal of pseudopotential theory and the scope of the book, let us ask the basic question: Why pseudopotential theory? Is there any appreciable advantage in using this model rather than conventional techniques?

Without attempting to answer this question in a comprehensive way, we would like to point out two aspects of pseudopotential theory which make the study and application of this method attractive. To the theoretical physicist we want to point out that, as what follows will show, pseudopotential theory is mathematically coherent and elegant. (By this we mean a formalism in which every step follows from the preceding steps in a logical fashion.) To the scientist who wants to apply the formalism to specific problems of atomic and molecular physics, we want to point out the simplicity of the theory, which is often combined with remarkable accuracy. It must be especially pointed out that modelization is much easier in this theory than in the conventional techniques. The great variety of potentials, presented in Sections 4.3 and 7.2, clearly demonstrates the unusual advantages of this theory in constructing simple and accurate model potentials.

1.4. NOTATION, NOMENCLATURE, AND REFERENCING

The notation used will be conventional, but consistency in the symbols used will be enforced only within each section. Thus the same symbol may mean different quantities in different sections. The Dirac notation will be used somewhat liberally, meaning that we shall simply put

$$\langle r \mid \psi \rangle \equiv \psi(r) \,, \tag{1.6}$$

and

$$\langle \phi | H | \psi \rangle \equiv \int \phi^* H \psi \, dv \,, \tag{1.7}$$

but it will not be implied, as in the original notation of Dirac, that the wave functions ϕ and ψ are eigenfunctions of an operator or that they are orthogonal. In our notation these may be unspecified wave functions.

The basic nomenclature of pseudopotential theory needs some explaining. A pseudo-orbital or pseudo-wave function is a wave function which has to satisfy only a reduced number of orthogonality conditions. As we have seen in Section 1.1, the pseudo-orbital which is the eigengunction of Eq. (1.2) does not need to satisfy any orthogonality conditions in the ground state. In the first excited state it will have to be orthogonal to the ground state, and so on. Thus in the $3s$, $3p$, $4s$, $4p$, and so on, states of the Na atom the pseudo-orbital will have to be orthogonal to the lower lying valence orbitals but not to the core orbitals. In contrast, the HF orbital in these states would have to be orthogonal to the core states as well as to the lower lying valence states.

The adjective *pseudo* in pseudo-orbital does not imply anything beyond the reduced number of orthogonality conditions. It does not imply that the wave function indicated is an approximation; neither does it imply a semiempirical character. Thus this adjective, which will be used in this work so often, is a *misnomer* insofar as it implies that such a wave function is somehow less "real" than the conventional wave functions. It should be clearly stated that measurable physical quantities can be obtained from exact pseudo-orbitals with the same degree of exactness as from conventional orbitals. It must be kept in mind, however, that the formulas giving the physical quantities in terms of pseudo-orbitals are generally different from the formulas giving the same quantities in terms of conventional orbitals.

Similarly, the designation *pseudopotential* is also a misnomer to a certain degree. This word simply means a potential (local or nonlocal) which represents the Pauli exclusion principle. The word does not imply an approximation; an equation with a pseudopotential in it is not necessarily more approximate than the corresponding conventional equation. As we shall see, there are exact and model pseudopotentials. The equations with exact pseudopotentials will be on the same level of accuracy as the conventional equations; the equations with model pseudopotentials will be approximate. The word *pseudopotential* does imply, quite accurately, that this quantity is not like the Coulomb potential which occurs in the wave equation. The Coulomb potential is a quantity defined in classical physics; the pseudopotential does not have a classical analog. We note that in some of the literature other quantities, such as the exchange interaction potential and the potential derived from density-dependent correlation-energy expressions, are also denoted as pseudopotentials. In this work such usage will not be followed.

Finally, in connection with the referencing we note that the following books will be referred to by abbreviations. Four textbooks of Slater will be cited as follows:

Slater I   J. C. Slater, *Quantum Theory of Atomic Structure, Vol. I* (McGraw-Hill, New York, 1960).

Slater II   J. C. Slater, *Quantum Theory of Atomic Structure, Vol. II* (McGraw-Hill, New York, 1960).

Slater III   J. C. Slater, *Quantum Theory of Molecules and Solids, Vol. I, Electronic Structure of Molecules* (McGraw-Hill, New York, 1963).

Slater IV   J. C. Slater, *Quantum Theory of Molecules and Solids, Vol. IV, The Self-Consistent Field for Molecules and Solids* (McGraw-Hill, New York, 1974).

Two older monographs which are convenient for the Thomas–Fermi model and for the density-dependent pseudopotentials are cited as follows:

Gombas I   P. Gombas, *The Statistical Theory of Atoms and its Applications* (Springer Verlag, Vienna, 1949).

Gombas II   P. Gombas, *Pseudopotentials* (Springer Verlag, Vienna and New York, 1967).

References to original publications follow the appendixes.

chapter 2

Exact Pseudopotentials for One-Valence-Electron Systems

2.1. THE METHOD OF SZEPFALUSY

An exact pseudopotential theory for atoms with one valence electron was first developed by Szepfalusy.[4] We present his work here in the somewhat more detailed formulation given by Szasz and McGinn.[5]

The theory is developed for an atom which has N core electrons in closed shells plus one valence electron. Let us consider first the atom with the valence electron removed. Let $\varphi_1, \varphi_2, \ldots, \varphi_N$ be the orthonormal spin-orbitals for the core electrons and let the wave function of the core be, in the HF approximation, the determinant,

$$\Phi = (N!)^{-1/2} \det[\varphi_1 \varphi_2 \cdots \varphi_N].$$ (2.1)

By varying the HF energy expression with respect to the spin-orbitals we obtain the HF equations,

$$H_F \varphi_i = \epsilon_i \varphi_i, \quad (i = 1, 2, \ldots, N),$$ (2.2)

where H_F is the HF Hamiltonian

$$H_F = t + g + U, \tag{2.3}$$

with t being the operator of kinetic energy, g the nuclear attraction, and U the total HF potential, which is defined as

$$U = \sum_{i=1}^{N} U_i, \tag{2.4}$$

where

$$U_i(1)f(1) = \int \frac{\varphi_i(2)\varphi_i^*(2)}{r_{12}} dq_2 f(1) - \int \frac{\varphi_i(1)\varphi_i^*(2)f(2)\, dq_2}{r_{12}}. \tag{2.5}$$

The last formula shows that U_i is the sum of the electrostatic and exchange potentials generated by the orbital φ_i. The integration is over spatial and spin coordinates which are jointly denoted by q. In Eq. (2.2) ϵ_i is the HF orbital energy. The HF equations do not contain any offdiagonal Lagrangian multipliers, since, the total wave function being the single determinant, Eq. (2.1), the multipliers can be eliminated by a unitary transformation [Slater II, p. 1].

Now let us add the valence electron to the core, and let us assume that the core orbitals do not change when the valence electron is added ("frozen core" approximation). The total wave function of the $(N + 1)$-electron atom will be

$$\Phi = \begin{vmatrix} \varphi_1(1) & \varphi_1(2) & \cdots & \varphi_1(N+1) \\ \varphi_2(1) & \varphi_2(2) & \cdots & \varphi_2(N+1) \\ \vdots & & & \vdots \\ \varphi_N(1) & \varphi_N(2) & \cdots & \varphi_N(N+1) \\ \psi(1) & \psi(2) & \cdots & \psi(N+1) \end{vmatrix}, \tag{2.6}$$

where ψ is the valence orbital. By putting Φ in an antisymmetric form, the Pauli exclusion principle is fully satisfied; there is no need to assume that the ψ is orthogonal to the core orbitals. Our goal is to derive the equation for the best ψ. Next we rewrite Eq. (2.6) as follows:

$$\Phi = \begin{vmatrix} \varphi_1(1) & \varphi_1(2) & \cdots & \varphi_1(N+1) \\ \varphi_2(1) & \varphi_2(2) & \cdots & \varphi_2(N+1) \\ \vdots & & & \vdots \\ \varphi_N(1) & \varphi_N(2) & \cdots & \varphi_N(N+1) \\ \varphi(1) & \varphi(2) & \cdots & \varphi(N+1) \end{vmatrix}, \tag{2.7}$$

where

$$\varphi = \psi + \sum_{i=1}^{N} c_i \varphi_i , \qquad (2.8)$$

with the c_i being constants. As is known, the replacement of ψ by φ does not change the determinant [Slater III, p. 283]. We determine the c_i in such a way as to Schmidt-orthogonalize φ to the core functions. We obtain

$$c_i = -\langle \varphi_i \mid \psi \rangle . \qquad (2.9)$$

Denoting $\alpha_i = -c_i$ we put,

$$\varphi = A_0 \left(\psi - \sum_{i=1}^{N} \alpha_i \varphi_i \right), \qquad (2.10)$$

where A_0 is a constant for the normalization of φ. Assuming that ψ is normalized, we get,

$$A_0 = \left\{ 1 - \sum_{i=1}^{N} |\alpha_i|^2 \right\}^{-1/2} . \qquad (2.11)$$

The normalization of the valence orbital with A_0 and the normalization of the total wave function [Eq. (2.7)] can be accomplished at the same time by multiplying Eq. (2.7) by the constant A where

$$A = \{(N+1)!\}^{-1/2} A_0 , \qquad (2.12)$$

so that our total wave function becomes

$$\Phi = \{(N+1)!\}^{-1/2} \det\{\varphi_1 \varphi_2 \cdots \varphi_N \varphi\} , \qquad (2.13)$$

where φ is given by Eq. (2.10).

The wave function Eq. (2.13) is the conventional, single-determinant, HF wave function. We have shown that the total wave function Eq. (2.6), in which the valence orbital is not orthogonal to the core orbitals, is equivalent to the HF wave function Eq. (2.13); the two differ by the multiplicative constant A given by Eq. (2.12).

Next we calculate the average value of the Hamiltonian with respect to the wave function Eq. (2.13), and vary the resulting expression with respect to the valence orbital φ while keeping the core orbitals fixed. In the variation we take into account the subsidiary conditions that φ must be orthogonal to the core orbitals,

$$\langle \varphi_i \mid \varphi \rangle = 0 , \quad (i = 1, 2, \ldots, N) , \qquad (2.14)$$

and that it must be normalized. We thus obtain the HF equation for the valence orbital:[6]

$$H_F \varphi = \epsilon \varphi, \tag{2.15}$$

where ϵ is the orbital energy of the valence electron. The orthogonality condition Eq. (2.14) will be satisfied by the solutions of Eq. (2.15), since φ and the core orbitals φ_i are the eigenfunctions of the same Hermitian operator H_F.

The HF equation Eq. (2.15) defines the best orthogonalized valence orbital φ with "frozen core." Returning now to our original task, the determination of the equation for the best nonorthogonal orbital ψ, we simply substitute Eq. (2.10) into Eq. (2.15) and obtain the equation for ψ,

$$(H_F + V_p)\psi = \epsilon \psi, \tag{2.16}$$

where V_p is the linear operator,

$$V_p = \sum_{i=1}^{N} (\epsilon - \epsilon_i)|\varphi_i\rangle\langle\varphi_i|. \tag{2.17}$$

Szepfalusy summarized the result contained in Eq. (2. 16) by saying that, if the valence orbital ψ is not orthogonal to the core orbitals, then the equation for the best ψ contains, besides the HF Hamiltonian H_F, also the (nonlocal) potential V_p. In current terminology this nonlocal potential is the pseudo-potential, and the wave function ψ is the pseudo-orbital (PO).

We have now two equivalent procedures for the determination of the valence electron orbital and orbital energy. Moving in the framework of conventional HF procedure we must solve Eq. (2.15), demanding that the solution be orthogonal to the core functions. In the pseudopotential method we must solve Eq. (2.16), in which case the pseudo-orbital does not have to satisfy orthogonality conditions with respect to the core functions. We can solve Eq. (2.16) "as if the core did not exist"—except, of course, the potentials will depend on the core functions and orbital energies.

From the derivation it is clear that the two procedures are equivalent. Both Eq. (2.15) and Eq. (2.16) are derived by minimizing the total energy of the atom (although the minimization is restricted by keeping the core functions frozen). The total energy depends only on the total wave function [Eq. (2.6)], and since the total wave function is the same in both procedures the energy minimization must lead to identical results.

Next we want to derive the form of V_p for central-field-type HF orbitals.[4] For the case under discussion, an atom with one valence electron outside closed shells, the core orbitals are of central field type

$$\varphi_i = (\hat{P}_{n_i l_i}(r)/r) Y_{l_i m_{li}}(\vartheta, \varphi) \eta_{m_{si}}(\sigma), \tag{2.18}$$

where $\eta_{m_{si}}$ is a spin function, $Y_{l_i m_{li}}$ is the normalized spherical harmonics, and $\hat{P}_{n_i l_i}$ is the radial part of the HF orbital.* Fock has shown[6] that if the core orbitals are of this form then the electrostatic potential in H_F [Eqs. (2.4)–(2.5)] is spherically symmetric and the kernel of the exchange operator depends on $P_l(\cos \gamma)$ where P_l is the Legendre polynomial and γ is the angle between r_1 and r_2. Fock pointed out that such an operator commutes with the square and z-component of the angular momentum operator, therefore, the valence solutions of Eq. (2.15) will also be of central field type:

$$\varphi = (\hat{P}_{nl}(r)/r) Y_{lm_l}(\vartheta, \varphi) \eta_{m_s}(\sigma). \tag{2.19}$$

Substitute Eq. (2.18) into the kernel of the pseudopotential operator Eq. (2.17). Then we obtain for the kernel,

$$K(r_1, r_2) = \sum_{i=1}^{N} (\epsilon - \epsilon_i) \varphi_i(1) \varphi_i^*(2)$$

$$= \sum_{n_i l_i m_{li} m_{si}} (\epsilon_{nl} - \epsilon_{n_i l_i}) \frac{\hat{P}_{n_i l_i}(r_1) \hat{P}_{n_i l_i}(r_2)}{r_1 r_2}$$

$$\times Y_{l_i m_{li}}(\vartheta_1 \varphi_1) Y_{l_i m_{li}}^*(\vartheta_2 \varphi_2) \eta_{m_{si}}(\sigma_1) \eta_{m_{si}}^*(\sigma_2). \tag{2.20}$$

The spherical harmonics satisfy the equation [Slater I, p. 182]

$$\sum_{m_{li}=-l_i}^{l_i} Y_{l_i m_{li}}(1) Y_{l_i m_{li}}^*(2) = \frac{2l_i + 1}{4\pi} P_{l_i}(\cos \gamma), \tag{2.21}$$

where γ is again the angle between r_1 and r_2. Substituting Eq. (2.21) into Eq. (2.20) we see that the kernel of the operator V_p will depend only on $P_l(\cos \gamma)$. Using Fock's result we see that the operator $(H_F + V_p)$ will commute with the relevant angular momentum operators; therefore, the solutions of Eq. (2.16) will be of central field type:

$$\psi = (P_{nl}/r) Y_{lm_l}(\vartheta, \varphi) \eta_{m_s}(\sigma). \tag{2.22}$$

Here we used the same quantum numbers as in Eq. (2.19). That we can do this is evident from Eq. (2.10). Multiply Eq. (2.10) from the left by φ^* and integrate. Then we get

$$\langle \varphi \mid \psi \rangle = 1/A_0, \tag{2.23}$$

from which we see that φ and ψ must have identical angular and spin parts.

---

* The word *orbital* is used to indicate either the total one-electron wave function or its radial part. The specific meaning will always be clear from the context.

Using Eq. (2.20) and Eq. (2.22) we obtain

$$
\begin{aligned}
V_p \psi &= V_p (P_{nl}/r) Y_{lm_l} \eta_{m_s} \\
&= \sum_{n_i} (\epsilon_{nl} - \epsilon_{n_i l})(\hat{P}_{n_i l}/r) Y_{lm_l}(\vartheta, \varphi) \eta_{m_s}(\sigma) \int \hat{P}_{n_i l}(r') P_{nl}(r')\, dr' ,
\end{aligned}
\tag{2.24}
$$

where the summation is over those electron states of the core which have the same azimuthal quantum number as the valence orbital.

Now let us suppose that we substituted the HF orbital [Eq. (2.19)] into Eq. (2.15) and eliminated the angular and spin parts. The result is a radial wave equation,

$$
H_R \hat{P}_{nl} = \epsilon \hat{P}_{nl} .
\tag{2.25}
$$

Let us substitute the PO [Eq. (2.22)] into Eq. (2.16). Then for $V_p \psi$ we obtain Eq. (2.24). Substituting that into Eq. (2.16) and using the notation of Eq. (2.25) we obtain that the radial part of the PO satisfies the equation,

$$
(H + V_p)P_{nl} = \epsilon P_{nl} ,
\tag{2.26}
$$

where V_p is now the linear operator

$$
V_p = \sum_{n_i} (\epsilon_{nl} - \epsilon_{n_i l}) |\hat{P}_{n_i l}\rangle \langle \hat{P}_{n_i l}| .
\tag{2.27}
$$

Our result is that, for central-field-type HF orbitals, the pseudopotential is *l*-dependent,* and is an operator which operates only on the radial part. The angular and spin parts are not affected by the introduction of the pseudopotential.

Putting Eq. (2.18) and Eq. (2.22) into Eq. (2.9) we see that α_i will be zero except for those core states which have the same l, m_l and m_s as the PO. Using this result we obtain from Eq. (2.10), after eliminating the spin and angular parts,

$$
\hat{P}_{nl} = A_0 \left(P_{nl} - \sum_{n_i} \alpha_{n_i l,\, nl} \hat{P}_{n_i l} \right),
\tag{2.28}
$$

where

$$
\alpha_{n'l',\, nl} \equiv \int \hat{P}_{n'l'} P_{nl}\, dr ,
\tag{2.29}
$$

---

* As we see from the presence of ϵ_{nl}, the potential is also *n*-dependent or energy-dependent. This fact has been recognized only in the later development of the theory and it is still not being explicitly indicated in many discussions, including some of the presentations which will follow in this book. For the (*nl*)-dependence of the pseudopotential, see Section 4.2.

and the summation in Eq. (2.28) is again over those core states which have the same azimuthal quantum number as the PO.

2.2. THE METHOD OF PHILLIPS AND KLEINMAN

A. The Formulation of the Method

Ever since its formulation by Herring[7] in 1940 the so-called OPW (Orthogonalized Plane Wave) method was considered to be an effective procedure for the calculation of the wave function of the valence electrons in crystals. In this method the Pauli exclusion principle was taken into account by (Schmidt-) orthogonalizing the valence-electron wave functions to the wave functions of the electrons in the atomic cores.

Starting from this method Phillips and Kleinman (PK) formulated an exact pseudopotential theory,[8] applicable to atoms, molecules, and solids, by showing that the orthogonalization procedure can be replaced by a pseudopotential.

Let H be the Hamiltonian operator of the system and let φ be the exact wave function of the valence electron, which we assume to be orthogonalized to the core orbitals $\varphi_1, \varphi_2, \ldots, \varphi_N$:

$$\varphi = \psi - \sum_{i=1}^{N} \alpha_i \varphi_i, \tag{2.30}$$

where

$$\alpha_i = \langle \varphi_i \mid \psi \rangle. \tag{2.31}$$

In these equations, ψ is the nonorthogonal part of the valence wave function. The Schroedinger equation for the valence orbital is

$$H\varphi = \epsilon\varphi. \tag{2.32}$$

On putting φ as given by Eq. (2.30) into Eq. (2.32) we obtain the equation for ψ:

$$(H + V_p)\psi = \epsilon\psi, \tag{2.33}$$

where V_p is the Phillips–Kleinman pseudopotential, defined as

$$V_p = \sum_{i=1}^{N} \alpha_i \frac{(\epsilon - \epsilon_i)\varphi_i}{\psi}, \tag{2.34}$$

and ψ is the pseudo-orbital. In deriving Eq. (2.33) we assumed that the core orbitals are eigenfunctions of H with the eigenvalues ϵ_i.

Arriving at Eq. (2.33) PK concluded that the wave equation satisfied by the nonorthogonal wave function ψ contains, besides H, the term V_p, which can be interpreted as a pseudopotential. The equations above comprise the basis of the exact pseudopotential theory of PK.

Comparing the Phillips–Kleinman theory with the formulation of Szepfalusy it is evident that, starting from different ideas, they have arrived at the same equation. Indeed, looking at Eqs. (2.16) and (2.17), which were derived by Szepfalusy, and at Eqs. (2.33) and (2.34), which are the Phillips–Kleinman equations, we see that the equation for the pseudo-orbital ψ is the same in both cases. This is not surprising, since both theories are exactly equivalent to the orthogonalized formulations. It follows from this that the physical interpretation which we have appended to Eq. (2.16) is equally valid for Eq. (2.33). Thus, the calculation of pseudo-orbitals and orbital energies will yield the same results in both theories.

When we look at the interpretation of the term which represents the Pauli principle in the wave equation, we see that this interpretation is different in the two theories. Szepfalusy defined the pseudopotential as a linear operator, while Phillips and Kleinman defined it as a local potential. The exact relationship between the two is as follows: let V_p^S represent the operator of Eq. (2.17) and let V_p^{PK} represent the local potential of Eq. (2.34). Then, denoting the pseudo-orbital by ψ, we have:

$$V_p^{PK} = \frac{(V_p^S \psi)}{\psi}. \tag{2.35}$$

Thus we may say that, although the two formulations lead to the same results, the physical interpretation is different. We want to discuss this difference here in some detail because it played an important role in the development of pseudopotential theory.

By defining an exact local pseudopotential, Phillips and Kleinman created a theory much wider in its scope and applicability than the formulation of Szepfalusy. The reason for this that an operator like the one in Eq. (2.17) cannot be given a direct physical interpretation, while the local potential in Eq. (2.34) can be plotted easily in a diagram and explained in plausible physical terms. This property of PK formulation ensures that the theory can be understood much easier. More importantly, the PK theory can be applied immediately to a much larger number of special cases; it is evident that it is much easier to work with a local potential than with an operator.

We want to demonstrate the importance of the local formulation with an example. Probably the most important application of the pseudopotential theory in atomic and molecular physics is the formulation of the molecular effective Hamiltonian [Eq. (8.92)]. Most of the applications of pseudopotential theory in molecular physics are based on that effective Hamiltonian. Reviewing Sections 8.3 and 8.4 below, the reader will recognize that the derivation of the effective Hamiltonian rests on the assumption that the pseudopotential operators representing the interaction between the core and valence electrons can exactly be transformed into local potentials. This step, the localization of the operators, is carried out in the same way as in the PK theory presented here; in fact, the localization is a straightforward application of the ideas of Phillips and Kleinman.

Another example of the usefulness of the local formulation is the case of one-electron model potentials. Model potentials are, by definition, local, and a sound theoretical justification can be given to them only after the exact pseudopotentials are put into local form. The conceptual difficulty inherent in the step of localization is demonstrated by Szepfalusy's work. In attempting to establish connection between the exact pseudopotential operator and the local, density-dependent pseudopotentials, Szepfalusy attempted to localize the operator of Eq. (2.17), but could do so only in an approximation (see Section 3.2.A). In connection with this, we may also recall the difficulties encountered by Slater in trying to localize the exchange potential in the HF equations.[9] Only after such a localization could the HF equations serve as the background for the development of the $X\alpha$ model.[20]

Thus we see that the most important feature of pseudopotential theory, emphasized in Section 1.2, the possibility of putting the theory in a computationally advantageous model form, rests, to a considerable degree, on the work of Phillips and Kleinman in which the operator representing the Pauli exclusion principle was exactly transformed into a local potential. Their work was the first presentation of an exact pseudopotential theory in which the Pauli exclusion principle was replaced by a local potential. The formulation of the pseudopotential in local form permitted a plausible physical interpretation and opened the way to wide-ranging applications.

B.   Exact Solutions of the Phillips–Kleinman Equation for Atoms

The first calculations of exact pseudopotentials and pseudo-orbitals for atoms were carried out by Szasz and McGinn.[11] We present here an overview of the calculations and a discussion of the results. (The results in their entirety are presented in Appendix A.)

Calculations were carried out for the valence electron pseudo-orbital and orbital energies of the monovalent atoms Li, Na, K, Rb, Be^+, Mg^+, Ca^+, Al^{++}, Cu, and Zn^+. Computed were all those low-lying states for which the pseudopotential is not zero. As we have seen from Eq. (2.27), the pseudopotential is l-dependent and zero for those l values which do not occur in the core; for example, for the Na valence electron we have s and p pseudopotentials but no d pseudopotentials, since the core does not contain d electrons. Calculations were done for the lowest s and p valence states.

The equation to be solved is Eq. (2.33) in which we identify H with the HF Hamiltonian [Eq. (2.3)]. Thus our equation becomes

$$(H_F + V_p)\psi = \epsilon\psi, \tag{2.36}$$

where V_p is the PK pseudopotential [Eq. (2.34)] and H_F is given by Eq. (2.3):

$$H_F = t + g + U, \qquad (2.37)$$

where

$$t = -\frac{1}{2}\Delta \qquad (2.38)$$

and

$$g = -\frac{Z}{r}. \qquad (2.39)$$

We reinterpret U in the following way: in Eq. (2.3) the U means the total HF potential operator defined by Eqs. (2.4) and (2.5). Let us denote the operator in Eq. (2.4) by U_{op}. Then our U in Eq. (2.37) is defined as

$$U = \frac{(U_{op}\psi)}{\psi}. \qquad (2.40)$$

In other words, our U is a Slater-type local potential rather than the original operator of Eq. (2.4).

Next we introduce the modified potential which is defined as

$$V_M = g + U + V_p. \qquad (2.41)$$

This is the HF potential modified by the addition of V_p and it is the sum of the electrostatic, exchange, and pseudopotentials. Using this quantity we get Eq. (2.36) in the form

$$\left(-\frac{1}{2}\Delta + V_M\right)\psi = \epsilon\psi. \qquad (2.42)$$

Explict formulas for the construction of V_M are given in Appendix A.

Equation (2.42) was solved by a self-consistent procedure. The calculations were started by constructing a first approximation to V_M. The core orbitals and orbital energies which are needed for V_M were taken from HF calculations. For ϵ and ψ, which are needed to construct the exchange potential and the pseudopotential, suitable initial expressions were found. Using the first approximation for V_M, Eq. (2.42) was solved by numerical integration. This yielded new ϵ and ψ. The procedure was repeated until self-consistency was achieved in the pseudo-orbital and orbital energy. In order to facilitate the easy applicability of the pseudo-orbitals, analytic fits were constructed for these; the numerical integration yields the pseudo-orbitals in the form of tables which are not so easy to use. Using the analytic fits for the pseudo-orbitals the pseudopotentials can also be constructed easily.

A crucial point of the numerical procedure is the choice of the first approximation for the pseudo-orbital. For the *s* states, rapid convergence was achieved by replacing V_M with the Hellmann potential [Eq. (1.1)], solving the resulting equation by numerical integration and using the resulting eigenfunction as the first approximation of the iteration. For the *p* and *d* states, this procedure failed to give convergence; for these states the final output of the lower lying *s* and *p* states was used for the construction of the initial approximations. With this alternative procedure rapid convergence was achieved.

As representative examples of the results we discuss here the 3*s* pseudo-orbital for the Na atom and the modified potentials for the Na and Cu atoms. In Figure 2.1 we have $R(3s)$ and $P(3s) = rR(3s)$ for the Na valence electron. The form of $P(3s)$ is very similar to the form of the corresponding Slater-type orbital; this is the reason why it is customary to refer to the pseudo-orbital as a "Slater-type orbital." This reference is correct insofar as the STO's are nodeless like the pseudo-orbitals in the ground state. The comparison is less accurate, however, when we look at the asymptotic behavior. At $r = 0$ the 3*s* pseudo-orbital goes like r^l while the 3*s* STO has the form of r^2. The fluctuations in $R(3s)$ correspond to the nodes of the 3*s* HF orbital; these fluctuations are the results of the presence of core contributions in the pseudo-orbital. We note that these fluctuations are "smoothed out" in the $P(3s)$.

In Figures 2.2 and 2.3 we have the modified potentials for all the computed states in the Na and Cu atoms, respectively, along with the HF potential for the lowest state. Looking at the diagrams we see that for the

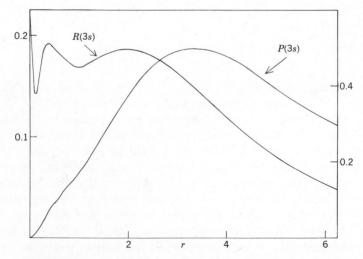

FIGURE 2.1.   The pseudo-orbital for the 3*s* state of the Na atom.

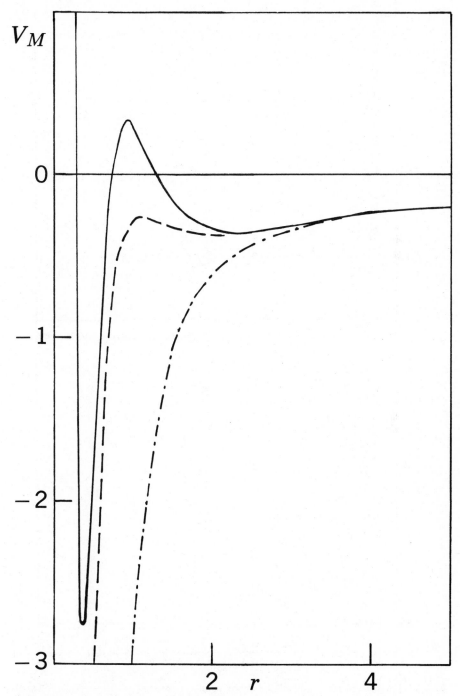

FIGURE 2.2. The modified potentials for the Na valence electron (————: the s potential; ----: the p potential; –·–·–·–: the HF potential in the 3s state).

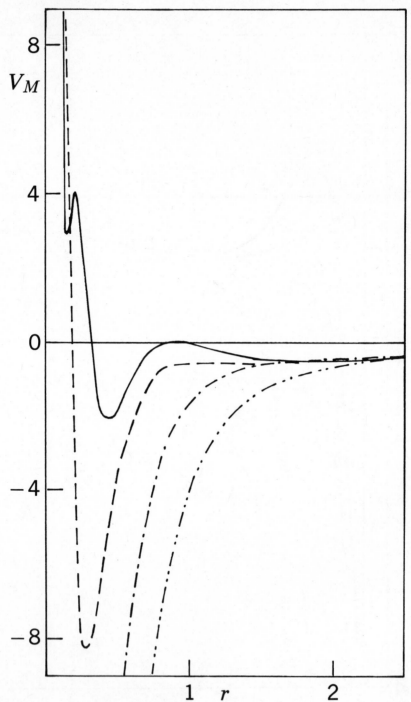

FIGURE 2.3. The modified potentials for the Cu valence electron (————: the *s* potential; – – – –: the *p* potential; – · – · – ·: the *d* potential; – · · – · · – · · –: the HF potential in the 4*s* state).

Na s state and for the Cu s and p states the potentials have the structure similar to the Hellmann potential which we described in Section 1.1; going from the nucleus outward the potentials show the positive barrier/negative well/Coulombic sequence. The barrier is missing in the p state of Na and in the d state of Cu. The modified potential should not be confused with the effective potential which is defined as

$$V_{eff} = V_M + \frac{l(l+1)}{2r^2}.$$ (2.43)

If the azimuthal term is added to V_M, all effective potentials will have high positive barriers in the core region. Thus the physical meaning of these diagrams is the same as the physical meaning of the Hellmann potential: the high barriers in V_M and in V_{eff}, respectively, are "preventing the electrons from falling into the core"; that is, the pseudopotentials are effectively playing the role of the Pauli exclusion principle. More accurately, as was pointed out by PK,[8] the pseudopotential in Eq. (2.34) provides the *radial* part of the orthogonalization energy of the valence electron, while the azimuthal part is provided by the azimuthal term in Eq. (2.43). These two potentials, the pseudopotential and the azimuthal term in Eq. (2.43), together play the role of the Pauli exclusion principle.

We note that while all potentials have the same overall structure, there are considerable differences in the details of the curves for different l values; that is, the l-dependence of V_M is quite pronounced.

The results of the PK-type pseudopotential calculations may be summarized as follows:

1. For each value of l, in the state with the lowest principal quantum number, the radial part of the valence electron PO is a nodeless function. (It is a Slater-type orbital, or STO.)
2. The structure of the modified potentials and/or the effective potentials show the typical barrier/well/Coulombic sequence.

The curves are considerably different for different l values.

Further discussion of these results may be found at the end of Section 2.3.B. The comparison of the calculated energies with the observed values will take place in Section 9.2.

2.3. THE METHOD OF COHEN AND HEINE

A. Indeterminacy of the PK Solutions

Having presented in the preceding sections the basic equations of exact pseudopotential theory, we now start to build up the formalism of the theory. It was observed by Cohen and Heine[12] that the solutions of the PK equation are not uniquely determined (PK-type indeterminacy). In order to show this let us consider an atom with one valence electron for which we have Eq. (2.16):

$$(H_F + V_p)\psi = \epsilon\psi, \tag{2.44}$$

where H_F is given by Eqs. (2.3)–(2.5) and V_p is the pseudopotential [Eq. (2.17)]:

$$V_p = \sum_{i=1}^{N} (\epsilon - \epsilon_i)|\varphi_i\rangle\langle\varphi_i| . \tag{2.45}$$

Let us assume that ψ and ϵ are solutions of Eq. (2.44) and let

$$\tilde{\psi} = \psi + \delta\psi, \tag{2.46}$$

where

$$\delta\psi = \sum_{j=1}^{N} \alpha_i\varphi_j \tag{2.47}$$

and the α_j are arbitrary constants. Substitution of $\tilde{\psi}$ into Eq. (2.44) gives

$$(H_F + V_p)\tilde{\psi} = (H_F + V_p)\psi + (H_F + V_p)\delta\psi$$

$$= \epsilon\psi + \sum_{j=1}^{N} \alpha_j H_F\varphi_j + \sum_{i,j=1}^{N} \alpha_j(\epsilon - \epsilon_i)|\varphi_i\rangle\langle\varphi_i \mid \varphi_j\rangle = \epsilon\psi + \sum_{j=1}^{N} \alpha_j\epsilon_j\varphi_j$$

$$+ \sum_{i=1}^{N} (\epsilon - \epsilon_i)\alpha_i\varphi_i = \epsilon\left(\psi + \sum_{i=1}^{N} \alpha_i\varphi_i\right) = \epsilon\tilde{\psi}, \tag{2.48}$$

where we have used the orthonormality of the HF core orbitals and Eq. (2.2). Our result is that if ψ is the solution of Eq. (2.44), then so is $\tilde{\psi}$ with the same eigenvalue. Therefore functions of the form

$$\psi = \varphi + \sum_{i=1}^{N} \alpha_i\varphi_i, \tag{2.49}$$

where the α_i's are arbitrary constants, are solutions of Eq. (2.44) with the same eigenvalue.

From this theorem Cohen and Heine (CH) concluded that the PO [Eq. (2.49)] can be subjected to a mathematical requirement in addition to the requirement that it be a solution of Eq. (2.44) with eigenvalue ϵ. An additional requirement will not disturb the exactness of the theory; that is, the pseudo-orbital which is the solution of Eq. (2.44) will represent the absolute energy minimum regardless of the additional condition imposed.

B. Kinetic Energy Minimization

It was suggested by Cohen and Heine that the most plausible additional condition is to demand that the PO [Eq. (2.49)] should minimize the expectation value of the kinetic energy operator; that is, it should be the "smoothest" PO.* The requirement means that the radial part of the PO will have the smallest amount of oscillations in the core region.

Let us investigate what happens to the pseudopotential equation if this requirement is imposed. We introduce here operator notation. Let the linear integral operator Ω be defined as

$$\Omega = \sum_{i=1}^{N} |\varphi_i\rangle\langle\varphi_i| \,, \tag{2.50}$$

or, in conventional notation, if f is an arbitrary function, then

$$\Omega(1)f(1) = \sum_{i=1}^{N} \varphi_i(1) \int \varphi_i^*(2)f(2)\, dq_2 \,. \tag{2.51}$$

Let

$$P = 1 - \Omega \,. \tag{2.52}$$

The operators Ω and P will be called orthogonality projection operators. Since they play a fundamental role throughout the pseudopotential theory, their properties are summarized in Appendixes B, D and E. Using these operators we can put Eq. (2.49) in the form

$$\varphi = (1 - \Omega)\psi \,. \tag{2.53}$$

Substituting this into the HF equation [Eq. (2.15)] we get

$$H_F (1 - \Omega)\psi = \epsilon(1 - \Omega)\psi \,, \tag{2.54}$$

which can be written as

$$(H_F + V_p)\psi = \epsilon\psi \,, \tag{2.55}$$

where

$$V_p = -(H_F - \epsilon)\Omega \,. \tag{2.56}$$

---

* This condition was suggested earlier by Szepfalusy[4] but was not followed up by a mathematical formulation.

Since the orbitals from which Ω is built are eigenfunctions of H_F, the Ω and P operators commute with H_F [Eq. (B.29)]. Thus we can write

$$V_p = -\Omega(H_F - \epsilon) . \tag{2.57}$$

We obtain the previously derived form of the pseudopotential, Eq. (2.45), by taking into account that the φ_i's are eigenfunctions of H_F.

Let E^K be the expectation value of the kinetic energy operator with respect to the PO [Eq. (2.49)]:

$$E^K = \langle \psi | t | \psi \rangle / \langle \psi | \psi \rangle . \tag{2.58}$$

Rearranging this equation and forming the variation we get

$$\langle \psi | \psi \rangle E^K = \langle \psi | t | \psi \rangle ,$$
$$\langle \delta\psi | \psi \rangle E^K + \delta E^K \langle \psi | \psi \rangle = \langle \delta\psi | t | \psi \rangle , \tag{2.59}$$

where we carry out the variation of E^K with respect to ψ^*. (It is known that the variation with respect to ψ leads to the same result [Slater I, p. 110].) According to our assumption ψ should minimize E^K; therefore we put

$$\delta E^K = 0 , \tag{2.60}$$

and obtain

$$\langle \delta\psi | t - E^K | \psi \rangle = 0 . \tag{2.61}$$

The PK-type indeterminacy permits variation only in the form of Eq. (2.47); therefore we put

$$\delta\psi = \sum_{j=1}^{N} \alpha_i \varphi_j , \tag{2.62}$$

and on putting this into Eq. (2.61) we get

$$\sum_{j=1}^{N} \alpha_j \langle \varphi_j | t - E^K | \psi \rangle = 0 . \tag{2.63}$$

This equation must be satisfied for any $\delta\psi$, which means that it must be satisfied for arbitrary α_j's. Thus we must have that

$$\langle \varphi_j | t - E^K | \psi \rangle = 0 , \quad (j = 1, 2, \ldots, N) . \tag{2.64}$$

Using Eq. (2.3) and reinterpreting U in such a way as to include g, the nuclear potential, we obtain from Eq. (2.44)

$$t\psi = (\epsilon - U - V_p)\psi, \tag{2.65}$$

and so

$$\langle \varphi_j | t - E^K | \psi \rangle = \langle \varphi_j | \epsilon - U - V_p - E^K | \psi \rangle. \tag{2.66}$$

Using Eq. (2.66) in Eq. (2.64) we get

$$\langle \varphi_j | V_p | \psi \rangle = \langle \varphi_j | \epsilon - U - E^K | \psi \rangle, \quad (j = 1, 2, \dots, N). \tag{2.67}$$

We multiply these equations from the left by φ_j and add them. Then we get

$$\sum_{j=1}^{N} \varphi_j \langle \varphi_j | V_p | \psi \rangle = \sum_{j=1}^{N} \varphi_j \langle \varphi_j | \epsilon - U - E^K | \psi \rangle. \tag{2.68}$$

Using the definition, Eq. (2.50), we can write

$$\Omega V_p \psi = \Omega(\epsilon - U - E^K)\psi. \tag{2.69}$$

Now operate on Eq. (2.55) from the left with the operator Ω. We obtain

$$\Omega(H_F - \epsilon)\psi = -\Omega V_p \psi, \tag{2.70}$$

and substituting this into Eq. (2.69) we get

$$-\Omega(H_F - \epsilon)\psi = \Omega(\epsilon - U - E^K)\psi. \tag{2.71}$$

The pseudopotential was given by Eq. (2.57). Using Eq. (2.71) we obtain from Eq. (2.57)

$$V_p \psi = \Omega(\epsilon - U - E^K)\psi. \tag{2.72}$$

The result of the derivation is that, if we demand that the pseudo-orbital should minimize the kinetic energy, then it must satisfy the equation

$$(H_F + V_p)\psi = \epsilon\psi, \tag{2.73}$$

where V_p is defined by Eq. (2.72). At the same time the solution of Eqs. (2.72)–(2.73) will reproduce the HF orbital and orbital energy, the former through Eq. (2.53). In order to see the precise form of the various operators we write down here Eq. (2.73) again, with the operators written out in detail:

$$\{t + U + \Omega(\epsilon - U - E^K)\}\psi = \epsilon\psi, \tag{2.74}$$

where

$$E^K = \langle \psi | t | \psi \rangle / \langle \psi \, | \, \psi \rangle . \tag{2.75}$$

The HF potential U, which now includes the nuclear potential, depends only on the core orbitals; in the pseudopotential we have U as well as Ω, which both depend on the core orbitals, as well as the eigenvalue ϵ and the constant E^K, where the latter depends on the pseudo-orbital ψ. Obviously the equation must be solved with a self-consistent procedure similar to the procedure outlined in Section 2.2.B.

It was pointed out by CH that conditions other than the kinetic energy minimization can be imposed. If we minimize the modified potential [Eq. (2.41)], which in our present notation will be

$$V_M = U + V_p , \tag{2.76}$$

that is, if we demand that the PO should make the expectation value of V_M to a minimum, then we obtain, with a derivation similar to that which led to Eq. (2.72), the potential

$$V_p \psi = \Omega(-U + V^M)\psi , \tag{2.77}$$

where

$$V^M = \langle \psi | V_M | \psi \rangle / \langle \psi \, | \, \psi \rangle . \tag{2.78}$$

In view of the indeterminacy of the PK solutions, it should be noted that the solutions of the PK equation which were presented in Section 2.2.B are not uniquely determined. We have seen that functions of the form of Eq. (2.49) are solutions of the PK equation with arbitrary α_i's. Although the pseudo-orbitals presented in Section 2.2.B are of the form of Eq. (2.49), they are not random solutions; they are wave functions which must be very close to the kinetic-energy-minimized pseudo-orbitals. That this is so is the result of the self-consistent procedure which is started with the PO obtained using the Hellmann potential. Solutions of the wave equation with the Hellmann potential are generally very smooth functions, besides being nodeless in the ground state. Starting the self-consistent procedure with a Hellmann solution ensures that the final PO will also be a smooth function, and thus it will be close to the kinetic-energy-minimized PO. (In connection with the quality of the PO's presented in Section 2.2.B see also the discussion at the end of Section 7.1.)

C.   General Form of the PK-type Potentials

It is evident from the preceding discussion that there is an arbitrarily large number of permissible pseudopotentials all leading to the correct orbital energy for the valence electron. It was suggested by Austin, Heine, and Sham[13] that the PK-type potentials can be written in the general form

$$V_p = \sum_{i=1}^{N} |\varphi_i\rangle\langle F_i|, \tag{2.79}$$

where F_i is an arbitrary function. All three pseudopotentials which we have written down thus far are of the form suggested by AHS. We had in the original PK potential

$$\langle F_i| = \langle\varphi_i|(\epsilon - \epsilon_i), \tag{2.80}$$

while in the kinetic-energy-minimized potential we had

$$\langle F_i| = \langle\varphi_i|(\epsilon - U - E^K), \tag{2.81}$$

and in the modified potential minimization

$$\langle F_i| = \langle\varphi_i|(-U + V^M). \tag{2.82}$$

Let us write down here again the HF equation for the valence electron, Eq. (2.15):

$$H_F\varphi = \epsilon\varphi, \tag{2.83}$$

and let us consider the equation:

$$(H_F + V_p)\psi = \epsilon\psi, \tag{2.84}$$

where V_p is given by Eq. (2.79). Let us assume that solutions of this equation exist. Let ψ be a wave function and ϵ an orbital parameter which are solutions of Eq. (2.84). Multiply Eq. (2.84) from the left by the operator P:

$$(PH_F + PV_p)\psi = \epsilon P\psi. \tag{2.85}$$

Using Eqs. (B.29) and (B.31) we obtain

$$H_F(P\psi) = \epsilon(P\psi). \tag{2.86}$$

On writing

$$\varphi = P\psi, \tag{2.87}$$

we get

$$H_F\varphi = \epsilon\varphi, \tag{2.88}$$

which is the HF equation [Eq. (2.83)]. Thus if ψ is a solution of Eq. (2.84) with the eigenvalue ϵ then $(P\psi)$ is a solution of the HF equation with the same eigenvalue. Therefore those solutions of Eq. (2.84) which are related to the HF valence orbital by Eq. (2.87) will reproduce the HF orbital parameters for the valence states regardless of the form of the F_i's in Eq. (2.79).* In other words, V_p in Eq. (2.79) is an exact pseudopotential, regardless of the form of the F_i's. (The General Pseudopotential Theorem of Austin, Heine, and Sham.[13])

We proceed now to show that solutions of Eq. (2.84) exist. The proof of existence will serve as a computational procedure. First we establish the spectrum of the orbital energies. Let us expand the solution of Eq. (2.84) in terms of the complete set described in appendix B:

$$\psi = \sum_{i=1}^{\infty} c_i \varphi_i . \tag{2.89}$$

On putting this into Eq. (2.84), multiplying from the left by φ_k^* and integrating we get,

$$\sum_{i=1}^{\infty} c_i(\epsilon_i \delta_{ik} + \langle \varphi_k | V_p | \varphi_i \rangle) = \epsilon c_k . \tag{2.90}$$

Substituting Eq. (2.79) we obtain

$$(\epsilon - \epsilon_k)c_k = \sum_{i=1}^{\infty} \sum_{j=1}^{N} c_i \langle \varphi_k | \varphi_j \rangle \langle F_j | \varphi_i \rangle = \sum_{i=1}^{\infty} \sum_{j=1}^{N} c_i \delta_{kj} \langle F_j | \varphi_i \rangle . \tag{2.91}$$

Now let φ_k be a valence state $k = v$. Then we get

$$(\epsilon - \epsilon_v)c_v = 0 . \tag{2.92}$$

Putting $c_v \neq 0$ we get

$$\epsilon = \epsilon_v . \tag{2.93}$$

---

* Two remarks are in order here. First, the argument above is valid only if ϵ is a nondegenerate eigenvalue. If ϵ is degenerate (i.e., if the orbitals $\varphi^{(1)}, \varphi^{(2)}, \ldots, \varphi^{(k)}$ belong to the same ϵ), then the argument shows that $(P\psi)$ will be a linear combination of these orbitals. This is sufficient to establish the identity of the eigenvalues of Eqs. (2.83) and (2.84). It is plausible to assume that, if the F_i's possess the correct symmetry, then the eigenvalues of Eq. (2.84) will show the same degree of degeneracy as Eq. (2.83). In this case there will be a set of PO's, $\psi^{(1)}, \psi^{(2)}, \ldots, \psi^{(k)}$, satisfying the relationships $\varphi^{(k)} = P\psi^{(k)}$. Second, in the argument we have also assumed that $P\psi \neq 0$. Equation (2.84) will also have solutions for which $P\psi = 0$. These will be the so-called core solutions which do not reproduce the valence levels.

If k is a core state, $k = c$, then

$$(\epsilon - \epsilon_c)c_c = \sum_{i=1}^{\infty} c_i \langle F_c \mid \varphi_i \rangle, \qquad (2.94)$$

and assuming that $c_c \neq 0$ we get

$$\epsilon = \epsilon_c + \sum_{i=1}^{\infty} \left(\frac{c_i}{c_c}\right) \langle F_c \mid \varphi_i \rangle. \qquad (2.95)$$

Thus the eigenvalues of Eq. (2.84) will be of the form of either Eqs. (2.93) or (2.95). The former are the valence solutions, the latter the core solutions. We note that if our potential is the PK expression, then we have Eq. (2.80); and putting that into Eq. (2.95) we obtain

$$\epsilon = \epsilon_c + \sum_{i=1}^{\infty} \left(\frac{c_i}{c_c}\right)(\epsilon_v - \epsilon_c)\langle \varphi_c \mid \varphi_i \rangle = \epsilon_c + (\epsilon_v - \epsilon_c) = \epsilon_v. \qquad (2.96)$$

The spectrum of the PK potential is identical with the valence levels.

Austin, Heine, and Sham have shown that unique solutions belonging to Eq. (2.93) as well as to Eq. (2.95), can be constructed. Their elegant proof will only be outlined here. It was shown that a pseudo-orbital of the form

$$\psi = \varphi_v + \sum_{i=1}^{N} \alpha_i \varphi_i \qquad (2.97)$$

can be constructed in such a way that ψ is a unique solution of Eq. (2.84) with the eigenvalue ϵ_v. This pseudo-orbital is a valence solution and the α_i's are the solutions of the linear, inhomogeneous equations

$$\sum_{j=1}^{N} M_{ij}\alpha_j = N_i, \quad (i = 1, 2, \ldots, N). \qquad (2.98)$$

In these equations we have

$$M_{ij} = \langle F_i \mid \varphi_j \rangle + (\epsilon_i - \epsilon_v)\delta_{ij} \qquad (2.99)$$

and

$$N_i = -\langle F_i \mid \varphi_v \rangle, \qquad (2.100)$$

where φ_v and ϵ_v are the valence orbital and eigenvalue respectively. The system in Eq. (2.98) will have unique solutions if the system determinant and the inhomogenity terms are not zero. Equation (2.98) must be solved in a self-consistent fashion, since the F_i's may depend on the pseudo-orbitals (i.e., on the α_i coefficients).

Next, it was shown by AHS that core solutions belonging to Eq. (2.95) and having the form

$$\psi = \sum_{i=1}^{N} \alpha_i \varphi_i, \tag{2.101}$$

can be constructed in such a way that ψ is a unique solution of Eq. (2.84) with the eigenvalue given by Eq. (2.95). In this case the α_i's are the solutions of the linear homogeneous equations

$$\sum_{j=1}^{N} M_{ij} \alpha_j = 0, \quad (i = 1, 2, \ldots, N), \tag{2.102}$$

where

$$M_{ij} = \langle F_i \mid \varphi_j \rangle + (\epsilon_i - \epsilon)\delta_{ij}. \tag{2.103}$$

The system in Eq. (2.102) will have solutions if the system determinant is zero. Since the F_i's may depend on the pseudo-orbitals, the solutions must be obtained by a self-consistent procedure.

The statement about the solutions being determined uniquely means that we obtain unique α_i's for a given set of F_i's. It is evident that there will be many physically meaningful choices for the F_i's, each of which will yield a unique pseudo-orbital which will be an exact solution of Eq. (2.84). Some exceptional cases should be mentioned. The procedure breaks down if there is an accidental degeneracy between Eqs. (2.93) and (2.95). The requirement that the inhomogeneity terms must not be zero also puts a restriction on the form of the F_i's. A forbidden form is, for example,

$$F_i = C\langle \varphi_i \mid, \tag{2.104}$$

where C is a constant.[14] In this case we get $N_i = 0$ in Eq. (2.100) and we do not get valence solutions. The PK expression Eq. (2.80) also leads to $N_i = 0$, which is just another way of saying that the PK equation does not have unique solutions. However, as we have seen in Section 2.2.B, meaningful exact solutions can be computed in this case.

D.   Cancellation Theorem

It was pointed out by Phillips and Kleinmann[8] and further elucidated by Cohen and Heine[12] that the modified potential of the pseudopotential theory has the interesting property of being constant, in a good approximation, over a considerable range of r (Cancellation Theorem[8,12,13]). This property is the result of the modified potential being the sum of the negative HF potential U and the positive pseudopotential V_p. It can be shown by analyzing the mathematical structure of the modified potential that there is a considerable cancellation between the negative and positive terms. Indeed, looking at the PK-type potentials in Figures 2.2 and 2.3, we see that these potentials could be replaced by a constant, in a good approximation, over a wide range of r. Even in the core region, where the modified potential shows strong fluctuations, it could be effectively replaced by an average—that is, by a constant.

2.4. THE METHOD OF WEEKS AND RICE

Although the pseudopotential of Eq. (2.79) is of very general form, it cannot be used in the case when H_F and P do not commute, that is, when the core orbitals are not eigenfunctions of the valence Hamiltonian; the step leading from Eq. (2.85) to Eq. (2.86) cannot be taken. Such a situation occurs, for example, if we are considering an atom in which the core electrons are in closed shells while the valence electrons are in an open shell. Even an atom with one valence electron belongs to this category if we remove the restriction of frozen core; that is, if we vary the $(N + 1)$ orbitals simultaneously, instead of varying the valence orbital with the core orbitals frozen.

The pseudopotential method developed by Weeks and Rice[14] set the goal of developing a pseudopotential formalism for an arbitrary valence Hamiltonian and an arbitrary set of one-electron core orbitals. The method can be formulated for one or for many valence electrons. In this section we discuss the one-valence-electron case; the application of the method to more than one valence electron will follow in Chapter 6.

Let φ and ϵ be the wave function and energy of the valence electron where

$$\epsilon = \langle \varphi | H' | \varphi \rangle . \tag{2.105}$$

The Hamiltonian H' should be Hermitian but otherwise unspecified. The φ is normalized and subjected to the condition that

$$\langle \varphi_i | \varphi \rangle = 0 , \quad (i = 1, 2, \ldots, N) . \tag{2.106}$$

The core orbitals are assumed to be known but they are not the eigenfunctions of H'. (Equation (B.29) is not valid.) On varying Eq. (2.105) under the subsidiary conditions of normalization and orthogonality we get

$$H'\varphi = \epsilon \varphi + \sum_{i=1}^{N} \lambda_i \varphi_i , \tag{2.107}$$

where the λ_i are Lagrangian multipliers for which we obtain

$$\lambda_i = \langle \varphi_i | H' | \varphi \rangle, \tag{2.108}$$

and using this expression, Eq. (2.107) becomes

$$H'\varphi = \epsilon\varphi + \sum_{i=1}^{N} \varphi_i \langle \varphi_i | H' | \varphi \rangle = \epsilon\varphi + \Omega H'\varphi. \tag{2.109}$$

Introducing the operator

$$H \equiv PH', \tag{2.110}$$

we obtain the wave equation for the valence electron in the form

$$H\varphi = \epsilon\varphi. \tag{2.111}$$

In the HF approximation with frozen core, the Lagrangian multipliers can be eliminated and we get Eq. (2.83) with $H = H_F$.

Adopting the idea from the theory of electron correlation, Weeks and Rice (WR) proceeded to solve Eq. (2.111) as follows. The wave equation [Eq. (2.111)] can be transformed into a variational equation [Slater I, p. 110]. In order to ensure orthogonality to the core orbitals, φ is put in the form

$$\varphi = (1 - \Omega)\psi = P\psi, \tag{2.112}$$

and we obtain from Eq. (2.111):

$$\epsilon = \text{Min} \frac{\langle P\psi | H | P\psi \rangle}{\langle P\psi \mid P\psi \rangle} = \text{Min} \frac{\langle \psi | PHP | \psi \rangle}{\langle \psi | P | \psi \rangle}. \tag{2.113}$$

where we used the Hermitian and idempotent character of P.*

We vary Eq. (2.113) unrestrictedly with respect to ψ^*. The procedure requires that (PHP) be Hermitian; this is proved in Appendix C. We obtain for ψ the wave equation

$$(H + V_p)\psi = \epsilon\psi, \tag{2.114}$$

where V_p is the Weeks–Rice pseudopotential

---

* The formulation of the problem in the form of a variational equation with the operator PHP was also given by the Author[82] but was not followed up by the formulation of the pseudopotential [Eq. (2.115)].

$$V_p = -\Omega(H - \epsilon) - (H - \epsilon)\Omega + \Omega(H - \epsilon)\Omega$$

$$= -\Omega(H - \epsilon) - P(H - \epsilon)\Omega$$

$$= -\Omega H - H\Omega + \Omega H\Omega + \epsilon\Omega, \qquad (2.115)$$

where we have written down three different forms which will be used alternatively, depending on which of the formulas is the most convenient in a particular application.

The potential in Eq. (2.115) reduces to the PK potential [Eq. (2.57)] if $H = H_F$. In that case we have in the second line

$$-P(H - \epsilon)\Omega = -(H - \epsilon)P\Omega = 0, \qquad (2.116)$$

where we used Eq. (B.19). On using Eq. (2.116) in the second line of Eq. (2.115) we get Eq. (2.57).

Next we show that the valence orbital φ as well as any of the core orbitals φ_i are eigenvunctions of Eq. (2.114). First consider

$$V_p\varphi = -\Omega(H - \epsilon)\varphi - P(H - \epsilon)(\Omega\varphi). \qquad (2.117)$$

The first term is zero because of Eq. (2.111) and the second because of Eq. (B.13). Therefore

$$V_p\varphi = 0. \qquad (2.118)$$

Next consider (φ_k = core orbital),

$$V_p\varphi_k = -\Omega(H - \epsilon)\varphi_k - P(H - \epsilon)(\Omega\varphi_k)$$

$$= -\Omega(H - \epsilon)\varphi_k - P(H - \epsilon)\varphi_k$$

$$= (-\Omega - 1 + \Omega)(H - \epsilon)\varphi_k$$

$$= -(H - \epsilon)\varphi_k, \qquad (2.119)$$

where we used Eq. (B.13).

Now put $\psi = \varphi$ in Eq. (2.114) and use Eq. (2.118). Then we get

$$(H + V_p)\varphi = H\varphi = \epsilon\varphi. \qquad (2.120)$$

Next put $\psi = \varphi_k$ in Eq. (2.114) and use Eq. (2.119):

$$(H + V_p)\varphi_k = (H - H + \epsilon)\varphi_k = \epsilon\varphi_k. \qquad (2.121)$$

Equations (2.120) and (2.121) show that both φ and φ_k, ($k = 1 \cdots N$), are solutions of Eq. (2.114) with the valence eigenvalue ϵ. From this it follows that we can form the linear combinations

$$\psi = \sum_{i=1}^{N} \alpha_i \varphi_i, \tag{2.122}$$

and

$$\psi = \varphi + \sum_{i=1}^{N} \alpha_i \varphi_i. \tag{2.123}$$

Both functions are solutions of Eq. (2.114) with an arbitrary set of α_i coefficients. The eigenvalue belonging to both functions is the valence level ϵ. The function in Eq. (2.122) is the core solution while ψ in Eq. (2.123), which is a pseudo-orbital, is the valence solution. The WR pseudopotential is subjected to the same kind of indeterminancy as the PK potential.

We are now able to generalize the WR potential in a way similar to Eq. (2.79).* Consider

$$V_p = \sum_{i=1}^{N} |\varphi_i\rangle\langle F_i| - P(H - \epsilon)\Omega. \tag{2.124}$$

where F_i is arbitrary. The potential suggested by AHS, Eq. (2.79) is obtained from Eq. (2.124) if the second term is zero. As we have seen, that term is zero if the core orbitals are eigenfunctions of H, in which case P and H commute and we obtain $P\Omega = 0$. The WR potential of Eq. (2.115) is obtained from Eq. (2.124) if we put

$$\langle F_i| = -\langle \varphi_i|(H - \epsilon). \tag{2.125}$$

We now prove the following theorem. Consider the wave equation

$$(H + V_p)\psi = \epsilon\psi, \tag{2.126}$$

where V_p is given by Eq. (2.124). We state that if ψ and ϵ are solutions of Eq. (2.126) then $\varphi = (P\psi)$ and ϵ are the solutions of Eq. (2.111), regardless of the form of the F_i's.

In order to prove the theorem let us multiply Eq. (2.126) from the left by P. We get

$$(PH + PV_p)\psi = \epsilon P\psi. \tag{2.127}$$

Using Eq. (2.110) we obtain

$$PH = P(PH') = P^2H' = PH' = H. \tag{2.128}$$

* The following argument is based on published[25] and unpublished work of the author.

On using Eq. (2.124) we get

$$PV_p = \sum_{i=1}^{N} P|\varphi_i\rangle\langle F_i| - P^2(H - \epsilon)\Omega \, . \qquad (2.129)$$

From Eq. (B.14) we know that P annihilates any core orbital; therefore, the first term will be zero. In the second term $P^2 = P$ and $P\Omega = 0$; thus we get

$$PV_p = -PH\Omega = -H\Omega \, , \qquad (2.130)$$

where we have again used Eq. (2.128). On using Eqs. (2.128) and (2.130) in Eq. (2.127) we obtain

$$(PH + PV_p)\psi = (H - H\Omega)\psi = H(1 - \Omega)\psi = HP\psi = \epsilon P\psi \, , \quad (2.131)$$

and using Eq. (2.112) we obtain

$$H\varphi = \epsilon\varphi \, , \qquad (2.132)$$

which proves the theorem.*

It is clear from the preceding discussion that the undetermined functions F_i in the potential of Eq. (2.124) can be fixed by the same method which we used to fix the undetermined functions in the potential of Eq. (2.79). We demand that the undetermined constants α_i in the PO should be fixed by satisfying a subsidiary mathematical condition. For such a condition, let us again use the kinetic energy minimization. Using the argument leading from Eq. (2.58) to Eq. (2.69) we obtain easily

$$\Omega V_p\psi = \Omega(\epsilon - U - E^K)\psi \, , \qquad (2.133)$$

where $U = U' - \Omega H'$ and U' is the total potential in H' including the nuclear attraction. Operating from the left on Eq. (2.124) by the operator Ω we get

$$\Omega V_p = \sum_{i=1}^{N} \Omega|\varphi_i\rangle\langle F_i| - \Omega P(H - \epsilon)\Omega$$

$$= \sum_{i=1}^{N} \sum_{j=1}^{N} |\varphi_j\rangle\langle\varphi_j \,|\, \varphi_i\rangle\langle F_i| = \sum_{i=1}^{N} |\varphi_i\rangle\langle F_i| \, , \qquad (2.134)$$

where we have used the condition that $\Omega P = 0$. Let us operate on ψ by ΩV_p;

---

* In this derivation, as in the derivation leading to Eq. (2.88), we have assumed that ϵ is nondegenerate and $(P\psi) \neq 0$. See the footnote following Eq. (2.88).

we get

$$\Omega V_p \psi = \sum_{i=1}^{N} \varphi_i \langle F_i \mid \psi \rangle . \tag{2.135}$$

Let us write down Eq. (2.133) again:

$$\Omega V_p \psi = \sum_{i=1}^{N} \varphi_i \langle \varphi_i | \epsilon - U - E^K | \psi \rangle . \tag{2.136}$$

Comparing the last two equations we get

$$\langle F_i | = \langle \varphi_i | (\epsilon - U - E^K) . \tag{2.137}$$

This is the same formula as Eq. (2.81), except that here U has a slightly different meaning. Therefore, if we demand that the PO should minimize the kinetic energy, the pseudopotential becomes

$$V_p = \Omega(\epsilon - U - E^K) - P(H - \epsilon)\Omega . \tag{2.138}$$

It is evident that the minimization of the modified potential would lead to an expression similar to Eq. (2.82); we would obtain the corresponding pseudopotential by substituting Eq. (2.82) into Eq. (2.124).

Next we show that for properly chosen F_i's the wave equation Eq. (2.126) will have a solution of the form of Eq. (2.123) with uniquely determined α_i coefficients.

Let us substitute Eq. (2.123) into the wave equation. We get $(i \leftrightarrow j)$

$$(H + V_p)\left(\varphi + \sum_{j=1}^{N} \alpha_j \varphi_j\right) = \epsilon\left(\varphi + \sum_{j=1}^{N} \alpha_j \varphi_j\right) . \tag{2.139}$$

Multiplying from the left by φ_i^* $(i = 1 \cdots N)$ and integrating, we obtain

$$\langle \varphi_i | H | \varphi \rangle + \langle \varphi_i | V_p | \varphi \rangle + \sum_{j=1}^{N} \alpha_j \langle \varphi_i | H | \varphi_j \rangle + \sum_{j=1}^{N} \alpha_j \langle \varphi_i | V_p | \varphi_j \rangle$$

$$= \epsilon \left\{ \langle \varphi_i \mid \varphi \rangle + \sum_{j=1}^{N} \alpha_j \langle \varphi_i \mid \varphi_j \rangle \right\} . \tag{2.140}$$

Now let

$$M_{ij} \equiv \langle \varphi_i | H | \varphi_j \rangle + \langle \varphi_i | V_p | \varphi_j \rangle - \epsilon \delta_{ij} , \tag{2.141}$$

and

$$N_i \equiv -\langle \varphi_i | H | \varphi \rangle - \langle \varphi_i | V_p | \varphi \rangle . \tag{2.142}$$

Using these notations we can write Eq. (2.140) in the form

$$\sum_{j=1}^{N} M_{ij}\alpha_j = N_i , \quad (i = 1, 2, \ldots, N) . \tag{2.143}$$

This is a set of linear inhomogeneous equations for the N coefficients α_j. We obtain a unique set of coefficients from these equations if the system determinant is not zero:

$$\det M_{ij} \neq 0 , \tag{2.144}$$

and if the inhomogeneity terms are not equal to zero,

$$N_i \neq 0 , \quad (i = 1, 2, \ldots, N) . \tag{2.145}$$

Let us substitute the operators H and V_p into the matrix components M_{ij} and N_i. We get

$$\langle \varphi_i | H | \varphi_j \rangle = \langle \varphi_i | PH' | \varphi_j \rangle = \langle P\varphi_i | H' | \varphi_j \rangle = 0 , \tag{2.146}$$

because of Eq. (B.14). Using Eq. (2.124) we obtain

$$V_p\varphi_j = \sum_{k=1}^{N} |\varphi_k\rangle\langle F_k | \varphi_j \rangle - P(H - \epsilon)(\Omega\varphi_j)$$

$$= \sum_{k=1}^{N} |\varphi_k\rangle\langle F_k | \varphi_j \rangle - P(H - \epsilon)\varphi_j , \tag{2.147}$$

where we used Eq. (B.13). Using Eq. (2.147) we obtain

$$\langle \varphi_i | V_p | \varphi_j \rangle = \sum_{k=1}^{N} \langle \varphi_i | \varphi_k \rangle\langle F_k | \varphi_j \rangle - \langle \varphi_i | P(H - \epsilon) | \varphi_j \rangle = \langle F_i | \varphi_j \rangle , \tag{2.148}$$

where the second term vanished since $P\varphi_i = 0$. Using Eqs. (2.146) and (2.148) we obtain

$$M_{ij} = \langle F_i | \varphi_j \rangle - \epsilon\delta_{ij} . \tag{2.149}$$

Now consider Eq. (2.142). The first term will be zero, similarly to Eq. (2.146). For the second term we get

$$\langle \varphi_i | V_p | \varphi \rangle = \sum_{k=1}^{N} \langle \varphi_i | \varphi_k \rangle\langle F_k | \varphi \rangle - \langle \varphi_i | P(H - \epsilon)\Omega | \varphi \rangle = \langle F_i | \varphi \rangle , \tag{2.150}$$

where we used that $\Omega \varphi = 0$. Substituting this into Eq. (2.142) we obtain

$$N_i = -\langle F_i \mid \varphi \rangle . \tag{2.151}$$

Comparing Eqs. (2.143), (2.149), and (2.151) with Eqs. (2.98), (2.99), and (2.100) we see that they are closely analogous. (The term with $\epsilon_i \delta_{ij}$ which appears in Eq. (2.99) would also appear in Eq. (2.149) if H would be identical with H_F.) From this it follows that the solutions of Eq. (2.143) may be obtained in the same way as we have described the calculation of the solutions of Eq. (2.98). We emphasize that, in general, the F_i's will depend on the pseudo-orbital; therefore, in order to get the solution, a self-consistent procedure will be needed. We note also that the remarks made at the end of Section 2.3.C, about the restrictions put on the form of the F_i's by Eq. (2.145) are equally valid here.

In summing up we observe that the method of Weeks and Rice is, in the slightly generalized form presented here, the most general pseudopotential formalism for one valence electron outside of an uncorrelated atomic core. In this method both the Hamiltonian H' and the core orbitals were completely arbitrary. The wave equation, Eq. (2.111), the solutions of which were orthogonal to the core orbitals, was exactly transformed into the wave equation, Eq. (2.114), which was not subjected to any orthogonality requirement. The general form of the pseudopotential was given by Eq. (2.124); the arbitrary functions F_i could be fixed by additional mathematical conditions. The valence solutions have the form of Eq. (2.123); the coefficients are uniquely determined after the F_i's are properly chosen. Finally, we note again that this method can be generalized to atoms with more than one valence electrons.

2.5. PSEUDOPOTENTIALS FOR CORRELATED CORE

A. Introduction

In the preceding four sections we have seen the forms which the pseudopotential takes for those cases in which the core electrons are represented by one-electron functions. Such cases are referred to as *uncorrelated* atomic cores.

By *electron correlation* we shall mean, somewhat loosely but very generally, those effects which cannot be treated in the framework of an independent particle model (i.e., in the framework of the HF approximation). The basic idea underlying the HF approximation, in its single-determinantal form, is that the wave function of the electrons is an (antisymmetrized) product of one-electron functions; that is, it is assumed that each electron moves independently from the others, feeling the presence of the other electrons only through an average potential. However, the HF determinant is only an approximation to the exact eigenfunction of the ab initio Hamiltonian. The exact eigenfunction will exhibit correlation between the electrons; that is, it will not be of such structure that the electrons can be said to move independently of each other. Rather, the structure of the exact eigenfunction will be such that the probability of finding the i-th electron in volume dv will depend on the momentary position of all other electrons.

In this section we shall discuss three cases in which electron correlation will be introduced into the wave function of the atomic core, and we shall show how the pseudopotential theory has to be modified in the case of correlated cores. The three correlation effects are as follows:

1.   In Section 2.5.B we discuss what happens to the pseudopotential if the electron correlation between the core electrons is taken into account; that is, if the wave function of the closed-shell-core cannot be described by a single determinantal wave function.

2.   In Section 2.5.C we discuss a special kind of core-valence correlation effect, consisting of the exchange interaction between the valence electron and the core electrons which have spins parallel to the spin of the valence electron. The radial part of those core electrons whose spin is parallel to the spin of the valence electron will be different from the radial parts of the core electrons with antiparallel spins. This effect, called spin-polarization, requires different orbitals for different spins, and the total wave function will not be a single determinant. We shall discuss what happens to the pseudo-potential in this case. It is evident that, this effect takes the discussion out of the framework of the "frozen core" approximation, since here the distribution of the core electrons will depend on the presence of the valence electron.

3.   The third correlation effect, which is also outside of the range of the frozen core approximation, is the so-called core polarization. Here it is taken into account that the presence of a point charge, the valence electron, outside of the core will destroy the original spherical symmetry of the atomic core with closed shells. As the result of this effect, the charge distribution of the atomic core will deviate from the spherically symmetric charge density of the HF approximation, and the potential in which the valence electron is moving will also be different from the potential of the core in the HF approximation. This difference is called the polarization potential. In Section 2.5.D we discuss this potential and the role it plays in pseudopotential theory. Section 2.5.D will close with a joint discussion of all three effects.

B.   The Method of Öhrn and McWeeny[15]

Since the material of this subsection is the generalization of the contents of Sections 2.1 and 2.2, the discussion will closely parallel those sections. Let us consider an atom with one valence electron, and let us first remove the valence electron and consider the core alone. Let the number of the core electrons be N and the Hamiltonian given as

$$H = \sum_{i=1}^{N} (t_i + g_i) + \frac{1}{2} \sum_{i,j=1}^{N} \frac{1}{r_{ij}}, \qquad (2.152)$$

where the notation is the same as in the preceding sections. Let the Schroedinger equation of the core be

$$H\Phi_c = E_c \Phi_c, \qquad (2.153)$$

where Φ_c is a fully antisymmetric, normalized, N-electron function

$$\Phi_c = \Phi_c(1, 2, \ldots, N), \qquad (2.154)$$

and E_c is the energy of the core. Let us assume that the exact or approximate solution of Eq. (2.153) has been obtained, and that Φ_c means this exact or approximate solution, in the discussion below. In any case Φ_c is such that the correlation between the core electrons (the core-core correlation) is taken into account in Φ_c. Thus Φ_c is a correlated wave function and not the single determinant of Eq. (2.1).

Now let us add the valence electron to the system and let the total wave function be

$$\Phi_T = \tilde{A}\{\Phi_c(1, 2, \ldots, N)\psi_v(N + 1)\}, \qquad (2.155)$$

where $\tilde{A}$ is an antisymmetrizer operator.

In order to determine ψ_v Öhrn and McWeeny proceeded as follows. First let us clarify the properties of the functions Φ_c and ψ_v. The core wave function Φ_c is correlated, and is the exact or approximate solution of Eq. (2.153). The valence orbital ψ_v is normalized but otherwise unspecified. It is not assumed that there is an orthogonality relationship between Φ_c and ψ_v. The application of the operator $\tilde{A}$ ensures that Φ_T will be fully antisymmetric, which is what the Pauli exclusion principle requires. The Pauli principle does not require orthogonality between ψ_v and Φ_c; this is a point that we have emphasized in connection with Eq. (2.6). We assume that the addition of the valence electron does not change Φ_c, thus keeping the frozen core approximation valid. Next let us form the expectation value of the $(N + 1)$-electron Hamiltonian with respect to Φ_T. On varying the resulting expression with respect to ψ_v^* while keeping Φ_c fixed, one can obtain the equation for the best ψ_v. The only subsidiary condition in the variation is the normalization of ψ_v. Öhrn and McWeeny have shown that the result is

$$H_{\text{eff}}\psi_v = \epsilon_v \psi_v, \qquad (2.156)$$

where

$$H_{\text{eff}} = t + g + \hat{U} + \hat{V}_p. \tag{2.157}$$

Here $\hat{U}$ is an operator expressing the electrostatic and exchange potential of the core in terms of the correlated core function Φ_c. The operator $\hat{V}_p$ is a pseudopotential, which keeps the valence electron out of the core. Like $\hat{U}$, the operator $\hat{V}_p$ depends on Φ_c; more accurately, it depends on the first-, second-, and third-order density matrices of Φ_c, and also on the total energy of the core, E_c, and on the valence energy ϵ_v.

Equation (2.156) is closely analogous to Eq. (2.16). In fact, it is easy to show that Eq. (2.16) is a special case of Eq. (2.156). We can see this without any calculations. The derivation of Öhrn and McWeeny becomes identical with the derivation of Szepfalusy if we put for Φ_c the HF wave function [Eq. (2.1)]:

$$\Phi_c = (N!)^{-1/2} \det[\varphi_1 \cdots \varphi_N]. \tag{2.158}$$

If Φ_c is given by this expression, then the wave function for the $(N+1)$-electron system, the function in Eq. (2.155), will become identical with Eq. (2.6). From Eq. (2.6) we obtain Eq. (2.16) directly. Thus if Φ_c is the HF wave function, the pseudopotential $\hat{V}_p$ will become identical with the potential of Eq. (2.17).

Now let us go one step further and change the operators of Eq. (2.157) into local potentials, in the manner of Phillips and Kleinmann. Let

$$\tilde{U} = \frac{(\hat{U}\psi_v)}{\psi_v}, \tag{2.159}$$

and

$$\tilde{V}_p = \frac{(\hat{V}_p\psi_v)}{\psi_v}. \tag{2.160}$$

Then Eq. (2.156) becomes

$$H_{\text{eff}}\psi_v = \{t + g + \tilde{U} + \tilde{V}_p\}\psi_v = \epsilon_v\psi_v. \tag{2.161}$$

The contents of this equation can be expressed in the form of a theorem: For an arbitrary, correlated core function it is possible to formulate a wave equation for the valence electron in which the Pauli exclusion principle is expressed in the form of a local potential.

The reason why we have not given the explicit expression for $\tilde{V}_p$ is that the main significance of this potential does not lie in its possible application in actual calculations, although there is nothing in the structure of $\tilde{V}_p$ that would prevent such calculations. The main significance of this discussion is that it will play an important role in the establishment of a theoretical foundation for the model pseudopotentials.

C. The G1 Method of Goddard

We turn now to the discussion of spin-polarization in pseudopotential theory. This effect can be incorporated into the pseudopotential theory by using the G1 method of Goddard.[16] Goddard developed a very general method for the calculation of a certain type of atomic or molecular wave functions. The spin-polarized wave functions, which we discuss here, are a special case of Goddard's method; consequently we shall not discuss the method in its full generality but only its application to the case of spin-polarization.

Let Φ be a spatial wave function representing the $(N + 1)$ electrons of an atom with N electrons in the closed-shell core plus one valence electron. Let χ be a spin function for the $(N + 1)$ electrons. Goddard considered the wave function

$$\Psi = G(\Phi\chi), \tag{2.162}$$

where G is an operator which, when operating on the product $(\Phi\chi)$, will generate from it a wave function which is an eigenfunction of the total spin square; since χ is assumed to be an eigenfunction of the z-component of the spin, Ψ will be an eigenfunction of S^2 and S_z. The spatial function Φ may be correlated or it may be an uncorrelated product. The operator G also ensures that Ψ will be fully antisymmetric, thus satisfying the Pauli principle.

Next we put Φ into the form of a product of one-electron functions. Goddard has shown that, after placing Ψ into the energy integral and varying the total energy of the atom with respect to the one-electron functions, one obtains a set of coupled integro-differential equations, one for each orbital, from which the best such orbitals can be determined.

In these derivations it is not assumed that the one-electron functions are orthogonal. This is done in order to avoid restricting the generality of the variation. We have seen in Section 2.1 that if the wave function is a single determinant, the orthogonalization of the one-electron orbitals is not a restriction. In this method, however, the total wave function Ψ is not a single determinant, even if Φ is a product of one-electron functions. Thus the orthogonalization would be a restriction here.

Applying the method to an atom with one valence electron outside closed shells one obtains the wave equation for the valence electron

$$H\psi_v = \epsilon_v \psi_v, \tag{2.163}$$

where H has the form

$$H = t + g + U^{G1}. \tag{2.164}$$

In this equation, U^{G1} is a potential operator which represents the complete interaction between the valence electron and the core electrons. This operator involves integrals over all core orbitals. Since those orbitals are not orthogonal, the operator U^{G1} is a rather complex expression. The point is that the U^{G1} will contain a pseudopotential which keeps the valence electron out of the core. The occurence of a pseudopotential is the result of the one-electron orbitals not being orthogonal to each other. Since the total wave function is antisymmetric, in the absence of orthogonality conditions, the Pauli principle will take the form of pseudopotentials.

The calculations carried out with this method for the Li, Be^+, and B^{++} atoms show that the orbital obtained for the valence electron of these atoms is nodeless (a pseudo-orbital), and the modified potential $g + U^{G1}$ shows the typical structure expected when U^{G1} contains a pseudopotential (see Figure 3 of the third paper of Ref. 16). Thus the method of Goddard clearly represents the extension of the pseudopotential theory to spin-polarized cores.

D. Core Polarization

As we mentioned in Section 2.5.A, core polarization is an effect which involves electron correlation between the core electrons and the valence electron. Strictly speaking, this effect is not a part of pseudopotential theory, and will not be treated here in detail. The significance of this effect for pseudopotential theory lies in the fact that this effect, like the other two correlation effects discussed in the preceding sections, is important for the theoretical justification of model potentials.

The theory of core polarization was developed, after some early work by Bethe, by Callaway.[17] In order to take into account the polarization of the core by the valence electron, Callaway started with a wave function of the type of Eq. (2.155), with Φ_c modified as follows:

$$\Phi_c = \det[\varphi_1(r_1, r_v)\varphi_2(r_2, r_v) \cdots \varphi_N(r_N, r_v)] . \qquad (2.165)$$

In this wave function, the one-electron core orbitals depend parametrically on the coordinate of the valence electron; thus core-valence correlation is introduced by making the distribution of the core electrons dependent on the momentary position of the valence electron. Callaway has shown that the core orbitals will be the solutions of HF-type equations in which the Coulomb potential of the valence electron will appear as a perturbation. Solving those equations in a plausible approximation, Callaway brought the wave equation of the valence electron to the form

$$\{t + g + U + V_c\}\psi_v = \epsilon_v\psi_v, \qquad (2.166)$$

where U is the HF potential and V_c is a potential function, called the polarization potential, which depends on the solutions of the above-mentioned HF-type equations. Thus the introduction of the valence electron coordinate into the core orbitals has the effect of an additional potential in the wave equation of the valence electron.

It is evident that Callaway's method does not contain the full core-valence correlation; the full treatment of this effect would make the valence electron position dependent on the momentary position of the core electrons. Callaway's results were incorporated into pseudopotential theory by Szasz and McGinn.[18] The polarization potentials were put into analytic forms, given in Appendix A.

We turn now to the joint discussion of the three correlation effects. Taken separately, the core-core correlation and the spin-polarization can be introduced into pseudopotential theory by making the appropriate changes in the HF potential and in the pseudopotential. On the other hand, the polarization potential is additive to the HF potential. Lacking a joint theory we shall postulate, on the basis of plausibility, that, the three effects can be taken into account simultaneously by an equation of the form

$$\{t + g + \tilde{U} + \tilde{V}_p + \tilde{V}_c\}\psi_v = \epsilon_v \psi_v, \tag{2.167}$$

where $\tilde{U}$ is a HF-type potential, $\tilde{V}_p$ is a pseudopotential, and $\tilde{V}_c$ is a polarization potential. We assume that the core-core correlation and the spin-polarization can be taken into account by the appropriate form of $\tilde{U}$, $\tilde{V}_p$, and possibly also of $\tilde{V}_c$.

chapter 3

Density-Dependent Pseudopotentials for One-Valence-Electron Systems

3.1. DERIVATION OF THE DENSITY-DEPENDENT PSEUDOPOTENTIAL FROM THE THOMAS–FERMI MODEL

A. Introduction

In Section 1.2 we outlined the difference between exact and model formulations of the pseudopotential method. Having presented the exact methods for one-valence-electron systems in Chapter 2, we now begin the presentation of methods which can be characterized as models. Actually, as the Table of Contents shows, the general presentation of models *per se* will take place in Chapter 4; this chapter describes density-dependent potentials. These potentials form a special group of model potentials; they are treated separately because, since they rest on the Thomas–Fermi (TF) model, they possess a better theoretical justification than most other models. It must be noted, however, that, in contrast to the methods presented in Chapter 2, this chapter describes approximate methods.

We have emphasized that a model should have a sound theoretical justification. It is time now to explain this a little more closely. By a sound theoretical justification we mean that the equations of a model should be related to the ab initio equations by a derivation. The word *derivation* here means something different from its strict mathematical sense, whereby we arrive at the result by exact mathematical operations. Derivation of a model from ab initio does not mean exact derivation: the concept of model precludes exactness. Derivation of a model from ab initio means that we arrive at the equations of the model by mathematical operations among which there are plausible approximate steps. It is evident that the soundness of such a derivation is a question of subjective value judgment. It is also evident that there is more than one way to establish a model.

The ultimate usefulness of a model is, of course, decided by its usefulness in applications. However, good results in applications do not remove the need for sound theoretical justification. It is the other way around: it is generally true that if the theoretical justification of a model is really sound then good results can and must be expected in applications.

B.   Hellmann's Method

Let us consider an atom or molecule with N core electrons in closed shells plus one valence electron. It was shown by Hellmann[19] that a pseudopotential can be derived using the TF model. (This method should not be confused with the Hellmann potential method described in Section 1.1.) First let us consider the core with the N electrons in the absence of the valence electron. Let the total energy of the core be represented by the density functional

$$\mathscr{E}_T = \int \mathscr{E}(\rho)\, dv,\tag{3.1}$$

where ρ is the electron density and $\mathscr{E}(\rho)$ is the energy-density defined as follows:

$$\mathscr{E}(\rho) = T(\rho) + V_n\rho + \frac{1}{2} V_e\rho + U_{X\alpha},\tag{3.2}$$

where $T(\rho)$ is the density of kinetic energy, V_n is the nuclear potential, V_e is the electrostatic potential

$$V_e = \int \frac{\rho(r')\, dv'}{|r - r'|},\tag{3.3}$$

and $U_{X\alpha}$ is the exchange energy

$$U_{X\alpha} = \alpha\rho^{4/3},\tag{3.4}$$

where we obtain Dirac's theoretical expression [Gombas I, p. 25] if we put

$$\alpha = -\frac{3}{4}\left(\frac{3}{\pi}\right)^{1/3} \text{a.u.} \qquad (3.5)$$

On putting the expression in Eq. (3.2) into Eq. (3.1) and varying $\mathcal{E}_T$ with respect to ρ under the subsidiary condition

$$\int \rho \, dv = N, \qquad (3.6)$$

we obtain the so called Thomas–Fermi–Dirac (TFD) statistical model [Gombas I, p. 76].

Let us assume that the density ρ_0 which minimizes $\mathcal{E}_T$ is available. In other words we assume that the total energy and density of the core are known. We now add the valence electron to the system and assume that the frozen core approximation is valid. Keeping the density of the core fixed we obtain for the total energy of the system

$$\mathcal{E}_T = \int \mathcal{E}(\rho_0 + \Delta\rho) \, dv. \qquad (3.7)$$

where $\Delta\rho$ is the valence electron density. Using Eqs. (3.2) and (3.3) we obtain easily

$$\int V_n(\rho_0 + \Delta\rho) \, dv = \int V_n \rho_0 \, dv + \int V_n \Delta\rho \, dv, \qquad (3.8)$$

and

$$\frac{1}{2}\int V_e(\rho_0 + \Delta\rho)[\rho_0 + \Delta\rho] \, dv = \frac{1}{2}\int\int \frac{[\rho_0(r') + \Delta\rho(r')][\rho_0(r) + \Delta\rho(r)] \, dv \, dv'}{|r - r'|}$$

$$= \frac{1}{2}\int V_e(\rho_0)\rho_0 \, dv + \int V_e(\rho_0) \Delta\rho \, dv, \qquad (3.9)$$

where we neglected the term quadratic in $\Delta\rho$. For the first and last term of Eq. (3.2) we use a Taylor expansion and we obtain, by keeping only the terms linear in $\Delta\rho$:

$$T(\rho_0 + \Delta\rho) = T(\rho_0) + \left(\frac{\partial T}{\partial \rho}\right)_{\rho_0} \Delta\rho, \qquad (3.10)$$

and

$$U_{X\alpha}(\rho_0 + \Delta\rho) = U_{X\alpha}(\rho_0) + \frac{3}{4}\alpha\rho_0^{1/3} \Delta\rho. \qquad (3.11)$$

Collecting Eqs. (3.8)–(3.11) we obtain

$$\mathscr{E}_T = \int \mathscr{E}(\rho_0 + \Delta\rho) \, dv = \int \mathscr{E}(\rho_0) \, dv + \int V_M \, \Delta\rho \, dv, \qquad (3.12)$$

where

$$V_M = V_n + V_e + \frac{3}{4}\alpha\rho_0^{1/3} + \left(\frac{\partial T}{\partial \rho}\right)_{\rho_0}. \qquad (3.13)$$

The second term of Eq. (3.12) is clearly the interaction energy between the core and the valence electron. Writing $\Delta\rho = |\psi|^2$ with ψ being the wave function of the valence electron we get

$$\mathscr{E}_T = \int \mathscr{E}(\rho_0) \, dv + \int \psi^* V_M \psi \, dv. \qquad (3.14)$$

The first three terms of V_M are the electrostatic and exchange potentials of the core. We postulate that the last term is the pseudopotential

$$V_p = \left(\frac{\partial T}{\partial \rho}\right)_{\rho_0}. \qquad (3.15)$$

Let us substitute for T the Fermi expression:

$$T = \frac{3}{10}(3\pi^2)^{2/3}\rho^{5/3}. \qquad (3.16)$$

Using this, we get from Eq. (3.15):

$$V_p = \frac{1}{2}(3\pi^2)^{2/3}\rho_0^{2/3}. \qquad (3.17)$$

In the TF model the connection between the Fermi sphere radius p_F and the density in given by

$$p_F = (3\pi^2)^{1/3}\rho^{1/3}. \qquad (3.18)$$

The maximum kinetic energy of the electron is

$$E_{\max} = \frac{p_F^2}{2m} = \frac{1}{2}(3\pi^2)^{2/3}\rho^{2/3}, \qquad (3.19)$$

where we have put $m = 1$. Comparing Eqs. (3.17) and (3.19) we get

$$V_p = E_{max} .$$ (3.20)

Hellmann's result[19] is that the pseudopotential for the valence electron of an atom or molecule is given by the expression

$$V_p = \left(\frac{\partial T}{\partial \rho}\right)_{\rho_0} .$$ (3.21)

where T is the density of the kinetic energy of the core. Using Eq. (3.16) for T we get the additional result that V_p is equal to the maximum kinetic energy of the core electrons.

Although this result was derived by Hellmann, the clear physical interpretation was given by Gombas[20] (see also [Gombas I, p. 150]). In the TF model the Pauli exclusion principle is taken into account by the requirement that each volume element of the size $h/2$ in the phase space can accomodate only one electron. The result of this will be that the electrons of the core will fill a sphere of radius p_F in the momentum space (Fermi sphere). Now if we add the valence electron to the core, in order to satisfy the Pauli exclusion principle the momentum of the valence electron must be greater than p_F. In terms of kinetic energy this means that the kinetic energy of the valence electron must be greater than $(p_F^2/2m)$. Using Gombas's phrase, the valence electron must be "raised" to an energy level greater than $(p_F^2/2m)$. This is approximately accomplished by adding to the electrostatic and exchange interaction potentials the expression given by Eq. (3.21) which, in the case when T is given by Eq. (3.16), is equal to $(p_F^2/2m)$. Therefore, since we have defined a pseudopotential as the replacement for the Pauli exclusion principle, it is justified to regard the expression given by Eq. (3.21) as a pseudopotential.

3.2. DERIVATION OF THE DENSITY-DEPENDENT PSEUDOPOTENTIAL FROM THE EXACT THEORY

A. Szepfalusy's Method

It is easy to show that density-dependent pseudopotentials can be derived from the exact theory.[4] Let us consider again an atom with N "frozen" core electrons plus one valence electron. The wave equation of the valence electron in the HF approximation is Eq. (2.15):

$$H_F \varphi = \epsilon \varphi ,$$ (3.22)

where H_F is given by

$$H_F = t + U, \tag{3.23}$$

with t being the operator of kinetic energy and U the HF potential. In the exact pseudopotential theory the wave equation of the valence electron is

$$(H_F + V_p)\psi = \epsilon\psi, \tag{3.24}$$

where V_p is the pseudopotential for which we chose the kinetic energy minimized expressed, Eq. (2.72). The connection between the HF orbital φ and the PO ψ is given by

$$\varphi = A_0\left(\psi - \sum_{i=1}^{N} \alpha_i\varphi_i\right), \tag{3.25}$$

where ψ and the φ_i's are normalized and A_0, which ensures the normalization of φ, is given by Eq. (2.11):

$$A_0 = \left\{1 - \sum_{i=1}^{N} |\alpha_i|^2\right\}^{-1/2}. \tag{3.26}$$

Multiply Eq. (3.22) from the left by φ^* and integrate; multiply Eq. (3.24) from the left by ψ^* and integrate. We obtain from the resultant expressions by subtraction

$$\langle V_p \rangle \equiv \langle \psi | V_p | \psi \rangle = \langle \varphi | H_F | \varphi \rangle - \langle \psi | H_F | \psi \rangle, \tag{3.27}$$

and taking into account that all orbitals are of central field type, we obtain, after integration over the angular parts,

$$\langle V_p \rangle = \int \hat{P}\left(-\frac{1}{2}\frac{d^2}{dr^2}\right)\hat{P}\, dr - \int P\left(-\frac{1}{2}\frac{d^2}{dr^2}\right)P\, dr$$

$$+ \int (\hat{P}^2 - P^2)\left(\frac{l(l+1)}{2r^2} + V_n + V_e\right) dr$$

$$+ \int \hat{P}V_X\hat{P}\, dr - \int P V_X P\, dr, \tag{3.28}$$

where V_n, V_e, and V_X are the nuclear, the electrostatic, and the exchange potentials of the core electrons, respectively. The radial parts of the HF orbital φ and of the pseudo-orbital ψ are denoted by $\hat{P}$ and P.

Let us define $\hat{\gamma}(r)$ and $\gamma(r)$ as follows:

$$\hat{\gamma}(r) \equiv -\frac{1}{2}\frac{d^2\hat{P}}{dr^2}\bigg/\hat{P}, \tag{3.29}$$

and

$$\gamma(r) \equiv -\frac{1}{2}\frac{d^2P}{dr^2}\bigg/ P.$$ (3.30)

Using these definitions we obtain

$$\langle V_p \rangle = \int \hat{P}^2 \hat{\gamma}\, dr - \int P^2 \gamma\, dr + \int (\hat{P}^2 - P^2) \left(\frac{l(l+1)}{2r^2} + V_n + V_e\right) dr$$

$$+ \int \hat{P}V_X\hat{P}\, dr - \int PV_XP\, dr.$$ (3.31)

This is still an exact expression. Now we introduce the approximation that V_X can be replaced by the statistical expression

$$V_{X\alpha} = \frac{4}{3}\alpha\rho^{1/3},$$ (3.32)

where the theoretical value of α is given by Eq. (3.5) and ρ is the density of the core electrons. Making the substitution* in Eq. (3.31) we obtain

$$\langle V_p \rangle = \int \hat{P}^2 \hat{\gamma}\, dr - \int P^2 \gamma\, dr$$

$$+ \int (\hat{P}^2 - P^2) \left(\frac{l(l+1)}{2r^2} + V_n + V_e + V_{X\alpha}\right) dr.$$ (3.33)

Szepfalusy[4] suggested that the structure of the pseudo-orbital is such that in a reasonable approximation one can put

$$P^2 \approx \hat{P}^2.$$ (3.34)

That this is indeed the case can be seen from Eq. (3.25). Eliminating the angular parts from Eq. (3.25) we obtain Eq. (2.28), which, in simplified notation, reads

$$\hat{P} = A_0\left(P - \sum_i \alpha_i \hat{P}\right).$$ (3.35)

---

* This step has only a slight effect on the argument. The replacement of the exact exchange operator by its statistical equivalent is probably a fairly good approximation considering the wide applications of the $X\alpha$ model. But even if it is not, the error committed is about the same in the two last terms of Eq. (3.31), so they cancel each other out to a considerable degree.

The bulk of both $\hat{P}$ and P will be outside of the core, in the valence area. In that area the core orbitals vanish and we have

$$\hat{P}^2 = A_0^2 P^2 . \tag{3.36}$$

As Appendix A shows the α_i coefficients are generally small, and therefore we can write

$$A_0^2 = \left\{1 - \sum_{i=1}^{N} |\alpha_i|^2\right\}^{-1} \approx \left\{1 + \sum_{i=1}^{N} |\alpha_i|^2\right\} \approx 1 , \tag{3.37}$$

and using this in Eq. (3.36) we get Eq. (3.34). Inside the core both $\hat{P}^2$ and P^2 are small, and qualitatively P^2 is an average of $\hat{P}^2$, from which we can again conclude the approximate validity of Eq. (3.34).

On using Eq. (3.34) in Eq. (3.33) we get

$$\langle V_p \rangle = \int (\hat{\gamma} - \gamma) P^2 \, dr = \langle \psi | \Phi_p | \psi \rangle , \tag{3.38}$$

where

$$\Phi_p(r) = \hat{\gamma}(r) - \gamma(r) . \tag{3.39}$$

Thus in the approximation in which the exchange operator can be replaced by the $X\alpha$ potential and in which Eq. (3.34) is valid, the pseudopotential can be written in the form of a local potential of the form of Eq. (3.39). From the definitions in Eqs. (3.29) and (3.30) we see that $\hat{\gamma}$ is the radial kinetic energy in the orthogonalized state and γ the radial kinetic energy in the nonorthogonal state. Indeed $\hat{\gamma}$ and γ are defined in such a way that the average values of these quantities with respect to $\hat{P}$ and P are exactly equal to the corresponding expectation values of the radial part of the kinetic energy operator.

In Eq. (3.39) the functions $\hat{\gamma}$ and γ are defined according to Eqs. (3.29) and (3.30) in terms of the exact $\hat{P}$ and P. Section 3.3 shows how these quantities can be transformed into density-dependent expressions.

In more recent work Topiol, Zunger, and Ratner[21] employ an argument which is similar to Szepfalusy's to define an exact pseudopotential. TZR replaced Eq. (3.22) by the valence electron equation of the Kohn–Sham theory[22] and set up the pseudopotential equation (Eq. (3.24)] in such a way that the density-dependent potentials in that equation depend on the pseudo-densities. The pseudopotential was defined by demanding that the eigenvalues of the exact equation, Eq. (3.22), and those of the pseudopotential equation, Eq. (3.24), be equal for all (n, l). Dividing Eq. (3.22) by φ and Eq. (3.24) by ψ and equating the eigenvalues, one obtains an explicit expression for V_p. The resulting expression is similar to Eq. (3.39), but it contains also the difference between the density-dependent potentials formed with the correct and pseudo-densities. Eliminating $\hat{\gamma}$ by using Eq. (3.22), one obtains an exact pseudopotential which will depend on the eigenvalue as well as on the pseudo-orbital. The resulting pseudopotential equation can be solved by iteration.

B.   The Author's Method

In the preceding section we have seen that, to the extent of a reasonable approximation, the pseudopotential can be expressed as a kinetic energy. We have also seen that for atoms the approximation of Eq. (3.34) is reasonably accurate; its validity, however, cannot be readily generalized for molecules. In order to obtain a density-dependent formula which is equally valid for atoms and molecules the author has shown[23] that one can arrive at a formula similar to Eq. (3.39) without the approximation of Eq. (3.34).

We assume now that the pseudopotential is a purely kinetic energy, basing this assumption on Hellmann's result, Eq. (3.21). We demand that the pseudopotential be defined in such a way that the total kinetic energy in the nonorthogonal state be approximately equal to the total kinetic energy in the orthogonalized state. Let the former be E_K and the latter $\hat{E}_K$. We demand that

$$E_K \approx \hat{E}_K. \tag{3.40}$$

Let $\hat{\gamma}$ and γ be the total kinetic energies in the orthogonalized and nonorthogonal states, respectively. We state that Eq. (3.40) will be satisfied if we put for the pseudopotential

$$\Phi_p = \hat{\gamma} - \gamma. \tag{3.41}$$

This can be proven easily. Put Eq. (3.41) into the wave equation Eq. (3.24):

$$(H_F + \Phi_p)\psi = (t + U + \hat{\gamma} - \gamma)\psi = \epsilon\psi. \tag{3.42}$$

Let us extract the kinetic energy from Eq. (3.42). We get

$$E_K = \langle\psi|t|\psi\rangle + \langle\psi|\Phi_p|\psi\rangle = \langle\psi|t|\psi\rangle + \langle\psi|\hat{\gamma}|\psi\rangle - \langle\psi|\gamma|\psi\rangle. \tag{3.43}$$

By definition of the γ we have

$$\langle\psi|\gamma|\psi\rangle \approx \langle\psi|t|\psi\rangle, \tag{3.44}$$

so we obtain

$$E_K \approx \langle\psi|\hat{\gamma}|\psi\rangle. \tag{3.45}$$

The kinetic energy of the orthogonalized orbital φ is

$$\hat{E}_K = \langle \varphi | t | \varphi \rangle . \tag{3.46}$$

If $\hat{\gamma}$ is properly constructed to represent the kinetic energy in the orthogonalized state, then

$$\langle \psi | \hat{\gamma} | \psi \rangle \approx \langle \varphi | t | \varphi \rangle = \hat{E}_K , \tag{3.47}$$

from which we obtain, using Eq. (3.45),

$$E_K \approx \langle \psi | \hat{\gamma} | \psi \rangle \approx \langle \varphi | t | \varphi \rangle = \hat{E}_K , \tag{3.48}$$

which proves our statement. Thus we have obtained for the atoms and molecules equally valid pseudopotential

$$\Phi_p = \hat{\gamma} - \gamma . \tag{3.49}$$

The previously derived expression [Eq. (3.39)] follows from Eq. (3.49) directly. For atoms the total kinetic energies appearing in Eq. (3.49) are equal to the sum of radial and azimuthal energies. The azimuthal energies are approximately equal in the orthogonalized and nonorthogonal states; therefore, they drop out from Eq. (3.49), leaving the expression of Eq. (3.39).

In connection with Eq. (3.49) we note that physical plausibility would require only the first term. We can see, however, that the second term must be subtracted if we consider again the wave equation Eq. (3.42):

$$\{t + U + \hat{\gamma} - \gamma\}\psi = \epsilon\psi . \tag{3.50}$$

If we demand that the kinetic energy in this equation be equal to the kinetic energy in the orthogonalized state, then the correct amount is given by $\hat{\gamma}$; the γ must be subtracted to compensate for the energy resulting from the operator t. The second term of Eq. (3.49) will be called the kinetic self-energy.[24] The expression Eq. (3.49) will be transformed into a density-dependent expression in Section 3.3.A.

3.3.   A SURVEY OF FORMULAS FOR DENSITY-DEPENDENT PSEUDOPOTENTIALS

A.   Pseudopotential for Atoms and Molecules

Starting from our three basic formulas, Eqs. (3.21), (3.39), and (3.49), we show in this section how various density-dependent expressions can be constructed. We have already seen how the pseudopotential in Eq. (3.17) can be obtained from Eq. (3.21). Next we turn to Eq. (3.49). We shall determine the form of $\hat{\gamma}$ and γ from the TF model.[23] Let the radius of the Fermi sphere which contains all electrons, core plus valence, be p and let the radius of the sphere containing only the core electrons be p_0. We define $\hat{\gamma}$ as

$$\hat{\gamma} = \frac{p^2}{2m}. \tag{3.51}$$

For the kinetic self-energy we put

$$\gamma = \frac{p_\varepsilon^2}{2m}, \tag{3.52}$$

where p_ε is the "momentum width" associated with the valence electron and is defined as

$$p_\varepsilon = p - p_0. \tag{3.53}$$

On using Eqs. (3.51) and (3.52) in Eq. (3.49) we obtain

$$\Phi_p = \hat{\gamma} - \gamma = \frac{1}{2m}(p^2 - p_\varepsilon^2). \tag{3.54}$$

Another useful form is obtained by substituting Eq. (3.53) into Eq. (3.54):

$$\Phi_p = \frac{1}{2m}(p_0^2 + 2p_0 p_\varepsilon). \tag{3.55}$$

The relationship between Fermi momentum and density is given by Eq. (3.18), from which we obtain

$$p_F^2 = (3\pi^2)^{2/3}\rho^{2/3}. \tag{3.56}$$

Let us define the densities ρ_0 and ρ in terms of the HF orbitals as follows:

$$\rho_0 = \sum_{i=1}^{N} |\varphi_i|^2, \tag{3.57}$$

and

$$\rho = \sum_{i=1}^{N} |\varphi_i|^2 + |\varphi|^2. \tag{3.58}$$

Using these densities we obtain from Eq. (3.56)

$$p^2 = (3\pi^2)^{2/3}\rho^{2/3} \,, \tag{3.59}$$

$$p_0^2 = (3\pi^2)^{2/3}\rho_0^{2/3} \tag{3.60}$$

and

$$p_\varepsilon^2 = (p - p_0)^2 = (3\pi^2)^{2/3}(\rho^{1/3} - \rho_0^{1/3})^2 \,. \tag{3.61}$$

On putting these into Eq. (3.54) we obtain

$$\Phi_p = \frac{1}{2}(3\pi^2)^{2/3}[\rho^{2/3} - (\rho^{1/3} - \rho_0^{1/3})^2] \,, \tag{3.62}$$

where we have put $m = 1$. Similarly we obtain from Eq. (3.55)

$$\Phi_p = \frac{1}{2}(3\pi^2)^{2/3}[\rho_0^{2/3} + 2\rho_0^{1/3}(\rho^{1/3} - \rho_0^{1/3})] \,. \tag{3.63}$$

Both expressions are dependent on the densities only. As we see, the density of the whole system ρ is needed to obtain Φ_p; that is, we need the valence orbital φ as well as the core orbitals. The situation is similar to the exact pseudopotential, where we likewise needed φ as well as the core orbitals for the calculation of the exact PO.

The two expressions, Eq. (3.62) and (3.63), are valid for atoms as well as molecules. We obtain Hellmann's formula, Eq. (3.17), by assuming that

$$p_\varepsilon \ll p_0 \,. \tag{3.64}$$

Omitting the second term in Eq. (3.55) as negligible relative to the first, we obtain

$$\Phi_p = \frac{1}{2}(3\pi^2)^{2/3}\rho_0^{2/3} \,, \tag{3.65}$$

which is identical with Eq. (3.17) since ρ_0 is the density of the core.

B. Pseudopotentials for Atoms

We consider now an atom with closed shells plus one valence electron. In this case we have a central field problem and can use Eq. (3.39), which we recall here:

$$\Phi_p = \hat{\gamma}(r) - \gamma(r) \,, \tag{3.66}$$

where $\hat{\gamma}$ and γ are radial kinetic energies. We transform this expression into a density-dependent potential using the author's derivation[25], which is based on two earlier derivations of Gombas.[26,27]

Let us consider the Fermi momenta p and p_0 which we have defined in the preceding section. Let p_r and p_φ be the radial and azimuthal components of p, and p_{0r} and $p_{0\varphi}$ the corresponding components of p_0. We put, in accordance with the definitions of $\hat{\gamma}(r)$ and $\gamma(r)$,

$$\hat{\gamma} = \frac{p_r^2}{2m},$$ (3.67)

and

$$\gamma = \frac{p_{\varepsilon r}^2}{2m} = \frac{(p_r - p_{0r})^2}{2m},$$ (3.68)

in complete analogy to Eqs. (3.51) and (3.52). $P_{\varepsilon r}$ is the "radial momentum width" associated with the valence electron. On using Eqs. (3.67) and (3.68) in Eq. (3.66) we obtain

$$\Phi_p = \frac{1}{2m}(p_r^2 - p_{\varepsilon r}^2).$$ (3.69)

The connection between the radial momenta p_r and p_{0r} and the density was established by Gombas.[26] In Figure 3.1 we have the subdivision of the momentum space according to Fermi.[28] Placed at the tip of the position vector r we have the Fermi sphere with the radius p_F. The radial and azimuthal components of p_F are p_r and p_φ. Fermi introduced angular momentum quantization into the TF model by putting

$$M = rp_\varphi = k\hbar,$$ (3.70)

where M is the angular momentum and k is the quantum number associated with M. (The exact quantum mechanical value is, of course, $k = \sqrt{l(l+1)}$.) From Eq. (3.70) we get

$$p_\varphi = k\frac{\hbar}{r}.$$ (3.71)

From this we see that the electrons with quantum number k will occupy in the momentum space the points on the surface of a cylinder the base of which is the circle with radius p_φ and the height of which is p_r. The electrons with quantum number between k and $k + dk$ will occupy the volume cut out of the Fermi sphere by the clinders with radius p_φ and $p_\varphi + dp_\varphi$. This volume

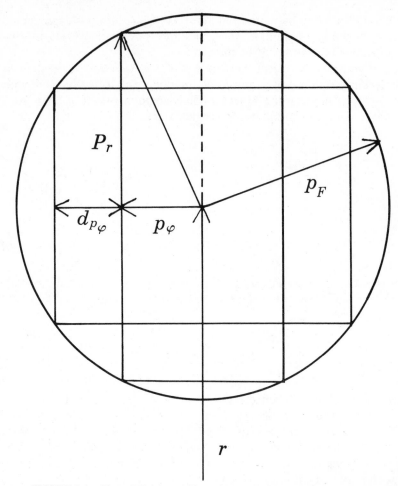

FIGURE 3.1.　The subdivision of the momentum space according to Fermi.

is given by

$$dv_p = (2\pi p_\varphi \, dp_\varphi)2p_r. \tag{3.72}$$

Let the volume in the coordinate space be dv and the number of electrons in it dN. Then we get, according to the Fermi–Dirac statistics,

$$dN = \frac{2 \, dv_p \, dv}{h^3}. \tag{3.73}$$

Let ρ be the electron density. Then

$$\rho = \frac{dN}{dv} = \frac{2 \, dv_p}{h^3} = \frac{p_\varphi(dp_\varphi)p_r}{\pi^2\hbar^3}. \tag{3.74}$$

Using Eq. (3.71) we get

$$\rho = k \frac{\hbar}{r} \, dk \, \frac{\hbar}{r} \frac{p_r}{\pi^2 \hbar^3} = k \, dk \, \frac{p_r}{\pi^2 r^2 \hbar} \, . \tag{3.75}$$

For k we put, according to Fermi,[28]

$$k = l + \frac{1}{2}, \quad (l = 0, 1, \ldots), \tag{3.76}$$

and putting $dk = 1$ we get

$$\rho = \frac{(2l+1)p_r}{2\pi^2 r^2 \hbar} \, . \tag{3.77}$$

Let D be the radial density $D = 4\pi r^2 \rho$; then we get

$$D = \frac{2}{\pi} \frac{(2l+1)p_r}{\hbar} \, , \tag{3.78}$$

and switching to atomic units we obtain Gombas's result:[26]

$$p_r = \frac{\pi}{2} \frac{D_l}{(2l+1)} \, , \tag{3.79}$$

where we have attached the index l to the D, indicating that this is the radial density of the electrons with quantum number l among those which fill the Fermi sphere with radius p_F.

We use now Eq. (3.79) to introduce the radial densities into Eq. (3.69). Let D_{0l} be the radial density of l-electrons in the core and D_l the radial density of l-electrons in the whole atom—that is, we define in accordance with Eq. (3.58)

$$D_l = D_{0l} + \hat{P}_{nl}^2 \, , \tag{3.80}$$

where $\hat{P}_{nl}$ is the radial part of the HF valence orbital. Then we have according to the definitions of p_r and p_{0r}.

$$p_{0r} = \frac{\pi}{2} \frac{D_{0l}}{(2l+1)} \, , \tag{3.81}$$

and

$$p_r = \frac{\pi}{2} \frac{D_l}{(2l+1)} = \frac{\pi}{2} \frac{(D_{0l} + \hat{P}_{nl}^2)}{(2l+1)} \, . \tag{3.82}$$

Using these expressions we get

$$p_r^2 = \frac{\pi^2}{4} \frac{(D_{0l}^2 + 2D_{0l}\hat{P}_{nl}^2 + \hat{P}_{nl}^4)}{(2l+1)^2}$$

(3.83)

and

$$p_{er}^2 = (p_r - p_{0r})^2 = \frac{\pi^2}{4} \frac{\hat{P}_{nl}^4}{(2l+1)^2}.$$

(3.84)

Substituting these into Eq. (3.69) we get the pseudopotential:[25]

$$\Phi_l = \frac{1}{2m}(p_r^2 - p_{er}^2) = \frac{\pi^2}{8} \frac{(D_{0l}^2 + 2D_{0l}\hat{P}_{nl}^2)}{(2l+1)^2}.$$

(3.85)

According to the derivation this is the pseudopotential which must be placed into the wave equation of the valence electron as a replacement for the Pauli exclusive principle. Since Φ_l is l-dependent, we can put the pseudopotential in the form of a semilocal potential.[12]

Let Ω_l be the angular momentum projection operator

$$\Omega_l = \sum_{m=-l}^{+l} |Y_{lm}\rangle\langle Y_{lm}|.$$

(3.86)

Using this operator, we can put

$$\Phi = \sum_{l=0}^{\infty} \Phi_l \Omega_l,$$

(3.87)

where $\Phi_l = 0$ for those l values which do not occur in the core. As we see from Eq. (3.85), the potential Φ_l is also n-dependent through the $\hat{P}_{nl}^2$. This feature, which is not indicated in Eq. (3.87), must be taken into account in such a way that, after the operator Ω_l has selected the proper Φ_l, the principal quantum number n must also be adjusted to the n value of the PO on which we operate.

The wave equation [Eq. (3.24)] now becomes

$$(H_F + \Phi)\psi = \epsilon\psi$$

(3.88)

where Φ is given by Eqs. (3.85), (3.86), and (3.87).

At this point, we elucidate the discussion by presenting the graph of the modified potential for the 3s state of the Na valence electron. In Figure 3.2 we have the modified potential constructed with the density-dependent

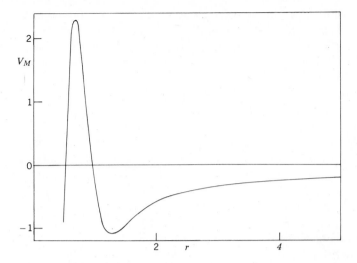

FIGURE 3.2. The density-dependent modified potential for the $3s$ state of the Na valence electron.

pseudopotential [Eq. (3.85)]. In the graph, the HF potential is the same as in Figure 2.2, while the $3s$ pseudopotential is constructed from Eq. (3.85), D_{0l} and $\hat{P}_{nl}^2$ being the HF radial densities.[3] As we see from comparing the diagram with Figures 2.2 and 2.3, the density-dependent modified potential has the same general form as the exact potentials. We have here again the typical barrier/well/Coulombic structure.

As we have mentioned above, prior to the derivation of Eq. (3.85) by the author,[25] two similar formulas were derived by Gombas. The first formula is[26]

$$G_l^{(1)} = \frac{\pi^2}{8} \frac{D_{0l}^2}{(2l+1)^2} + \frac{1}{8r^2} \tag{3.89}$$

and the second is[27]

$$G_l^{(2)} = \Phi_l + \frac{1}{8r^2} \tag{3.90}$$

[See Gombas II, p. 66]. At the time of the derivation of $G_l^{(1)}$ it was not yet recognized that, according to Eq. (3.66), the kinetic self-energy γ must be subtracted from the orthogonalization energy $\hat{\gamma}$. Hence the missing second term of Eq. (3.85). The potential $G_l^{(2)}$ was derived by taking the self-energy into account. In both potentials we have the term $(1/8r^2)$ which is an azimuthal-type kinetic energy. It was pointed out by Phillips and Kleinmann[8] that the inclusion of a term of this type in the pseudopotential is a mistake. The correctness of this statement is evident for at least two reasons. First, we have emphasized in the discussion of the exact pseudopotentials that, for a valence electron outside closed shells, the pseudopotential is a radial kinetic energy. Second, the derivation of Eq. (3.85) which we have presented here clearly shows that there is no necessity for the inclusion of an azimuthal term.

Comparing Eqs. (3.89) and (3.90) with Eq. (3.85), it is clear that, except for the azimuthal term, the formulas are identical. The main strength of the author's derivation, however, is that, in constrast to the work of Gombas, the author tied the density-dependent potential to the exact theory through Eq. (3.66) which, as we have seen in Section 3.2.A, is derived from the exact theory. Only with this step has this model acquired a sound theoretical justification.

The pseudopotential of Eq. (3.85) can be brought to a form in which only the total (as opposed to radial) densities occur.[23] In order to show this, we refer to Figure 3.1, from which we see that the momenta p_F, p_r, and p_φ satisfy the relationship

$$p_F^2 = p_r^2 + p_\varphi^2 . \tag{3.91}$$

From this we get

$$p_r^2 = p_F^2 - p_\varphi^2 . \tag{3.92}$$

The radial "momentum width" $p_{\varepsilon r}$ was defined as

$$p_{\varepsilon r} = p_r - p_{0r} \tag{3.93}$$

while the "total momentum width" was, according to Eq. (3.53),

$$p_\varepsilon = p - p_0 . \tag{3.94}$$

It is physically plausible as well as provable by a simple derivation that we can put

$$p_\varepsilon \approx p_{\varepsilon r} . \tag{3.95}$$

Using Eqs. (3.92) and (3.95) in Eq. (3.69) we obtain

$$\Phi_p = \frac{1}{2m} [p_F^2 - p_\varphi^2 - p_\varepsilon^2] . \tag{3.96}$$

Using Eq. (3.59) for p_F^2 and Eq. (3.61) for p_ε^2 we obtain

$$\Phi_p = \frac{1}{2} (3\pi^2)^{2/3} [\rho^{2/3} - (\rho^{1/3} - \rho_0^{1/3})^2] - \frac{p_\varphi^2}{2m} . \tag{3.97}$$

Apart from the last term, this is already a total-density-dependent expression. For the azimuthal term we could use Eq. (3.71), but that would bring into Φ_p an undesirable long-range term proportional to $1/r^2$. The difficulty can be overcome, as always, by falling back on the exact theory.[23] The presence of this azimuthal term is the consequence of the PO having the same azimuthal part as the HF valence orbital; that is, it is the consequence of the PO being orthogonal to certain core orbitals because of its angular part.

To formulate the argument in general terms, let ψ be orthogonal to the lowest m core orbitals $\varphi_1 \cdots \varphi_m$ because of symmetry; that is, let

$$\langle \psi \mid \varphi_k \rangle = 0, \quad (k = 1, 2, \ldots, m). \tag{3.98}$$

Translating this into the language of the TF model, we may say that the electron does not need to be raised from the origin of the momentum space to p_F, but only from the Fermi sphere which contains the lowest m orbitals to p_F. Let p_{00} be the Fermi momentum associated with the orbitals $\varphi_1 \cdots \varphi_m$. Then we put

$$\frac{p_\varphi^2}{2m} \approx \frac{p_{00}^2}{2m}, \tag{3.99}$$

which means that we replace the left side by the right side on the basis of physical plausibility. Now put

$$\rho_{00} \equiv \sum_{i=1}^{m} |\varphi_i|^2, \tag{3.100}$$

and using Eq. (3.18) we get

$$\frac{p_\varphi^2}{2m} \approx \frac{1}{2} (3\pi^2)^{2/3} \rho_{00}^{2/3}. \tag{3.101}$$

On putting this into Eq. (3.97) we obtain:[23]

$$\Phi_p = \frac{1}{2} (3\pi^2)^{2/3} [\rho^{2/3} - \rho_{00}^{2/3} - (\rho^{1/3} - \rho_0^{1/3})^2], \tag{3.102}$$

which is now an expression dependent only on (total) densities.

We get another useful form by rewriting Eq. (3.96) with the aid of Eq. (3.94):

$$\Phi_p = \frac{1}{2m} [p_0^2 - p_\varphi^2 + 2p_0 p_\varepsilon]. \tag{3.103}$$

Using Eq. (3.60) for p_0, Eq. (3.101) for the azimuthal term, and Eq. (3.61) for the p_ε, we obtain

$$\Phi_p = \frac{1}{2}(3\pi^2)^{2/3}[\rho_0^{2/3} - \rho_{00}^{2/3} + 2\rho_0^{1/3}(\rho^{1/3} - \rho_0^{1/3})] \,. \qquad (3.104)$$

If we assume that $p_\varepsilon \ll p_0$, we can omit the third term as negligible relative to the first two and obtain

$$\Phi_p = \frac{1}{2}(3\pi^2)^{2/3}[\rho_0^{2/3} - \rho_{00}^{2/3}] \,. \qquad (3.105)$$

This expression was derived first in the pioneer work of Gombas.[20] [See also Gombas I, p. 150.] In order to see the significance of this formula we recall that one of the results of the exact pseudopotential theory was that, for a valence electron of an atom outside of closed shells, the pseudopotential is *l*-dependent. The expression in Eq. (3.105), which was derived by Gombas long before the development of exact theory, was the first pseudopotential which shows the correct *l*-dependence. Indeed, looking at the definition of ρ_{00}, Eq. (3.100), we can deduce how Φ_p depends on the azimuthal quantum number. Let the valence electron state be (nl). Let $(n'l)$ be the core state with the lowest energy among the *l* electrons. Then we can write the summation in Eq. (3.100), consistently with the definition of ρ_{00}, as follows:

$$\rho_{00} = \sum_{l''=0}^{l-1} \sum_{n''=l''+1}^{n'} q(n''l'')\hat{P}_{n''l''}^2 \,, \qquad (3.106)$$

where $q(n''l'')$ is the occupation number of the core electron state $(n''l'')$. Since evidently $n' = l + 1$, we can write

$$\rho_{00} = \sum_{l''=0}^{l-1} \sum_{n''=l''+1}^{l+1} q(n''l'')\hat{P}_{n''l''}^2 \,. \qquad (3.107)$$

If the valence electron is in an s state we have $l = 0$. Then $\rho_{00} = 0$. For a p state $l = 1$ and $l'' = 0$, with n'' taking the values 1 and 2. For a d state $l = 2$ and l'' will take the values 0 or 1. We get $n'' = 1, 2, 3$ for $l'' = 0$ and $n'' = 2, 3$ for $l'' = 1$. Thus we obtain the densities as follows: for an s-electron $\rho_{00} = 0$; for a p electron ρ_{00} contains $(1s)^2(2s)^2$; and for a d electron ρ_{00} contains all electrons up to $3p$. Therefore ρ_{00} will contain all orbitals among the lowest-lying states to which the valence electron PO is orthogonal by symmetry.

Since Eq. (3.107) clearly shows that ρ_{00} depends on l we can put Eq. (3.105) in the form of Eq. (3.87) where now

$$\Phi_l = \frac{1}{2}(3\pi^2)^{2/3}[\rho_0^{2/3} - \rho_l^{2/3}],\qquad (3.108)$$

and ρ_l is the expression in Eq. (3.107).

While the significance of Gombas's formula for the development of the pseudopotential theory is very clear, we note that the author's derivation leading to Eqs. (3.102) and (3.104) has the merit of tying the density-dependent potentials to the exact theory and strengthening thereby their theoretical justification.

We note that for s states of an atom, Eq. (3.104) reads

$$\Phi_{l=0} = \frac{1}{2}(3\pi^2)^{2/3}[\rho_0^{2/3} + 2\rho_0^{1/3}(\rho^{1/3} - \rho_0^{1/3})].\qquad (3.109)$$

This formula, which is also obtainable from Eq. (3.63) if we apply the latter to the s state of an atom, was obtained by Gombas,[29] for this special case, prior to the authors derivations.[23,25]

An interesting form of a density-dependent pseudopotential operator was suggested by Gombas and Kisdi.[30] We recall that according to Eq. (2.17) the exact pseudopotential can be written as the operator V_p, the kernel of which is

$$K(r_1, r_2) = \sum_{i=1}^{N} (\epsilon - \epsilon_i)\varphi_i(r_1)\varphi_i^*(r_2).\qquad (3.110)$$

If we replace the core orbitals by plane waves, the orbital energies by $p/2m$, and the valence energy ϵ by $p_F/2m$, the summation becomes an integration in the momentum space with the upper limit p_F. The result of this integration is

$$K(r_1, r_2) = \frac{3}{\pi^2}\frac{p_F^5}{x^5}\left[\sin x - x\cos x - \frac{1}{3}x^2\sin x\right],\qquad (3.111)$$

where

$$x = p_F|r_1 - r_2|.\qquad (3.112)$$

We recall that according to Eq. (3.18),

$$p_F = (3\pi^2)^{1/3}\rho^{1/3}.\qquad (3.113)$$

Thus K is the kernel of a density-dependent operator and V_p operates on the PO as follows:

$$V_p(\boldsymbol{r}_1)\psi(\boldsymbol{r}_1) = \int K(\boldsymbol{r}_1, \boldsymbol{r}_2)\psi(\boldsymbol{r}_2)\, dv_2 \,, \qquad (3.114)$$

where K is given by Eq. (3.111).

C. Pseudopotential for Cylindrical Symmetry

Let us consider a molecule with cylindrical symmetry which has one electron outside closed shells. We want to derive a pseudopotential which, when placed in the wave equation, replaces the Pauli exclusion principle. We present here a modified version of the work of Szondi,[31] who has shown that a density-dependent potential can be formulated for this case.

Let R, φ, z be the cylindrical coordinates, with the z axis coinciding with the axis of symmetry. Let p_R, p_φ, and p_z be the cylindrical coordinates of the momentum. In a potential field which has cylindrical symmetry, only the z-component of the angular momentum is quantized. We have

$$L_z = \lambda\hbar\,, \quad (\lambda = 0, \pm 1, \pm 2, \ldots)\,. \qquad (3.115)$$

Since $L_z = Rp_\varphi$ we get the quantization of the azimuthal component of the momentum

$$p_\varphi = \lambda\hbar/R\,. \qquad (3.116)$$

We postulate that if we omit the kinetic self-energy then the pseudopotential is

$$\Phi_p = \frac{1}{2m}\,(p_F^2 - p_\varphi^2)\,, \qquad (3.117)$$

which is set up in analogy to Eqs. (3.69) and (3.92).

In Figure 3.3 we have the Fermi sphere with radius p_F and the momentum components p_R, p_φ, and p_z forming the coordinate axes. From Eq. (3.116) we see that, for a given R, the electrons with quantum number λ will occupy points of a plane perpendicular to the p_φ axis at a distance of p_φ from the origin. The electrons with quantum number between λ and $\lambda + d\lambda$ will occupy the volume cut out of the Fermi sphere by two planes which are at the distance p_φ and $p_\varphi + dp_\varphi$ from the origin (shaded area). (The electrons with quantum number $-\lambda$ will occupy the corresponding points on the opposite side of the Fermi sphere.) The volume of the shaded area is

$$dv_p = (p_{zm}^2\,\pi)\, dp_\varphi\,, \qquad (3.118)$$

where p_{zm} is the maximum of the z component for the shaded area. Let dv be the volume in the coordinate space and dN the number of λ-electrons in

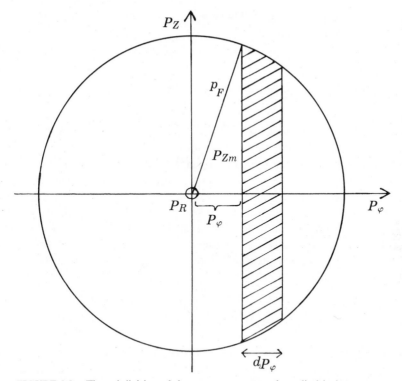

FIGURE 3.3.   The subdivision of the momentum space for cylindrical symmetry.

it. Then, according to the TF model, we have

$$dN = \frac{dv_p \, dv}{(h^3/2)},$$ (3.119)

and we get for the electron density

$$\rho_\lambda = \frac{dN}{dv} = \frac{2v_p}{h^3} = \frac{2}{h^3} p_{zm}^2 \pi \, dp_\varphi = \frac{p_{zm}^2 \, d\lambda}{4\pi^2 \hbar^2 R},$$ (3.120)

where we used Eq. (3.116). Putting $d\lambda = 1$, we get the connection between p_{zm} and ρ_λ:

$$p_{zm} = (4\pi^2 R)^{1/2} \rho_\lambda^{1/2},$$ (3.121)

where we have put $\hbar = 1$. This relationship is analogous to Eq. (3.79). On using Eq. (3.121) in Eq. (3.117) we obtain the result for the pseudopotential

$$\Phi_p = \frac{1}{2m} (p_F^2 - p_\varphi^2) = \frac{1}{2m} p_{zm}^2 = (2\pi^2 R)\rho_\lambda,$$ (3.122)

where we have put $m = 1$, and where ρ_λ is the density of all core electrons with the quantum number λ.

Our result differs from Szondi's in that he obtained the expression[31]

$$\Phi_p = (2\pi^2 R)\rho_\lambda + \frac{\lambda^2 + (1/12)}{2R^2}. \qquad (3.123)$$

The difference between our expression [Eq. (3.122)] and Szondi's potential [Eq. (3.123)] is the additional $1/R^2$ term in the latter. Our derivation is consistent with the derivation of Eq. (3.85), while Eq. (3.123) is analogous to Eq. (3.89). In discussing Eq. (3.89) we pointed out that a consistent derivation does not yield the $1/r^2$ term. The same remark applies here; a derivation which is constructed in such a way as to be consistent with the derivation of Eq. (3.85) will not contain the additional $1/R^2$ term. The decisive requirement is consistency with the exact theory; since Eq. (3.85) was derived from the exact theory, our derivation leading to Eq. (3.122) is also indirectly tied to the exact theory.

3.4. THE VIRIAL THEOREM

In this section we show that properly constructed density dependent pseudopotentials satisfy the virial theorem. As a representative example, valid for molecules as well as atoms, we have chosen Eq. (3.62); from among those which are valid only for spherically symmetric atoms, we have chosen Eq. (3.85). (The discussion would be similar for the other potentials.) We start with the wave equation [Eq. (3.42)], which we recall here:

$$(H_F + \Phi_p)\psi = (t + U + \Phi_p)\psi = \epsilon\psi. \qquad (3.124)$$

The operators t and U are the kinetic energy and the HF potential, respectively. We assume that depending on the choice of Φ_p, Eq. (3.124) is valid for an atom or a molecule.

Let us consider first a molecule, and let Φ_p be the potential of Eq. (3.62). We assume that Eq. (3.124) has been solved with fixed core orbitals $\varphi_1 \cdots \varphi_N$ and orbital parameters $\epsilon_1 \cdots \epsilon_N$. Let ψ and ϵ be the solutions of Eq. (3.124), that is, the quantities which make the expression

$$\epsilon = \langle\psi|H_F|\psi\rangle + \langle\psi|\Phi_p|\psi\rangle \qquad (3.125)$$

to an absolute minimum. Let (xyz) be the coordinates of the valence electron, and $X_1 Y_1 Z_1 \cdots X_p Y_p Z_p$ be the coordinates of the fixed nuclei. Since we are working in the framework of the Born–Oppenheimer approximation, all orbitals, the core orbitals as well as ψ, will depend on the nuclear coordinates: for example, we shall have

$$\psi = \psi(x, y, z, X_1, Y_1, Z_1, \ldots, X_p, Y_p, Z_p) \,. \qquad (3.126)$$

We assume that all orbitals are normalized regardless of the values of the nuclear coordinates; that is, we have

$$\int |\psi|^2 \, dv = 1 \,, \qquad (3.127)$$

and similar equations for the core orbitals.

We apply the scaling method of Fock.[32] [Slater IV, p. 289]. Let the scaled PO be

$$\psi^\lambda = \lambda^{3/2} \psi(\lambda r, \lambda R) \,, \qquad (3.128)$$

where r stands for the triplet (xyz) and R indicates all nuclear coordinates. If Eq. (3.127) is valid, then ψ^λ is normalized for any λ. Let us now subject all orbitals which occur in Eq. (3.125) to scaling, and let the resulting expression be ϵ^λ. We obtain

$$\epsilon^\lambda = \epsilon_F^\lambda + \epsilon_p^\lambda \,, \qquad (3.129)$$

where

$$\epsilon_F = \langle \psi | H_F | \psi \rangle \,, \qquad (3.130)$$

and

$$\epsilon_p = \langle \psi | \Phi_p | \psi \rangle \,. \qquad (3.131)$$

Both ϵ_F and ϵ_p depend on the nuclear coordinates

$$\epsilon_F = \epsilon_F(R) \,, \qquad (3.132)$$

and

$$\epsilon_p = \epsilon_p(R) \,. \qquad (3.133)$$

If the orbitals are subjected to scaling then it is known from the HF theory that

$$\epsilon_F^\lambda = \lambda^2 E_k(\lambda R) + \lambda E_p(\lambda R) \,, \qquad (3.134)$$

where E_K and E_p are the kinetic and potential part of the energy; since they are, in general, functions of R, the scaling makes them functions of (λR).

Next consider Eq. (3.131) with Φ_p being given by Eq. (3.62):

$$\Phi_p = \frac{1}{2}(3\pi^2)^{2/3}[\rho^{2/3} - (\rho^{1/3} - \rho_0^{1/3})^2] \,, \tag{3.135}$$

where, according to Eqs. (3.57) and (3.58), we have

$$\rho_0 = \sum_{i=1}^{N} |\varphi_i|^2 \,, \tag{3.136}$$

and

$$\rho = \rho_0 + |\varphi|^2 \,. \tag{3.137}$$

These densities are functions of $\boldsymbol{R}$ and will be normalized for any values of $\boldsymbol{R}$. From Eq. (3.128) it is evident that the scaled densities will be

$$\rho_0^\lambda = \lambda^3 \rho_0(\lambda r, \lambda \boldsymbol{R}) \,, \tag{3.138}$$

and

$$\rho^\lambda = \lambda^3 \rho(\lambda r, \lambda \boldsymbol{R}) \,. \tag{3.139}$$

Use Eqs. (3.138) and (3.139) in Eq. (3.135), and put the resulting expression into ϵ_p^λ. We obtain

$$\begin{aligned}
\epsilon_p^\lambda &= \langle \psi^\lambda | \Phi_p(\lambda) | \psi^\lambda \rangle \\
&= A \int \psi^{\lambda *} \{(\rho^\lambda)^{2/3} - [(\rho^\lambda)^{1/3} - (\rho_0^\lambda)^{1/3}]^2\} \psi^\lambda \, dv \,,
\end{aligned} \tag{3.140}$$

where A is the constant in Eq. (3.135). Using Eqs. (3.128), (3.138), and (3.139), we obtain

$$\begin{aligned}
\epsilon_p^\lambda = \lambda^2 A \int \psi^*(\lambda r, \lambda \boldsymbol{R}) \{&[\rho(\lambda r, \lambda \boldsymbol{R})]^{2/3} \\
&- ([\rho(\lambda r, \lambda \boldsymbol{R})]^{1/3} - [\rho_0(\lambda r, \lambda \boldsymbol{R})]^{1/3})^2\} \\
&\times \psi(\lambda r, \lambda \boldsymbol{R}) \, d(\lambda r) = \lambda^2 \epsilon_p(\lambda \boldsymbol{R}) \,.
\end{aligned} \tag{3.141}$$

On using Eqs. (3.141) and (3.134) in Eq. (3.129), we obtain

$$\epsilon^\lambda = \epsilon_F^\lambda + \epsilon_p^\lambda = \lambda^2(E_k + \epsilon_p) + \lambda E_p \,. \tag{3.142}$$

Now differentiate ϵ^λ with respect to λ:

$$\frac{\partial \epsilon^{\lambda}}{\partial \lambda} = 2\lambda (E_k + \epsilon_p) + E_p + \sum_k \frac{\partial \epsilon^{\lambda}}{\partial (\lambda X_k)} X_k, \tag{3.143}$$

where we have denoted one of the nuclear coordinates by X_k so that $k = 1, 2, \ldots, 3p$. In Eq. (3.143) the ϵ^{λ} in the last term is given by Eq. (3.129) and, by definition of that expression, it depends on $(\lambda \boldsymbol{R})$.

Now since ψ is making ϵ to an absolute minimum we must have

$$\lim_{\lambda=1} \left(\frac{\partial \epsilon^{\lambda}}{\partial \lambda} \right) = 0, \tag{3.144}$$

and we obtain from Eq. (3.143)

$$2(E_k + \epsilon_p) + E_p = -\sum_k X_k \frac{\partial \epsilon}{\partial X_k}, \tag{3.145}$$

which is the virial theorem. At the position where the ϵ has its minimum in terms of the nuclear positions, we have

$$2(E_k + \epsilon_p) = -E_p, \tag{3.146}$$

and since

$$E_p = \epsilon - (E_k + \epsilon_p), \tag{3.147}$$

we obtain

$$\epsilon = -(E_k + \epsilon_p) = \frac{E_p}{2}. \tag{3.148}$$

Thus our result is that the kinetic energy associated with the pseudo-orbital state ψ will be $(E_k + \epsilon_p)$; and this quantity satisfies the virial theorem. In Section 3.2.B we have seen that the density-dependent pseudopotentials are constructed in such a way that the quantity $(E_k + \epsilon_p)$ should accurately approximate the kinetic energy in the original orthogonalized HF state φ. The HF state φ satisfies the virial theorem; that is, the kinetic energy in the orthogonalized state is the negative of the orbital energy. Equation (3.148) says that if $(E_k + \epsilon_p)$ is a good approximation to the kinetic energy of the HF state, then the orbital energy ϵ computed from Eq. (3.124) will be a good approximation to the HF orbital energy. Therefore it is accurate to say that the link between the exact and density-dependent pseudopotentials is strengthened by the latter's satisfying the virial theorem.

Turning to the potential in Eq. (3.85) we observe that that expression is valid only for atoms; that is, the nuclear coordinates do not occur in the derivation. Let ϵ_p and ϵ_p^λ be the unscaled and scaled expectation values of Φ_l. In order to show that the virial theorem is satisfied it is enough to show that

$$\epsilon_p^\lambda = \lambda^2 \epsilon_p . \tag{3.149}$$

The orbital $\hat{P}_{nl}$ and the radial density D_{0l} will be scaled as follows:

$$\hat{P}_{nl}^\lambda = \lambda^{1/2} \hat{P}_{nl}(\lambda r) , \tag{3.150}$$

and

$$D_{0l}^\lambda = \lambda D_{0l}(\lambda r) . \tag{3.151}$$

Using these expressions we obtain for the pseudopotential [Eq. (3.85)]:

$$
\begin{aligned}
\epsilon_p^\lambda = \langle \psi^\lambda | \Phi_l^\lambda | \psi^\lambda \rangle &= B \int \psi^{\lambda *} \{ (D_{0l}^\lambda)^2 + 2(D_{0l}^\lambda)(\hat{P}_{nl}^\lambda)^2 \} \psi^\lambda \, dv \\
&= \lambda^2 B \int \psi^*(\lambda r) \{ [D_{0l}(\lambda r)]^2 + 2[D_{0l}(\lambda r)][\hat{P}_{nl}(\lambda r)]^2 \} \psi(\lambda r) \, d(\lambda r) \\
&= \lambda^2 \epsilon_p ,
\end{aligned}
\tag{3.152}
$$

where B is the constant in Eq. (3.85). Thus we see that Eq. (3.149), and with it the virial theorem, are again satisfied.

3.5.   TEST CALCULATIONS

Since the density-dependent pseudopotentials are models, we now ask: how accurate are they? Calculations for answering this question were carried out by Gombas and his collaborators[33] [See also Gombas I, p. 206], and more recently by Schwarz.[34] In these calculations the energy of the valence electron of an atom was calculated in the frozen core approximation, using a modified potential similar to that given by Eq. (3.13); that is, the potential used was

$$V_M = V_n + V_e + \Phi_l + V_{X\alpha} , \tag{3.153}$$

where V_n is the Coulomb potential of the nucleus, V_e is the electrostatic potential [Eq. (3.3)], Φ_l is the pseudopotential, and $V_{X\alpha}$ is the $X\alpha$-type exchange potential [Eq. (3.32)], with α given by Eq. (3.5). For Φ_l, the expression [Eq. (3.105)] was used by Gombas, and in the calculations of Schwarz the expressions given by Eqs. (3.105), (3.89), and (3.90) were tested. In addition, Schwarz used a modified form of Eq. (3.90) in which an

TABLE 3.1. The Results of the Calculations with Density-Dependent Pseudopotentials. (All Energies in eV Units.)

		Eq. (3.105)	Eq. (3.105)	Eq. (3.89)	Eq. (3.90)	Eq. (3.90)+ Eq. (3.154)	
Atom	State	Single Zeta	STO	STO	STO	STO	HF
Li	2s		6.91	6.25	6.10	5.36	5.34
	2p		3.59	3.59	3.59	3.59	3.50
	3s		2.37	2.20	2.18	2.01	2.01
	3p		1.52	1.52	1.52	1.52	1.54
Na	3s	4.92	5.09	6.08	5.91	5.19	4.95
	3p	3.02	3.17	3.17	3.14	2.98	2.97
	3d	1.505					1.51
	4s	1.92	1.96	2.15	2.12	1.96	1.90
	4p	1.33	1.44	1.44	1.44	1.38	1.37
K	4s	4.21	4.33	5.01	4.82	4.22	4.01
	4p	2.76	2.78	2.93	2.88	2.68	2.60
	4f	0.85					
	5s		1.79	1.90	1.86	1.71	1.66
	5p		1.30	1.35	1.33	1.27	1.24
	3d	1.59					1.58
Ca$^+$	4s	12.41					11.26
Al^{++}	3s	28.83					28.35
Cu	4s		4.33	9.44	9.16	7.51	6.44
	4p		2.79	4.05	3.96	3.52	3.32
	5s		1.80	2.54	2.49	2.35	2.18
	5p		1.33	1.68	1.65	1.53	1.47

inhomogeneity correction was introduced.[35] This amounts to replacing D_{0l} in Eq. (3.90) by

$$\hat{D}_{0l} = AD_{0l}, \tag{3.154}$$

where

$$A = 1 + (n_l - l + 1)/4(n_l - l), \tag{3.155}$$

and n_l is the highest principal quantum number in the core among the states with quantum number l. The calculations of Gombas were single Zeta calculations, and in Schwarz's work an STO basis was used. (The details of the valence orbitals were not published.) The expectation value of the energy in the potential field V_M was first calculated without $V_{X\alpha}$; afterwards the $V_{X\alpha}$ was treated as perturbation.

The results of the calculations are summarized in Table 3.1. The equation number at the head of the columns indicates the pseudopotential used. There are some conclusions which clearly emerge from the calculations. The density-dependent pseudopotentials are generally too small; that is, the computed levels are too deep. The results improve as we move along the sequence Eq. (3.89) → Eq. (3.90) → Eq. (3.90) + Eq. (3.154), the last being the best. The results obtained with Eq. (3.105) are almost as good as those obtained with Eq. (3.90) + Eq. (3.154). The deviations from the HF value are a few percent for the best calculations.

Similar calculations were carried out by Csavinsky and Hucek[36] using the potential of Eq. (3.105). These calculations were carried out for the first few s levels of the K atom, and results of the same quality were obtained.

chapter 4

Model Pseudopotentials
For One-Valence-
Electron Systems

4.1. INTRODUCTION

In Chapters 2 and 3 the pseudopotential theory was presented in its exact and Thomas–Fermi model-based forms. It is evident from the presentation that both the exact and TF formulations have reached a high degree of mathematical coherence and completeness. At this point one might ask: can the method be formulated in a simpler way?

We have, in fact, seen that it can. In Section 1.1 we presented Hellmann's ideas, which were very simple mathematically. The early calculations of Hellmann have shown that useful results can be obtained even with such a simple formalism.

In this chapter we start to build up the model formulation of the pseudopotential method. We have seen in Chapter 3 that the density-dependent potentials form a special class of models; here we attack the problem of modelization in its general form. The purpose of a model formulation is, in general, to provide a formalism which is simple mathematically but not significantly less accurate than the original formulation. The emphasis is on simplicity; a model is of little use if it is as complicated, or almost as complicated, mathematically as the exact formalism.

The presentation of models is organized as follows. In this chapter we deal with one-electron model potentials for the valence electron of one-valence-electron atoms. In Section 7.2 we discuss one-electron model potentials for the valence electron of atoms with several valence electrons. Both of these are one-electron model potentials. In building up the pseudo-potential theory we also attack the problem of modelization of many-electron effective Hamiltonians. That task is carried out in Sections 6.3, 7.4, and 8.3. Modelisation is a pyramid-like process; the one-electron model potentials are the basic building stones of the many-electron model Hamiltonians.

The first systematic effort to give one-electron pseudopotential theory a model formulation properly based on the exact theory was carried out by Abarenkov and Heine.[37] The formal part of this chapter (as opposed to the presentation of actual model potential formulas) rests mainly on the work of Abarenkov and Heine,[37] Hazi and Rice,[38] Szasz and McGinn,[39] Schwarz,[40] and Simons.[41]

4.2. GUIDELINES FOR THE CONSTRUCTION OF MODEL POTENTIALS

We now outline the basic guidelines which must be followed in the construction of model pseudopotentials. We consider two basic types of pseudopotential equations. The first type is Eq. (2.42), which is the exact pseudopotential equation for an electron outside closed shells. The second type is Eq. (2.167), which is also for an electron outside closed shells, but contains (almost) all correlation effects in the potential. First let us write down Eq. (2.42):

$$\{t + V_M\}\psi = \epsilon\psi.\tag{4.1}$$

Here the modified potential V_M is defined as

$$V_M = g + U + V_p,\tag{4.2}$$

where the quantities U and V_p are given by Eqs. (2.40) and (2.34), respectively. Next let us write down Eq. (2.167):

$$\{t + \tilde{V}_M\}\psi = \epsilon\psi,\tag{4.3}$$

where the modified potential is now

$$\tilde{V}_M = \tilde{U} + \tilde{V}_p + \tilde{V}_c.\tag{4.4}$$

The meaning of the individual potentials in this expression was given in connection with Eq. (2.167).

We define now model equations for both cases. For Eq. (4.1) we put

$$(t + V_m)\psi = \epsilon\psi, \qquad (4.5)$$

and for Eq. (4.3) we put

$$(t + \tilde{V}_m)\psi = \epsilon\psi. \qquad (4.6)$$

In these equations V_m and $\tilde{V}_m$ are model potentials replacing V_M and $\tilde{V}_M$.

The general definition of V_m and $\tilde{V}_m$ is that these potentials should replace, for all practical purposes, the potentials V_M and $\tilde{V}_M$. In other words, we shall assume that all properties of the valence electron can be computed with almost the same accuracy from the model equations as from the original, exact equations. A salient feature of these model equations is that we can work with them as if the core did not exist.

In *the most ideal case* the model potentials should satisfy the following conditions:

1. V_m and $\tilde{V}_m$ should be Hermitian and of the simplest analytic structure possible. They should have the same overall properties—asymptotic behavior, (nl)-dependence, and so on—as the exact potentials.

2. The eigenvalues of Eq. (4.5) should match the HF energy levels; the eigenvalues of Eq. (4.6) should match the empirical energy levels. (See below for a discussion of this point.)

3. The eigenfunctions of both equations should approximate the form of an exact pseudo-orbital as closely as possible.

In order to construct model potentials which satisfy these points, we must now establish the properties of V_M and $\tilde{V}_M$. Taking V_M first, we turn to Appendix A, where we have the explicit formulas for the modified potential. The reader should, at this point, check the derivation from Eqs. (A.1) to (A.11). Putting the core orbitals and the valence electron PO which occur in V_M in the central field form, one obtains Eq. (A.9) for U and Eq. (A.11) for V_p. The first term in U is the electrostatic potential of the core electrons. This potential is independent of the (nl) quantum numbers of the valence electron. The second term is the exchange potential, which is clearly (nl)-dependent. This term will introduce a weak (nl)-dependence into V_M, since the exchange potential is weak relative to the electrostatic potentials of the nucleus plus core. The pseudopotential V_p is given by Eq. (A.11). This is clearly (nl)-dependent. The dependence on l will be strong, since the summation over n_i is for those core electrons whose azimuthal quantum number is the same as the valence electron's. Thus the summation will be a different expression for each l. Within a given l the n-dependence is manifested through the presence of the parameter ϵ_{nl} in the numerator and the pseudo-orbital P_{nl} in the denominator.

We have several arguments to support the assertion that this n-dependence will be weak. The valence electron orbital energy in the numerator is much smaller than the core energies which stand behind it. About the pseudo-orbital in the denominator Phillips and Kleinmann remarked[8] that the pseudopotential will not depend strongly on the detailed form of this PO. Furthermore, exact, PK-type calculations for the excited states (Section 9.2) show that the pseudopotential is only weakly n-dependent. Finally, the structure of the density-dependent pseudopotentials also supports the assertion. The fact that the pseudopotential is (nl)-dependent is clearly seen on Eq. (3.85). That the n-dependence is weak can be deduced from the alternative form [Eq. (3.108)] which is n-independent altogether.

Thus we have at least four arguments in favor of the assertion that, while the PK-type pseudopotential is strongly l-dependent, it depends only weakly on the principal quantum number n. In view of the structure of Eq. (A.9), we can make the same statement about the whole modified potential V_M. Thus we can write this potential in the semilocal form:[12,37,38]

$$V_M = \sum_{l=0}^{\infty} V^l(r)\Omega_l, \tag{4.7}$$

where Ω_l is the angular momentum projection operator

$$\Omega_l = \sum_{m=-l}^{+l} |Y_{lm}\rangle\langle Y_{lm}|, \tag{4.8}$$

with Y_{lm} being the normalized spherical harmonics. We have assumed, but not explicitly indicated, that each V^l is weakly n-dependent; that is, operating with V_M on a central-field-type orbital with (nl), the potential with the correct l will be selected automatically by the projection operator after which the potential function with the proper n value must be substituted. This last step need not be done if the n-dependence is omitted from the formulation; that is, if V^l is supposed to be the same for all n values with the same l.

By definition, the pseudopotential vanishes for those l values which do not occur in the core. For these l values V^l will contain only $(g + U)$. According to our analysis, this will be a weakly (nl)-dependent expression.

About the asymptotic form of V^l we can say this. As we see in Appendix A, the analytic fits for the s states of Li, Na, and K have the following form:

$$V^{l=0} = -\frac{Z-N}{r} + V' + V_p \tag{4.9}$$

where Z is the nuclear charge, N is the number of core electrons, and

$$V' = \sum_i \alpha_i e^{-\beta_i r}, \tag{4.10}$$

and

$$V_p = \sum_i \alpha_i' e^{-\beta_i' r} . \tag{4.11}$$

These expressions give an analytic fit to the PK-type potentials. The α's and β's are constants providing an accurate fit to the numerical expressions. The formula in Eq. (4.9) shows that the modified potential consists of the Coulomb potential of the ionic charge of the core plus two exponentially declining expressions representing the electrostatic, exchange, and pseudo-potentials of the core. Although these formulas were obtained for the *s* states, it is evident from Eqs. (A.9) and (A.11) that the asymptotic form for large *r* will be the same for all *l* values.

We are now able to formulate the conditions which must be satisfied by the model potential V_m if it is to replace effectively the modified potential V_M. We put the model potential in the semilocal form

$$V_m = \sum_{l=0}^{\infty} V_m^l(r)\Omega_l , \tag{4.12}$$

where the model potential V_m^l replaces the exact V^l. The function V_m^l should be a weakly *n*-dependent function for those *l* values which occur in the core and a weakly (nl)-dependent function for those *l* values which do not occur in the core. For large *r* values we expect the asymptotic form

$$\lim_{r \to \infty} V_m^l(r) = -\frac{Z-N}{r} , \tag{4.13}$$

and the terms representing the electrostatic, exchange, and pseudopotentials should rapidly (exponentially) decline outside of the core.

We turn now to the model potential $\tilde{V}_m$ which replaces the $\tilde{V}_M$ of Eq. (4.3). As we outlined in Section 2.5.D, the expression in Eq. (4.4) contains correlation effects. We saw that the core-core correlation and the spin-polarization are included in the electrostatic, exchange, and pseudopotentials which are denoted by $\tilde{U} + \tilde{V}_p$. The last term in $\tilde{V}_M$, the potential $\tilde{V}_c$, is the core polarization. This last term does not fully contain the core-valence correlation. As a result, an exact agreement between the eigenvalues of Eq. (4.3) and the empirical energies is not to be expected.

The question arises here whether it is justifiable to replace Eq. (4.4) with a model potential and then demand that the eigenvalues of Eq. (4.6) match exactly the empirical levels. We shall assume that such a procedure is permissible. We argue as follows. It is true that the polarization potential does not contain the core-valence correlation effects fully. Missing is an electron correlation effect which, if properly taken into account, would lower the energy parameter ϵ of Eq. (4.3). Thus, even if a more detailed theory would show that this effect cannot be expressed as a potential in the valence electron wave equation, it certainly can be *simulated* by an attractive potential.

At this point we want to mention an argument presented by Freed.[42] Freed investigated the question of whether it is possible, for the valence electron of an atom or molecule, to define a model Hamiltonian which has the property that its parameters can be consistently determined from empirical data. Freed discussed this problem in its full generality and concluded that it is possible. By applying the theory to atoms with one valence electron it is possible to show that there exists for the valence electron an exact, one-electron wave equation in which the effective potential contains all core-core and core-valence correlation effects. Thus the theory of Freed strengthens our argument for demanding that the eigenvalues of Eq. (4.6) match the empirical energy levels.

Therefore we put the model potential in the form[43],

$$\tilde{V}_m = \sum_{l=0}^{\infty} \tilde{V}_m^l(r)\Omega_l + V_m^c , \qquad (4.14)$$

where the first term replaces $\tilde{U} + \tilde{V}_p$, and the model polarization potential, V_m^c, replaces $\tilde{V}_c$. This last term is supposed to contain all core-valence correlation (with the exception of spin polarization, which is supposed to be included in the first term). By writing Eq. (4.14) we have assumed that the electrostatic, exchange, and pseudopotentials have the same general form for correlated core as for uncorrelated. Thus the general properties of $\tilde{V}_m^l$ in Eq. (4.14) are the same as the properties of the corresponding function in Eq. (4.12).

We now place Eq. (4.14) into the wave equation and demand that the model potential be determined in such a way that the eigenvalues of Eq. (4.6) match the empirical energy values of the valence electron spectrum.

The general properties of V_m and $\tilde{V}_m$ are now established. It is evident that this analysis still leaves a great deal of freedom in the choice of the actual, mathematical form of the potential functions V_m^l. This is a special feature of the pseudopotential method. When we stated in Section 1.3 that modelization is especially easy in pseudopotential theory, it was this property of the pseudopotential that we had in mind. We shall now demonstrate that a model psuedopotential may have an analytic form significantly different from the exact potentials and still satisfy the conditions laid down above for model potentials.

Let $\hat{V}_m$ be a model potential which has been determined by some unspecified method; for example, $\hat{V}_m$ may be a density-dependent potential of the sort we discussed in Chapter 3. We want to check whether this potential satisfies the second condition—that is, whether the eigenvalues match the HF levels. Instead of using the direct method of obtaining the exact solution of the wave equation, we shall use, for the purpose of a qualitative demonstration, perturbation theory. Thus let $\hat{\psi}$ and $\hat{\epsilon}$ be the exact solutions of the wave equation

$$(t + \hat{V}_m)\hat{\psi} = \hat{\epsilon}\hat{\psi}. \tag{4.15}$$

Let us consider the wave equation with the exact potential, and let us write

$$(t + V_M)\psi = (t + \hat{V}_m + (V_M - \hat{V}_m))\psi = (t + \hat{V}_m + V_{pt})\psi = \epsilon\psi. \tag{4.16}$$

We can solve this equation approximately by perturbation method; treating V_{pt} as a perturbation, we get for the energy in the first approximation

$$\epsilon = \hat{\epsilon} + \int \hat{\psi}^*(V_M - \hat{V}_m)\hat{\psi}\,dv. \tag{4.17}$$

We now formulate a qualitative argument. In a qualitative fashion we may say that $\hat{\epsilon}$ will be a good approximation to ϵ if the integral in Eq. (4.17) is small. This integral may be made small in several ways. One way is to demand that the model potential $\hat{V}_m$ should approximate the exact V_M as closely as possible. By this we mean that all potential functions $\hat{V}_m^l$ should approximate the corresponding exact potential V^l as closely as possible for all r. Thus in this case

$$\hat{V}_m \approx V_M, \tag{4.18}$$

and

$$\int \hat{\psi}^*(V_M - \hat{V}_m)\hat{\psi}\,dv \approx 0. \tag{4.19}$$

The structure of the exact modified potential is such that the integral can be made small in a different way. We have seen that the modified potential shows the barrier/well/Coulombic structure. Inside of the barrier, for small r, the potential fluctuates between positive and negative values. Thus we can choose the model potential in such a way that it forms a kind of average of the exact potential. Then the difference $(V_M - \hat{V}_m)$ would not be small at all r values, but the integral of Eq. (4.17) would be small because, in that integral, the difference $(V_M - \hat{V}_m)$ is multiplied by the pseudo-orbitals, which are positive everywhere (in the lowest l states of the valence electron); and if $(V_M - \hat{V}_m)$ consists of positive and negative parts, then the integral of this function, multiplied with a function that is positive everywhere, can be very small or vanishing.

The cancellation theorem, discussed above, provides an example. We saw that over a considerable part of the core the exact potential can be approximated by a constant. In that area we have of course $\hat{V}_m = \text{const} \approx$

V_M. Inside of the positive barrier of V_M the exact potential fluctuates between positive and negative values. In that area the difference $(V_M - \hat{V}_m)$ will also fluctuate, since $\hat{V}_m$ is a constant. Thus the integral of Eq. (4.17) would be very small in this case.

Summing up this argument we may say that the fluctuating character of the pseudopotential makes it easier to construct model potentials than, for example, in the HF theory, where the effective potential is a monotonic, everywhere negative function.

Finally we note that the problem of core solutions, discussed in Chapter 2, can be eliminated by modelization. This is accomplished by simply constructing the model potentials in such a way that they will not have core solutions.

4.3. A SURVEY OF MODEL POTENTIALS

In this section we review the main types of model potentials for one-valence-electron atoms. The goal is to give the reader an overview of the developments, without attempting to present every model potential which has been suggested.

We begin with the Hellmann expression [Eq. (1.1)]. This potential function occupies a significant place in pseudopotential theory, not only because it was a pioneer effort, but also because it gives remarkably accurate results despite its simplicity. We now present Hellmann's method in detail.

We write down here our two basic equations, Eqs. (4.12) and (4.14). For potentials which should match the HF energy spectrum we postulated the form

$$V_m = \sum_{l=0}^{\infty} V_m^l(r)\Omega_l, \tag{4.20}$$

while for those which should match the empirical spectrum we suggested the formula

$$V_m = \sum_{l=0}^{\infty} V_m^l(r)\Omega_l + V_m^c. \tag{4.21}$$

In Eq. (4.21) we have omitted the tildes of Eq. (4.14); the conceptual difference between the two expressions having been clarified, there is no need to complicate the notation.

Hellmann postulated that V_m should be written in the form

$$V_m = -\frac{(Z-N)}{r} + A\frac{e^{-\kappa r}}{r}, \tag{4.22}$$

and suggested that A and κ be used to match the empirical spectrum. We put for the lowest s state, the lowest p state, and the first excited s state, the wave functions

$$R_{1s} = e^{-\epsilon r}, \tag{4.23}$$

$$R_{2p} = re^{-\omega r}, \tag{4.24}$$

and

$$R_{2s} = \left(1 - \frac{1}{3}(\epsilon + \eta)r\right)e^{-\eta r}, \tag{4.25}$$

where ϵ, ω, and η are variational parameters. These functions are H-like orbitals with the $2s$ orthogonalized to the $1s$. Hellmann used these functions regardless of the true quantum numbers of the ground state of the valence electron; that is, they are used even if the ground state is $5s$ or $6s$. On putting these orbitals into the energy expression

$$\epsilon = \langle \psi | t + V_m | \psi \rangle, \tag{4.26}$$

which corresponds to Eq. (4.5), we obtain $\epsilon(1s)$, $\epsilon(2p)$ and $\epsilon(2s)$ as functions of A, κ, ϵ, ω, η. Differentiating Eq. (4.26) with respect to the variational parameters ϵ, ω, and η, we get three more equations by demanding that the derivatives be zero; that is, that the energies of the valence electron be minimized with the H-like orbitals. If we demand that the minima of the energies be at the empirical levels—that is, if we substitute for the $\epsilon(nl)$'s the empirical values—then we have six equations for the determination of the constants. Hellmann fitted the ground state exactly and adjusted the computed $2p$ and $2s$ levels as closely as possible to the empirical levels.

As we see from the description, this is a very crude approximation. The potential is (nl)-independent, and it does not have a correlation potential like V_m^c in Eq. (4.21). The empirical levels are fitted with primitive H-like functions.

Several modifications have been introduced into the determination of the potential without changing its basic analytic form. Szasz and McGinn[39] and Iafrate[44] fitted the ground state and the lowest p state exactly with A, κ, ϵ, ω. Ladanyi suggested the extended form

$$V_m = -\frac{Z - N}{r} + A\frac{e^{-\kappa r}}{r} + B\frac{e^{-\lambda r}}{r}, \tag{4.27}$$

for the heavier atoms like Cu, Ag, and Au, where the original formula does not give a deep enough well. Here the constants can be determined in such a way that κ and λ are chosen arbitrarily and then A and B used to fit $\epsilon(1s)$

and $\epsilon(2s)$.[45] An alternative procedure is to choose λ arbitrarily and then use A, B, κ, ε, ω, and η to fit all three energy levels.[39] The parameters determined by these procedures are listed in Table 4.1. The values for Na, K, and so on, beginning with Na(I) belong to this group; those beginning with Na(II) are discussed below.

Callaway and Laghos[46] and McGinn[47] modified the procedure as follows. They kept Eq. (4.22) but fitted the lowest two states by solving the wave equation, with the Hellmann potential in it, exactly. These are the parameter values indicated by Na(II) and so on in Table 4.1.

Schwarz improved the determination of the parameters considerably.[34] He wrote the potential in the form

$$V_m = \sum_{l=0}^{\infty} \left\{ -\frac{Z-N}{r} + A_l \frac{e^{-\kappa_l r}}{r} \right\} \Omega_l, \qquad (4.28)$$

where the basic analytic form is maintained but there is a set of coefficients defined for each l. Schwarz then fitted the lowest l state with each of the V_m's in such a way that A_l was chosen arbitrarily and κ_l used to fit the

TABLE 4.1. The Parameters of the Hellmann Potentials

Atom	A	κ	B	λ	Atom	A	κ
Na(I)	1.826	1.072			Al^{2+}	7.486	2.146
K(I)	1.989	0.898			Si^{3+}	10.009	2.494
Rb(I)	1.640	0.716			P^{4+}	12.580	2.838
Cs(I)	1.672	0.666			S^{5+}	15.190	3.182
Mg^+	4.656	1.688			Cl^{6+}	17.864	3.534
Ca^+	3.653	1.012			A^{7+}	20.498	3.870
Sr^+	2.877	0.730			Ga^{2+}	7.084	2.260
Ba^+	2.679	0.596			Ge^{3+}	8.338	2.174
Ra^+	2.371	0.522			As^{4+}	9.725	2.176
Cu	5.353	1.020	−3.178	0.680	Se^{5+}	11.297	2.254
Ag	5.644	1.200	−3.139	0.800	In^{2+}	5.710	1.676
Au	9.195	0.960	−6.232	0.640	Sn^{3+}	6.840	1.574
Zn^+	4.376	1.800	−0.114	0.020	Sb^{4+}	8.486	1.674
Cd^+	4.283	1.712	−0.062	0.020	Te^{5+}	9.811	1.692
Hg^+	5.995	2.784	0.009	0.020	Tl^{2+}	5.001	1.606
Na(II)	10.090	2.004			Pb^{3+}	5.986	1.446
K(II)	2.8451	0.940			Bi^{4+}	7.101	1.388
Rb(II)	1.8423	0.690					
Cs(II)	1.5840	0.562					

empirical energy level. The wave function used in the fitting was a linear combination of four STO's.

A further improvement was introduced by Bardsley.[43] The potential was now written in the theoretically correct form of Eq. (4.21). For V_m^c Bardsley used

$$V_m^c = -\frac{\alpha_d}{2(r^2 + d^2)^2} - \frac{\alpha_q}{2(r^2 + d^2)^3}.$$ (4.29)

This is a polarization potential in which α_d and α_q are the dipole and quadrupole polarizabilities of the core, and the cutoff radius d is the radius of the core. The polarization potential of Eq. (4.29) has the correct asymptotic behavior for large r.

The potential of Bardsley is then

$$V_m = \sum_{l=0}^{\infty} \left\{ -\frac{Z-N}{r} + A_l \frac{e^{-\kappa r}}{r} \right\} \Omega_l - \frac{\alpha_d}{2(r^2 + d^2)^2} - \frac{\alpha_q}{2(r^2 + d^2)^3},$$ (4.30)

and for each l the two lowest terms are fitted with A_l and κ_l by solving the wave equation exactly. This V_m is l-dependent, has the correct asymptotic behavior for large r, and fits the two lowest levels of the empirical spectrum exactly. As we see from Eq. (4.30), this is still a Hellmann-type potential.

The parameter values of Schwarz's and Bardsley's potentials are collected in Table 4.2. As we see from the table, parameter values were determined for $l = 0, 1, 2$. Thus we have potentials for those l values which occur in the core we well as for some of those which do not occur. All the parameters were computed by Schwarz, except that the lines which are indicated by (II) are the values computed with Bardsley's method. In these last cases we also have the parameters of the polarization potential in the table.

The many applications which have been carried out with these potentials attest to the popularity of these expressions. The reason for this popularity is, of course, the extreme simplicity of these expressions relative to the ab initio potentials. But how accurate are these expressions? Good results obtained in a particular application do not prove the accuracy of a potential; there are always a number of other approximations in any application which might compensate for the inaccuracy of the model potential. In order to form a judgment we must look at the three conditions laid down in the preceding section. These were simplicity, matching of the energy spectrum, and reproducing of the pseudo-orbitals.

The Hellmann potentials certainly satisfy the first condition; they are simple. If we look at the potentials whose parameters are listed in Table 4.1, we see that they do not satisfy the second and third conditions. In addition, it is clear that by fitting the s and p levels simultaneously, we have obtained a potential which is some kind of an average between the correct s and p potentials. The potentials determined by McGinn, which are also in Table

TABLE 4.2. The Parameters of the Hellmann Potentials

Atom	A_0	κ_0	A_1	κ_1	A_2	κ_2	α_d	α_q	d
Li(I)	10	2.202	-2	2.330	-2	1.855			
Li(II)	26.762	2.896	-1.855	2.676	0.696	2.612	0.192	0.112	0.75
Na(I)	14	2.267	14	2.231	-10	1.953			
Na(II)	164.896	3.858	26.368	2.403	-2.788	1.794	0.945	1.520	1.10
K	18	1.866	18	1.635	-18	1.973			
Rb	26	1.940	26	1.568	-6	0.984			
Cu	30	6.134	-10	3.600	-10	1.709			
Be$^+$	20	3.362	-2	3.347	-2	3.031			
Mg$^+$	30	2.855	30	2.895	-10	2.604			
Ca$^+$	50	2.336							
Sr$^+$	60	2.205							
Zn$^+$	60	4.505							

4.1, are somewhat better in that they satisfy the second condition, but they are still averages between the *s* and *p* potentials. These observations must be taken into account in any application of these potentials.

Turning to Table 4.2, we consider first the potentials of Schwarz. Here the ground state is fitted. Schwarz calculated the first and second excited states for each *l* and obtained good results. Similarly, Bardsley has shown that the potentials in Eq. (4.30), which are fitted using the lowest two *l* states, give good results for the two next higher *l* states. Thus we can say that both potentials satisfy the second condition in the lower range of the energy spectrum.

We turn now to other model potentials. The formulas are collected in Table 4.3. On the top of each section of the table we have a serial number for future reference, the name of the author(s), and the reference of the publication. The serial numbers reflect roughly the chronological order of publication. The phrase "fitting parameters" identifies those parameters which were used to adjust the spectrum of the potentials. As stated before, we do not attempt to list all the potentials that were suggested. We have omitted, for example, some expressions which are clearly unphysical; also,

TABLE 4.3.

---

1. Abarenkov and Heine[37]

$$V_m = \sum_{l=0}^{\infty} V_m^l \Omega_l$$

$$V_m^l = -A_l, r < R_c; \quad V_m^l = -\frac{Z-N}{r}, r \geq R_c$$

Fitting parameters: A_l, R_c.

---

2. Bingel, Koch, and Kutzelnigg[107]

$$V_m = \sum_{l=0}^{\infty} V_m^l \Omega_l$$

$$V_m^l = -\frac{Z-N}{R_c^l}, \quad r \leq R_c^l; \qquad V_m = -\frac{Z-N}{r}, \quad r \geq R_c^l.$$

Fitting parameter: R_c^l.

---

3. Bardsley[43]

$$V_m = \sum_{l=0}^{\infty} \left\{ -\frac{Z-N}{r} + A_l e^{-\kappa_l r} \right\} \Omega_l + V_m^c$$

V_m^c given by Eq. (4.29); Fitting parameters: A_l, κ_l.

TABLE 4-3 (*continued*)

---

4. Bardsley[43]

$$V_m = \sum_{l=0}^{\infty} \left\{ -\frac{Z-N}{r} + A_l r^2 e^{-\kappa_l r^2} \right\} \Omega_l + V_m^c$$

V_m^c given by Eq. (4.29); Fitting parameters: A_l, κ_l.

---

5. Simons[106]

$$V_m = \sum_{l=0}^{\infty} \left\{ -\frac{Z-N}{r} + \frac{A_l}{r^2} \right\} \Omega_l$$

Fitting parameter: A_l.

---

6. Bardsley[69]

$$V_m = \sum_{l=0}^{\infty} \left\{ -\frac{Z-N}{r} + A_l e^{-\kappa_l r^2} \right\} \Omega_l + V_m^c$$

V_m^c given by Eq. (4.29); Fitting parameters: A_l, κ_l.

---

7. Switalski and Schwartz[108]

$$V_m = \sum_{l=0}^{\infty} \left\{ -\frac{Z-N}{r} + A_l \frac{e^{-\kappa_l r^2}}{r} \right\} \Omega_l$$

Fitting parameters: A_l, κ_l.

---

8. Chang, Habitz, Pittel, and Schwarz[90]

$$V_m = -\frac{Z-N}{r} - A \frac{e^{-\lambda r}}{r} + \sum_{l=0}^{l_m} (B_l e^{-\kappa_l r^2}) \Omega_l$$

l_m is the largest l in the core; Fitting parameters: A, λ, B_l, κ_l.

---

9. Dolgarno, Bottcher, and Victor[109]

$$V_m = -\frac{Z}{r} + V(r) + (A + Br)e^{-\kappa r} - \frac{\alpha_d}{2r^4}(1 - e^{-6\kappa r}) - \frac{\alpha_q}{2r^6}(1 - e^{-8\kappa r})$$

$V(r)$ is the first term of Eq. (2.5);
α_d = dipole polarizability; Fitting parameters: A, B, κ, α_q.

TABLE 4.3 (*continued*)

10. Flad, Stoll, and Preuss[91]

$$V_m = -\frac{Z-N}{r} + \frac{Be^{-\beta r^2} + rCe^{-\gamma r^2}}{r}$$

Fitting parameters: B, C, β, γ.

11. Preuss, Stoll, Wedig, and Krueger[93]

$$V_m = -\frac{Z-N}{r} - A\frac{e^{-\lambda r^2}}{r} + \sum_{l=0}^{l_m} (B_l e^{-\kappa_l r^2})\Omega_l$$

l_m is the largest l in the core;
Fitting parameters: A, λ, B_l, κ_l.

12. Schwerdtfeger, Stoll, and Preuss[110]

$$V_m = -\frac{Z-N}{r} + A\frac{e^{-\lambda r^2}}{r} + \sum_{l=0}^{l_m} (C_l e^{-\lambda r^2} + B_l e^{-\kappa_l r^2})\Omega_l$$

Fitting parameters: A, λ, C_l, B_l, κ_l.

13. Fuentealba, Preuss, Stoll, and Szentpaly[111]
   (for alkali compounds with single valence electron)

$$V_m = -\frac{Z-N}{r} + \sum_{l=0}^{2} (B_l e^{-\kappa_l r^2})\Omega_l - \frac{\alpha_d}{2r^4}(1 - e^{-(r/\rho)^2})^2 - \frac{\alpha_d}{2R^4} + \frac{\alpha_d(\mathbf{r}\cdot\mathbf{R})}{r^3 R^3}(1 - e^{-(r/\rho)^2})$$

Fitting parameters: B_l, κ_l.
α_d = dipole polarizability
R = internuclear distance
ρ = cutoff parameter

we omitted some which are in contradiction to the exact pseudopotential theory. (We note that the model potentials for many-valence-electron atoms will be discussed in Section 7.2.)

Looking at the Hellmann potentials which were given in the text and at the potentials in the table, we see that, mathematically, they range from the very simple to the fairly complicated. For a scientist who wants to use such models for calculations, these potentials provide formulas for a wide variety of atoms and molecular situations. We note also that, while the early potentials were mostly of the exponential type, the formulas developed more recently are mostly of Gaussian type. The reason for this is, of course, that the latter are more convenient in molecular calculations.

It is not our goal to give a detailed evaluation of the model potentials. Nevertheless, we can make some general observations. We have seen that the exact pseudopotential theory provides a framework for model potentials, which we discussed in Section 4.2. The guidlines which we discussed there found expression in Eqs. (4.20) and (4.21), which were the formulas for model potentials matching the (HF) and empirical spectrum, respectively. Now, as we see from the text and from Table 4.3, all models are adjusted to the empirical spectrum but only about half have the form which is prescribed for such potentials by the exact theory.

This clear contradiction between theory and models was discussed by Kahn, Baybutt, and Truhlar.[48] In those cases in which a potential of the form of Eq. (4.20) is used to match the empirical spectrum, the potential is assumed to contain core-valence polarization despite the fact that, as we have seen, the core-valence correlation effects are described by a potential function which has a different analytic form. Thus in these cases the total potential is *simulated* by the expression in Eq. (4.20). Such potentials still can be useful and accurate, but KBT have rightly pointed out that, in such cases, the errors inherent in this choice of analytic form are usually concealed by the procedure with which the parameters are determined. It is generally true for all model potentials, even for those which contain core-valence correlation in the correct form, that the method of the adjustment of the parameters *might* compensate for the inconsistencies inherent in the analytic form of the model.

In order to compare various forms of model potentials Schwarz made atomic calculations using some of the early models.[34] He concluded that, in atomic calculations, the potentials which have the barrier/well/Coulombic structure perform better than those which are constant within a certain radius.

chapter 5

The All-Electron Pseudopotential Model (APM)

5.1. INTRODUCTION

By an *All-Electron Pseudopotential Theory* we mean a quantum-mechanical formalism in which the Pauli exclusion principle is taken into account in the form of pseudopotentials rather than orthogonality conditions, and one in which this is done not only for the valence electrons but for *all electrons*.

When we outlined the basic idea of pseudopotentials in Section 1.1, we have talked exclusively of valence electrons. Up to this point the pseudopotential theory has been presented as a valence-only theory, and up to the time of this writing it has been applied almost exclusively to valence-electron problems. But in some recent work of the author[49] it is shown that the theory can be extended to all electrons of an atom or molecule, and that this extension results in the same kind of conceptual and computational advantages as in the case of valence electrons. The extension to all electrons, which enlarges considerably the scope of the theory, will be presented in this chapter; the rest of the book deals with valence-electrons-only formulations.

The idea that the total radial density of an atom can be built up entirely from PO's is implicitly contained in Slater's early work, in which he introduced the Slater-orbitals. [Slater I, p. 368] Then Gombas and Gaspar showed[50] that the orthogonality constraints can be replaced by density-dependent pseudopotentials for all electrons of an atom. Gombas and Ladanyi developed a statistical model[51] in which the Pauli exclusion principle was represented by density-dependent pseudopotentials for all electrons. On the basis of this model, calculations were carried out for all atoms of the periodic system.[52]

In contrast to these developments, which were of a semiquantitative character, Szepfalusy attempted[53] to build up an exact all-electron pseudopotential theory, starting from the HF equations for a closed shell atom and using the pseudopotential transformation which we discussed in Section 2.1. Using some of Szepfalusy's ideas and the techniques which were developed for systems with one valence electron, the author has shown[49] that a mathematically consistent all-electron theory can be formulated.

We present the author's work in this chapter, following closely the strategy outlined in Section 1.2. We start with an arbitrary form of one-electron approximation, which may be a HF model for closed shells, for nonclosed shells, or for the average of a configuration; the discussion is not restricted to a HF model, however, but can be applied to other types of one-electron approximations as well, for example, the $X\alpha$ model. The equations of this one-electron approximation will be transformed exactly into pseudopotential equations and the equivalence of the resulting formalism to the original will be demonstrated. Then it will be shown how the exact formalism can be replaced by a much simpler model. The general method is applied to two special cases: for the weighted average of an atomic configuration an APM is presented in Section 5.4, along with the results of the calculations carried out using this model for atoms between Li and Kr. In Section 5.5 it is discussed how the $X\alpha$ method can be transformed into an all-electron pseudopotential formalism and how a simplified model formulation can be obtained.

5.2. EXACT ALL-ELECTRON PSEUDOPOTENTIAL THEORY

Let us consider an atom or molecule with N electrons and let $\varphi_1, \varphi_2, \ldots, \varphi_N$ be the spin-orbitals assigned to the electrons. Let E_T be the total energy of the system. The detailed form of E_T will not be specified; we may be talking about any of the approximations listed at the end of the preceding section. Variation of E_T yields the one-electron equations

$$H'_s\varphi_s = \epsilon_s\varphi_s + \sum_{\substack{t=1 \\ (t \neq s)}}^{N} \lambda_{st}\varphi_t, \quad (s = 1, 2, \ldots, N).\tag{5.1}$$

Here H'_s is a one-electron Hamiltonian operator which, in general, will be different for each orbital and has the form

$$H'_s = t + U'_s, \tag{5.2}$$

where t is the kinetic energy operator and U'_s is an arbitrary composition of electrostatic and exchange potentials (operators) including the nuclear attraction. λ_{st} is a Lagrangian multiplier which is introduced in the process of variation to insure the orthogonality of one-electron orbitals. Multiplying Eq. (5.1) from the left by $\varphi_t^*(t \neq s)$ and integrating we get

$$\lambda_{st} = \langle \varphi_t | H'_s | \varphi_s \rangle. \tag{5.3}$$

Let us introduce the projection operators

$$\Omega^s(1)f(1) = \sum_{\substack{t=1 \\ (t \neq s)}}^{N} \varphi_t(1) \int \varphi_t^*(2) f(2) \, dq_2, \tag{5.4}$$

and

$$P^s = 1 - \Omega^s. \tag{5.5}$$

Using Eqs. (5.3), (5.4), and (5.5) we can rewrite Eq. (5.1) as follows

$$H'_s \varphi_s = \epsilon_s \varphi_s + \sum_{t=1}^{N} \varphi_t \langle \varphi_t | H'_s | \varphi_s \rangle = \epsilon_s \varphi_s + \Omega^s H'_s \varphi_s, \tag{5.6}$$

and this equation can be written in the form

$$H_s \varphi_s = \epsilon_s \varphi_s, \tag{5.7}$$

where

$$H_s = (1 - \Omega^s)H'_s = P^s H'_s. \tag{5.8}$$

In Eq. (5.7) we obtained a conventional HF-type equation after the Lagrangian multipliers were replaced by the projection operator [Eq. (5.4)] and the latter incorporated into the Hamiltonian according to Eq. (5.8). We note that, in general, H_s is not Hermitian, since H'_s and P^s do not commute. The solutions of Eq. (5.7) are orthonormal:

$$\langle \varphi_s | \varphi_t \rangle = \delta_{st}, \quad (s, t = 1, 2, \ldots, N). \tag{5.9}$$

Multiplying Eq. (5.7) from the left by φ_s^* and integrating we obtain

$$\epsilon_s = \langle \varphi_s | H_s | \varphi_s \rangle. \tag{5.10}$$

Let us define E_0, a functional of the orbitals $\varphi_1 \cdots \varphi_N$ as follows:

$$E_0[\varphi_i] \equiv E_T - \sum_{s=1}^{N} \langle \varphi_s | H_s | \varphi_s \rangle . \qquad (5.11)$$

Using Eq. (5.10) we obtain

$$E_T = E_0[\varphi_i] + \sum_{s=1}^{N} \epsilon_s . \qquad (5.12)$$

Our next goal is to transform Eq. (5.7) into a pseudopotential equation. Such a transformation can be carried out exactly in the same way as in the case of systems with one valence electron. In the case of Eq. (5.7) the orbital φ_s plays the role of the valence orbital, and the orbitals $\varphi_1 \cdots \varphi_{s-1}$ play the role of core orbitals. We want a pseudopotential which would replace the orthogonality requirement placed on φ_s with respect to the orbitals $\varphi_1 \cdots \varphi_{s-1}$. Reviewing the exact pseudopotentials presented in Chapter 2 we observe that since the "core orbitals" $\varphi_1 \cdots \varphi_{s-1}$ are not eigenfunctions of the "valence" Hamiltonian H_s, the procedure that we must use is the Weeks–Rice method which was presented in Section 2.4.

In order to apply the WR method let us introduce the projection operators

$$\Omega_s = \sum_{t=1}^{s-1} |\varphi_t\rangle\langle\varphi_t| , \qquad (5.13)$$

and

$$P_s = 1 - \Omega_s . \qquad (5.14)$$

These operators are the generalizations of Eqs. (2.50) and (2.52). The properties of the operators [Eqs. (5.4), (5.5), (5.13), and (5.14)] are summarized in Appendix D.

Let us consider the pseudopotential given by Eq. (2.138), which is a WR potential derived by demanding that the PO of the valence electron should minimize the expectation value of the kinetic energy. Replacing the valence-electron quantities in Eq. (2.138) with their analogs for the core orbitals φ_s we obtain

$$V_s = \Omega_s(\epsilon_s - U_s - E_s^K) - P_s(H_s - \epsilon_s)\Omega_s , \qquad (5.15)$$

where the quantities which have not yet been mentioned, U_s and E_s^K, are defined as follows:

$$U_s = U_s' - \Omega^s H_s' , \qquad (5.16)$$

and

$$E_s^K = \frac{\langle \psi_s | t | \psi_s \rangle}{\langle \psi_s | \psi_s \rangle}, \tag{5.17}$$

where ψ_s is the PO for the state s. Using Eq. (5.15) we obtain the equations

$$(H_s + V_s)\psi_s = \epsilon_s \psi_s, \quad (s = 1, 2, \ldots, N). \tag{5.18}$$

We now prove the following theorem. If ψ_s and ϵ_s are solutions of Eq. (5.18), then $\varphi = P_s \psi_s$ is the solution of Eq. (5.7) with the same eigenvalue ϵ_s.

In order to prove the theorem we multiply Eq. (5.18) from the left by P_s and obtain

$$(P_s H_s + P_s V_s)\psi_s = \epsilon_s P_s \psi_s. \tag{5.19}$$

Taking into account Eq. (5.8) and Eq. (D.14), we get

$$P_s H_s = P_s P^s H_s' = P^s H_s' = H_s. \tag{5.20}$$

Using Eq. (5.15) we obtain

$$P_s V_s = P_s \Omega_s (\epsilon_s - U_s - E_s^K) - P_s^2 (H_s - \epsilon_s)\Omega_s. \tag{5.21}$$

Since $P_s \Omega_s = 0$ we obtain

$$P_s V_s = -P_s H_s \Omega_s = -H_s \Omega_s, \tag{5.22}$$

where we have used Eq. (5.20) again. On using Eqs. (5.20) and (5.22) in Eq. (5.19), we obtain

$$(H_s - H_s \Omega_s)\psi_s = H_s P_s \psi_s = \epsilon_s P_s \psi_s, \tag{5.23}$$

from which it follows that

$$H_s \varphi = \epsilon_s \varphi, \tag{5.24}$$

where

$$\varphi = P_s \psi_s. \tag{5.25}$$

Since Eq. (5.24) is the same equation as Eq. (5.7), we conclude that φ is indeed the eigenfunction of H_s with the eigenvalue ϵ_s, which proves our theorem. We can put*

---

* See the footnotes following Eqs. (2.88) and (2.132).

$$\varphi = \varphi_s = P_s \psi_s . \tag{5.26}$$

Writing out this equation in detail,

$$\varphi_s = \psi_s - \sum_{t=1}^{s-1} |\varphi_t\rangle\langle\varphi_t | \psi_s\rangle = \psi_s - \sum_{t=1}^{s-1} \alpha_{st}\varphi_t , \tag{5.27}$$

where

$$\alpha_{st} = \langle\varphi_t | \psi_s\rangle . \tag{5.28}$$

The explicit formula for the pseudo-orbital ψ_s is

$$\psi_s = \varphi_s + \sum_{t=1}^{s-1} \alpha_{st}\varphi_t . \tag{5.29}$$

Comparing Eq. (5.29) with Eq. (2.123) we see that the former is the analog of the latter. Equation (5.18) will also have "core-type" solutions analogous to Eq. (2.122), but we can safely ignore those; what is important for the mathematical consistency of the theory is to show that solutions of the type of Eq. (5.29) exist.

In Sections 2.3 and 2.4 we have shown that for the PK-type equation [Eq. (2.84)] and for the WR-type equation [Eq. (2.126)] there are many meaningful choices for the form of the F_i functions which occur in the pseudopotentials. We have seen that the calculation of valence solutions for these equations led to the system of inhomogeneous equations [Eqs. (2.98) and (2.143)]. Meaningful choice of F_i was one for which those equations had unique solutions. We have seen that unique solutions existed if the system determinant and the inhomogeneity terms were not zero.

The situation is perfectly analogous for the calculation of ψ_s in Eq. (5.29). For the pseudopotential V_s in Eq. (5.18) we have chosen the expression of Eq. (5.15) which minimizes the kinetic energy associated with each PO. We show now that, similarly to the case of a valence electron, this choice leads to uniquely determined α_{st} coefficients; that is, to uniquely determined pseudo-orbitals ψ_s.

Let us assume that the eigenfunctions and eigenvalues of Eq. (5.7) have been computed and that they are uniquely determined. Along with the eigenfunctions, the operators in H_s are also uniquely determined. Substitute now Eq. (5.29) into the pseudopotential equation, Eq. (5.18), multiply from the left by the lower-lying orbital $\varphi_k^*(k < s)$, and integrate. We obtain

$$\sum_{l=1}^{s-1} \alpha_{sl}M_{kl}^s = -N_{ks}^s , \quad (s = 1, \dots, N; k = 1, \dots, s-1), \tag{5.30}$$

where

$$M^s_{kl} = \langle \varphi_k | H_s + V_s - \epsilon_s | \varphi_l \rangle, \tag{5.31}$$

and

$$N^s_{kl} = \langle \varphi_k | H_s + V_s | \varphi_l \rangle. \tag{5.32}$$

For each s we have, in Eq. (5.30), a set of $(s-1)$ linear, inhomogeneous algebraic equations for the determination of the coefficients $\alpha_{s1}, \alpha_{s2}, \ldots, \alpha_{s,s-1}$. Nontrivial solutions will exist if the system determinant and the inhomogeneity terms are not zero. As we have seen in Section 2.3, the last condition imposes requirements on the pseudopotential; for example, it excludes the PK-potential [Eq. (2.80)].

We want to investigate the form of the inhomogeneity terms for our choice of the pseudopotential. Substituting Eq. (5.15) into Eq. (5.32) we obtain

$$\begin{aligned} N^s_{ks} &= \langle \varphi_k | H_s + V_s | \varphi_s \rangle = \epsilon_s \delta_{ks} \\ &+ \langle \varphi_k | \Omega_s (\epsilon_s - U_s - E^K_s) - P_s (H_s - \epsilon_s) \Omega_s | \varphi_s \rangle. \end{aligned} \tag{5.33}$$

The first term is zero since $k \neq s$. In the second term we take into account Eqs. (D.5) and (D.6) and obtain

$$\begin{aligned} N^s_{ks} &= \langle \varphi_k | \Omega_s (\epsilon_s - U_s - E^K_s) | \varphi_s \rangle \\ &= \langle \Omega_s \varphi_k | \epsilon_s - U_s - E^K_s | \varphi_s \rangle \\ &= \langle \varphi_k | \epsilon_s - U_s - E^K_s | \varphi_s \rangle. \end{aligned} \tag{5.34}$$

Taking into account Eqs. (5.2), (5.8), and (5.16), we can write

$$H_s = H'_s - \Omega^s H'_s = t + U'_s - \Omega^s H'_s = t + U_s, \tag{5.35}$$

and using Eq. (5.7) we obtain, for the combination which occurs in Eq. (5.34),

$$(\epsilon_s - U_s)\varphi_s = t\varphi_s. \tag{5.36}$$

Putting this into Eq. (5.34) and taking into account that E^K_s is a constant, we obtain

$$N^s_{ks} = \langle \varphi_k | t | \varphi_s \rangle, \quad (s = 1, \ldots, N; k = 1, \ldots, (s-1)). \tag{5.37}$$

Our result is that the inhomogeneity terms are the matrix components of the kinetic energy operator with respect to the eigenfunctions of H_s. For a given orbital φ_s such a matrix component will be zero for those orbitals φ_k

which are of different symmetry from φ_s. From the form of Eq. (5.29) we see that the PO ψ_s will have the same symmetry as φ_s.

Now let us look at Eq. (5.30). For a given s, there will be no nontrivial solutions if all inhomogeneity terms $N_{1s}^s, N_{2s}^s, \ldots, N_{s-1,s}^s$ are zero. But according to our analysis this means that the orbitals $\varphi_1, \varphi_2, \ldots, \varphi_{s-1}$ are all of different symmetry from ψ_s. But in such a case ψ_s will be orthogonal to those orbitals anyhow, and there is no need to have a pseudopotential in the equation of this particular ψ_s. Thus for this particular s, the pseudopotential V_s will be zero, and all the α_{st} coefficients in Eq. (5.29) will be zero. The PO ψ_s will be the same as φ_s.

Reversing the argument we may say that if at least one of the orbitals $\varphi_1 \cdots \varphi_{s-1}$ is of the same symmetry as ψ_s, then one of the inhomogeneity terms will be different from zero and one uniquely determined α_{st} will be determined by Eq. (5.30). In general, if some or all the orbitals $\varphi_1 \cdots \varphi_{s-1}$ will be of the same symmetry as ψ_s, then the corresponding α_{st} coefficients will be uniquely determined by Eqs. (5.30).

As an example for the case when there is no pseudopotential in the equation of a PO, let us consider the $3d$ orbital of an atom. In that case the lower orbitals are of s and p symmetry. The pseudo-orbital will be of d symmetry; thus it will be orthogonal to all lower orbitals by virtue of the angular part. All inhomogeneity terms will be zero; consequently all α_{st} coefficients will also be zero; and from Eq. (5.29) we get $\psi_{3d} = \varphi_{3d}$. There will be no pseudopotential in the equation for the radial part of ψ_{3d}.

One omission in the argument above must be corrected. In order to get a set of coefficients from Eq. (5.30) we need not only the eigenfunctions and eigenvalues of H_s but also the quantity E_s^K which itself depends on the pseudo-orbital; that is, it depends on the α_{st} coefficients. It is obvious that Eq. (5.30) must be solved in a self-consistent fashion: start with an initial set $\alpha_{st}^{(0)}$, compute E_s^K, and, substituting it into Eq. (5.30), obtain a new set of coefficients $\alpha_{st}^{(1)}$; repeat the procedure until.

$$\alpha_{st}^{(n)} = \alpha_{st}^{(n-1)} . \tag{5.38}$$

The argument about the inhomogeneity terms is not affected by this self-consistent procedure; Eq. (5.37) shows that the inhomogeneity term does not depend on the choice of E_s^K.

We transform now the expression, Eq. (5.12), for the total energy in a way that will be useful later. If ψ_s is normalized we obtain from Eq. (5.18)

$$\epsilon_s = \langle \psi_s | H_s | \psi_s \rangle + \langle \psi_s | V_s | \psi_s \rangle . \tag{5.39}$$

Putting this into Eq. (5.12) we get

$$E_T = E_0[\varphi_i] + \sum_{s=1}^{N} \langle \psi_s | H_s | \psi_s \rangle + \langle \psi_s | V_s | \psi_s \rangle . \tag{5.40}$$

For comparison, we write down here Eq. (5.12) again, but with the ϵ_s replaced according to Eq. (5.10):

$$E_T = E_0[\varphi_i] + \sum_{s=1}^{N} \langle \varphi_s | H_s | \varphi_s \rangle . \tag{5.41}$$

We see now that, for a given set of eigenfunctions and eigenvalues of H_s, the pseudopotential equations, Eqs. (5.18), will have a unique set of solutions which reproduce exactly the eigenfunctions of H_s through Eq. (5.29):

$$\varphi_s = \psi_s - \sum_{t=1}^{s-1} \alpha_{st} \varphi_t . \tag{5.42}$$

If the orbitals $\varphi_1, \ldots, \varphi_N$ and the parameters $\epsilon_1, \ldots, \epsilon_N$ are uniquely determined, then the pseudo-orbitals $\psi_1, \ldots, \psi_N$ will also be uniquely determined. The exact pseudopotential formalism will exactly reproduce the orbitals $\varphi_1 \cdots \varphi_N$, and since the total energy of the system E_T depends only on these orbitals, it will also be exactly reproduced. The equivalence of the pseudopotential formalism to the conventional formalism is thereby proved.

We have now two equivalent procedures for the determination of the properties of an atomic or molecular system. The conventional procedure consists of solving the Eqs. (5.7) by a self-consistent procedure and obtaining the orthogonal orbitals $\varphi_1 \cdots \varphi_N$. The expectation values of all operators can be computed using these orbitals; for example, the total energy of the system is given by Eq. (5.12). The new procedure presented here consists of solving the pseudopotential equations, Eqs. (5.18), by a self-consistent procedure and obtaining the PO's $\psi_1 \cdots \psi_N$. From these the orbitals $\varphi_1 \cdots \varphi_N$ must be formed according to Eq. (5.42), and the expectation values of all operators are then computed as before. The two procedures are exactly equivalent. This is a clear demonstration of the fact that the introduction of pseudo-orbitals does not in any way imply that new approximations have been introduced.

Thus we see that the pseudopotential formalism is equivalent to the conventional. If this is the case, then in what does its usefulness lie? We pointed out in Section 1.3 that the advantage of pseudopotential theory is that it can be easily modelized. Thus the exact formalism forms the basis for the formulation of a model. We show in the rest of this chapter that the formalism just presented can be used to formulate the simple all-electron pseudopotential model which yielded good results for atoms. It is erroneous to say, however, that the exact formalism can serve only as the basis of a model. (We discuss this point because, in the literature, one encounters remarks to this effect.)

At the time of writing the pseudopotential method has been developed both in its exact and model form. The ab initio applications are not extensive enough, however, to form a judgment about their usefulness relative to the conventional ab initio methods. A case in point is the system Eqs. (5.18). If H_s is a HF Hamiltonian, then Eqs. (5.18) are equivalent to this HF procedure. There are numerous accurate HF calculations for atoms; the exact solutions of Eqs. (5.18) are not yet available. Physically, Eqs. (5.18) would be equivalent to the atomic HF procedure, but the computational work might be reduced. This might be the case, for example, if, instead of solving Eqs. (5.18) numerically, one obtained accurate approximation solutions by using Roothaan's expansion method.[54] The lack of orthogonality constraints might make the calculation of PO's easier than the calculation of HF orbitals.

Finally we note that the exact formalism developed here is based on the kinetic-energy-minimized pseudopotential. Choosing other expressions for the F_i functions which occur in Eq. (2.124), other exact formalisms can be developed. They will be equivalent to the conventional formalism as long as the choice of the F_i's leads to unique pseudo-orbitals.

5.3. THE ALL-ELECTRON PSEUDOPOTENTIAL MODEL

Let us write down here again our exact pseudopotential equations. The pseudo-orbitals $\psi_1 \cdots \psi_N$ must be determined from Eqs. (5.18):

$$(H_s + V_s)\psi_s = \epsilon_s \psi_s , \quad (s = 1, 2, \ldots, N) , \tag{5.43}$$

where H_s is the original one-electron Hamiltonian given by Eqs. (5.2) and (5.8), while V_s is the pseudopotential,

$$V_s = \Omega_s(\epsilon_s - U_s - E_s^K) - P_s(H_s - \epsilon_s)\Omega_s . \tag{5.44}$$

We now replace the exact formalism by a model. This will be done in two steps: first we replace Eq. (5.44) by a density-dependent expression, then we replace the orbitals $\varphi_1 \cdots \varphi_N$ by the PO's $\psi_1 \cdots \psi_N$.

A. Density-Dependent Pseudopotentials

The potential of Eq. (5.44) is a rather complicated operator. In order to obtain simpler expressions we follow the example of the $X\alpha$ model, in which the exchange interaction operator is replaced by a density-dependent potential. As we have seen in Chapter 3, for a valence electron the pseudopotential can be put in density-dependent form. We have derived expressions for atoms as well as molecules.

The considerations of Chapter 3 can be transferred here directly. We can treat the Eqs. (5.43) in the same way as we treated the HF equation of the valence electron. Thus, for an atom, we can introduce the density-dependent pseudopotential in the same way as we did in Section 3.2.A; and, for a molecule, as we did in Section 3.2.B. Let us denote the density-dependent pseudopotential for the state s by Φ_s. Replacing Eq. (5.44) by Φ_s, Eqs. (5.43) become

$$(H_s + \Phi_s)\psi_s = \epsilon_s \psi_s, \quad (s = 1, 2, \dots, N). \tag{5.45}$$

For Φ_s we choose the theoretically best established forms. Thus for an atom we choose Eq. (3.87) with Eq. (3.85) and Eq. (3.86). These expressions were derived for a valence electron. We transfer these results to Eq. (5.45), where s is an arbitrary state, valence or core, by identifying the state s with the valence electron state of Chapter 3, and the states below s with the core states of Chapter 3. In other words, the transfer of the pseudopotential concept from the discussions of Chapter 3 is done in such a way that the orbital ψ_s plays the role of the valence electron, and the orbitals below play the role of core electrons. Just as in the exact theory the potential V_s replaced the Pauli exclusion principle with respect to the orbitals $\varphi_1 \cdots \varphi_{s-1}$, in the APM model the density-dependent potential Φ_s will replace the exclusion principle with respect to the same lower-lying orbitals.

Accordingly we put

$$\Phi_s = \Phi_{nl} = \sum_{l'=0}^{\infty} \Phi_{nl'} \Omega_{l'}, \tag{5.46}$$

where $\Omega_{l'}$ is the angular momentum projection operator whose sole task is to select the correct l value for the potential, and is given by Eq. (3.86):

$$\Omega_{l'} = \sum_{m'=-l'}^{+l'} |Y_{l'm'}\rangle\langle Y_{l'm'}|, \tag{5.47}$$

where $Y_{l'm'}$ is the spherical harmonics. Φ_{nl} is given by Eq. (3.85):

$$\Phi_{nl} = \frac{\pi^2}{8(2l+1)^2} [D_{0l}^2 + 2D_{0l}\hat{P}_{nl}^2]. \tag{5.48}$$

In this formula we have emphasized the n-dependence of the potential, which is not displayed in Eq. (3.85). The reason for this is that in transferring Eq. (3.85) to a core state, the n-dependence of D_{0l} must be emphasized. According to the definition, this is the radial density of all "core" electrons with the quantum number l. If the state s has the quantum numbers (nl), as we have already assumed in writing down Eq. (5.46), then we get

$$D_{0l} = \sum_{n'=l+1}^{n-1} q(n'l)\hat{P}_{n'l}^2, \tag{5.49}$$

where $\hat{P}_{n'l}$ is the radial part of one of the lower-lying orbitals $\varphi_1 \cdots \varphi_{s-1}$, and

$q(n'l)$ is the occupation number of the state $(n'l)$. The n-dependence is displayed in the limit. For a valence electron, the upper limit of the summation in D_{0l} is always the core shell with the highest n; in the present discussion the upper limit is the principal quantum number of the shell below the selected (nl) state. The $\hat{P}_{nl}$ in Eq. (5.48) is the radial part of φ_s.

Considering now a molecule, we use for Φ_s the expression of Eq. (3.63):

$$\Phi_s = \frac{1}{2}(3\pi^2)^{2/3}[\rho_0^{2/3} + 2\rho_0^{1/3}(\rho^{1/3} - \rho_0^{1/3})], \qquad (5.50)$$

where we obtain, in a straightforward fashion, from Eqs. (3.57) and (3.58),

$$\rho_0 = \sum_{i=1}^{s-1} |\varphi_i|^2, \qquad (5.51)$$

and

$$\rho = \rho_0 + |\varphi_s|^2. \qquad (5.52)$$

Thus for an atom our pseudopotential is given by Eq. (5.46), and for a molecule by Eq. (5.50).

B. Pseudo-Orbital Condition

Having replaced the exact pseudopotential by the density-dependent expression, the equation for the PO's is Eq. (5.45). Equation (5.40) for the total energy must also be modified, since it contains the exact V_s. Replacing V_s by Φ_s we obtain

$$E_T = E_0[\varphi_i] + \sum_{s=1}^{N} \langle \psi_s|H_s|\psi_s\rangle + \langle \psi_s|\Phi_s|\psi_s\rangle. \qquad (5.53)$$

We want to clarify the mathematical structure of this expression. In order to do so we write down here again the equations for the PO's:

$$(H_s + \Phi_s)\psi_s = \epsilon_s\psi_s, \quad (s = 1, 2, \ldots, N), \qquad (5.54)$$

where the Hamiltonian operator is given by Eq. (5.35):

$$H_s = t + U_s = t + U'_s - \Omega_s H'_s, \qquad (5.55)$$

and the pseudopotential is either by Eq. (5.46) or by Eq. (5.50).

Equations (5.54) must be solved with a self-consistent procedure. The results of the calculations are the PO's $\psi_1 \cdots \psi_N$, but we must realize that the orthogonal orbitals have not disappeared; the potentials in the U'_s and H'_s as well as the pseudopotentials are defined in terms of the orthogonal orbitals $\varphi_1 \cdots \varphi_N$, which are related to the PO's by Eq. (5.42). Consequently the expression E_T in Eq. (5.53) will be a functional of the orthogonal orbitals $\varphi_1 \cdots \varphi_N$, as well as the PO's $\psi_1 \cdots \psi_N$. Let us indicate this functional dependence by the following notation:

$$E_T = E_T[\varphi_i, \psi_i] . \tag{5.56}$$

We introduce now an approximation, which will be called, for the sake of easier reference, the *pseudo-orbital condition*. This approximation consists of replacing the orthogonal orbitals everywhere by the PO's, putting

$$E_T[\varphi_s, \psi_s] \rightarrow E_T[\psi_s, \psi_s] . \tag{5.57}$$

This step means that the functional form of the energy expression as well as of the various potentials remains unchanged, but the place of the orbitals $\varphi_1 \cdots \varphi_N$ is everywhere taken by the PO's $\psi_1 \cdots \varphi_N$. With this step the E_T becomes solely the functional of the pseudo-orbitals.

An example here will be useful. Let us consider a system with closed shells. The exchange interaction between the orbital ψ_s and the rest of the orbitals is given, in Eq. (5.54), by the expression

$$U_e \psi_s = \int \frac{1}{r_{12}} \sum_{t=1}^{N} \varphi_t(1) \varphi_t^*(2) \psi_s(2) \, dq_2 . \tag{5.58}$$

After the introduction of the PO condition this becomes

$$U_e \psi_s = \int \frac{1}{r_{12}} \sum_{t=1}^{N} \psi_t(1) \psi_t^*(2) \psi_s(2) \, dq_2 . \tag{5.59}$$

The total density of the system in the exact pseudopotential theory is given by

$$\rho(r) = \sum_{s=1}^{N} |\varphi_s(r)|^2 , \tag{5.60}$$

where φ_s is given by Eq. (5.42). After the replacement of the exact potentials by the density-dependent expressions, Eq. (5.60) is still valid. After the introduction of the PO condition we have

$$\rho(r) = \sum_{s=1}^{N} |\psi_s(r)|^2 . \tag{5.61}$$

In order to elucidate the physical meaning of the PO condition we observe that after the replacement of the exact pseudopotentials by the density-dependent expressions, all potentials, with the exception of the exchange interactions, are functionals of the (one-electron or total) densities only. Thus, apart from the exchange potentials, the PO condition actually means the transition

$$|\varphi_s|^2 \rightarrow |\psi_s|^2 . \tag{5.62}$$

This is a noteworthy point because the approximation of Eq. (5.62) is more plausible than the transition $\varphi_s \rightarrow \psi_s$. In Section 3.2.A we used the condition of Eq. (5.62) for justifying the introduction of the density-dependent potentials for the valence electron of an atom [see Eq. (3.34)]. After Eq. (3.34) was postulated, we showed that the structure of the computed exact PO's shows the plausibility of this approximation. Thus, apart from the exchange potentials, the PO condition means an approximation which has been shown to be plausible in the atomic case.

In the very general form postulated here, the PO condition was introduced by the author.[49] Earlier developments, however, clearly pointed in the direction of this approximation. As we have mentioned above, the replacement of the density formed from orthogonal orbitals by the density formed from PO's was implicitly contained in Slater's early work [Slater I, p. 368]. In formula, Slater's idea amounts to the approximation

$$\sum_{s=1}^{N} |\varphi_s|^2 \rightarrow \sum_{s=1}^{N} |\psi_s|^2 . \tag{5.63}$$

This approximation was adopted by Gombas and Gaspar.[50] In addition to the approximation, Eq. (5.62), which was discussed in Section 3.2.A, Szepfalusy introduced (for atoms with closed shells) the transformation

$$\sum_{s=1}^{N} \varphi_s(1)\varphi_s^*(2) \rightarrow \sum_{s=1}^{N} \psi_s(1)\psi_s^*(2) , \tag{5.64}$$

where the first-order density matrix was approximated by pseudo-orbitals.[53] From this it is only one step to the general formulation of the author.

C. Formulation of the APM

Our goal here is to give a simple prescription for how to set up an APM as the replacement for an ab initio formalism. Let us consider an ab initio one-electron formalism—for example, any form of the HF approximation, the $X\alpha$ model, and so on.

Let us assume that the formalism is based on the one-electron orbitals $\varphi_1 \cdots \varphi_N$ which are solutions of the equations

$$H_s \varphi_s = \epsilon_s \varphi_s , \quad (s = 1, 2, \ldots, N) . \tag{5.65}$$

The Pauli exclusion principle is taken into account by requiring that the orbitals $\varphi_1 \cdots \varphi_N$ be orthogonal. The total energy of the system is given by

$$E_T = E_0[\varphi_i] + \sum_{s=1}^{N} \langle \varphi_s | H_s | \varphi_s \rangle . \qquad (5.66)$$

The total density of the system is

$$\rho = \sum_{s=1}^{N} |\varphi_s|^2 . \qquad (5.67)$$

The prescription for setting up the APM model is as follows. Introduce the pseudo-orbitals $\psi_1 \cdots \psi_N$. The only difference between the orbitals $\varphi_1 \cdots \varphi_N$ and these PO's is that the latter do not have to obey any orthogonality conditions. The PO's are the solutions of the equations

$$(H_s + \Phi_s)\psi_s = \epsilon_s \psi_s, \qquad (s = 1, 2, \ldots, N) . \qquad (5.68)$$

The Pauli exclusion principle is taken into account by the density-dependent pseudopotentials Φ_s. The total energy of the system is now given by

$$E_T = E_0[\psi_i] + \sum_{s=1}^{N} \langle \psi_s | H_s | \psi_s \rangle + \sum_{s=1}^{N} \langle \psi_s | \Phi_s | \psi_s \rangle , \qquad (5.69)$$

and the total density is

$$\rho = \sum_{s=1}^{N} |\psi_s|^2 . \qquad (5.70)$$

Although the notation does not show it, the operator H_s in Eqs. (5.68)–(5.70) is not the same as the operator H_s in Eqs. (5.65)–(5.67). The former is obtained from the latter by replacing everywhere the orthogonal orbitals $\varphi_1 \cdots \varphi_N$ by the pseudo-orbitals $\psi_1 \cdots \psi_N$. Especially simple is the juxta-position of Eqs. (5.69) and (5.66). From these two formulas we see that the APM energy expression is obtained from the conventional by replacing the set $\varphi_1 \cdots \varphi_N$ by the PO set $\psi_1 \cdots \psi_N$ and adding to the resultant expression the expectation value of the pseudopotentials with respect to the PO's, which is the last term in Eq. (5.69).

5.4. SINGLE ZETA CALCULATIONS FOR THE AVERAGE ENERGY OF ATOMIC CONFIGURATIONS

To test the ideas presented in this chapter, the author carried out cal-culations of total energies and densities of atoms between Li (Z = 3) and Kr

($Z = 36$). The approximation chosen was the HF model for the average energy of atomic configurations, developed by Slater [Slater I, p. 322, and Slater II, p. 27]. In a certain sense this approach occupies a central position in atomic physics. The correct energy of any atomic configuration differs from the average by only a few terms. In presenting this approach Slater remarked that "the weighted average of a configuration has a fundamental meaning since it is determined by a simple rule." The approach is obviously related to the $X\alpha$ model and, through that, to the density-functional formalism. The exact, numerical solutions of the HF equations of this model for neutral atoms have been obtained by Froese–Fischer[3] and provide us with material against which we can check our results (*ab initio* versus *model* comparison).

In setting up an APM model for the average of a configuration we follow the prescription in Section 5.3.C. Looking at the formulas for the energy expression, Eqs. (5.66) and (5.69), we see that for our case the two terms on the right side of Eq. (5.66) become the HF energy, which we denote by $E_{HF}[\varphi_i]$. Using this notation we get from Eq. (5.66)

$$E_T = E_{HF}[\varphi_i]. \tag{5.71}$$

For the energy expression of the APM model we obtain, using Eqs. (5.69) and (5.71),

$$E_T = E_{HF}[\psi_i] + \sum_{s=1}^{N} \langle \psi_s | \Phi_s | \psi_s \rangle. \tag{5.72}$$

Thus what we must do is take the HF energy expression in its original form, replace the orthogonal orbitals in it by the PO's, and add to the resulting expression the last term on the right side of Eq. (5.72).

The expression for E_{HF} is given by Slater [Slater I, Sec. 14.2, Eqs. (14.19), (14.21), and (14.23)]. Using those formulas we obtain,

$$E_{HF} = \sum_{n,l} q(n,l) I(nl) + \sum_{n,l} \left[\frac{q(n,l)}{2} \right] [q(n,l) - 1]$$

$$\times \left\{ F_0(nl, nl) - \frac{1}{(4l+1)} \sum_k c^k(l,l) F_k(nl, nl) \right\} + \sum_{n,l} \sum_{n',l'} q(n,l) q(n'l')$$

$$\times \left\{ F_0(nl, n'l') - \frac{[(2l+1)(2l'+1)]^{-1/2}}{2} \sum_k c^k(l,l') G_k(nl, n'l') \right\}. \tag{5.73}$$

In this energy expression $q(nl)$ is the number of electrons in the (nl) group; the summation over (nl) is for all filled states in the configuration; in the last term either $n' \neq n$ or $l' \neq l$ or both; the c^k are angular coefficients tabulated by Slater [Slater II, p. 281]. The quantities I, F, and G are the usual Slater integrals given by the formulas

$$I(nl) = -\frac{1}{2} \int_0^\infty P_{nl}(r) \left[\frac{d^2}{dr^2} + \frac{2Z}{r} - \frac{l(l+1)}{r^2} \right] P_{nl}(r) \, dr \; ; \tag{5.74}$$

$$F_k(nl, n'l') = \int P_{nl}^2(r_1) P_{n'l'}^2(r_2) L_k(r_1, r_2) \, dr_1 \, dr_2 \; ; \tag{5.75}$$

$$G_k(nl, n'l') = \int P_{nl}(r_1) P_{n'l'}(r_1) P_{nl}(r_2) P_{n'l'}(r_2) L_k(r_1, r_2) \, dr_1 \, dr_2 \, , \tag{5.76}$$

with

$$L_k(r_1, r_2) = \frac{r_<^k}{r_>^{k+1}} . \tag{5.77}$$

Equation (5.73) is Slater's original expression; that is, the orbitals in this expression are the radial parts of the orthogonal HF orbitals. In accordance with our discussion in Section 2.1 and in subsequent sections, we should have used the symbol $\hat{P}_{nl}$ to denote these orbitals. As the expressions stands it means $E_{HF}[\psi_i]$; that is, we have skipped the step of writing down $E_{HF}[\varphi_i]$, and have written down $E_{HF}[\psi_i]$ directly. The orbitals P_{nl} in Eq. (5.73) are the radial parts of the pseudo-orbitals; that is, the relationship between our ψ_i and the P_{nl} is given by the formula

$$\psi_i = \psi_{n_i l_i m_{l_i} m_{s_i}} = \frac{P_{n_i l_i}(r)}{r} Y_{l_i m_{l_i}}(\vartheta, \varphi) \eta_{m_{s_i}}(\sigma) \, , \tag{5.78}$$

where we have used the same symbols as in Eqs. (2.18) and (2.19).

Having written down the first term of Eq. (5.72), we turn now to the second term. In $\langle \psi_s | \Phi_s | \psi_s \rangle$ we use Eq. (5.78) for ψ_s and Eq. (5.48) for Φ_s. The radial density appearing in Eq. (5.48) is given by Eq. (5.49). We replace again the orthogonal radial orbitals $\hat{P}_{nl}$ by the non-orthogonal P_{nl} and obtain, after the spin and angular integration,

$$\langle \psi_s | \Phi_s | \psi_s \rangle = E_p[P_{nl}]$$

$$= \frac{\pi^2}{8(2l+1)^2} \sum_{n'=l+1}^{n-1} q(n'l) \left\{ q(n'l) \int_0^\infty P_{nl}^2(r) P_{n'l}^4(r) \, dr \right.$$

$$+ 2 \sum_{n''=n'+1}^{n-1} q(n''l) \int_0^\infty P_{nl}^2(r) P_{n'l}^2(r) P_{n''l}^2(r) \, dr + 2 \left. \int_0^\infty P_{nl}^4(r) P_{n'l}^2(r) \, dr \right\} . \tag{5.79}$$

The physical meaning of this quantity is that this is the expectation value of the pseudopotential Φ_s with respect to the PO ψ_s; that is, this is the orthogonalization energy (resulting from the Pauli exclusion principle) of the non-orthogonal electron state ψ_s.

We note that the mathematical structure of Eq. (5.79) is very simple: the density-dependent pseudopotential leads to integrals which are even more elementary than the otherwise rather simple integrals in Slater's expression. From the point of view of simplicity Eq. (5.79) compares favorably with the energy expression which we would obtain by averaging the exact PK pseudopotential [Eq. (A.11)].

On putting Eqs. (5.73) and (5.79) into Eq. (5.72), we obtain the APM model energy expression for the average energy of a configuration. Before writing this expression down let us introduce a constant factor η which is similar to the α constant of the $X\alpha$ model. This constant regulates the "strength" of the density-dependent pseudopotential and serves to compensate partly for the approximations which were made by the replacement of the exact pseudopotential by the density-dependent. Thus we obtain

$$E_T = E_{\mathrm{HF}}[P_{nl}] + \sum_{n,l} q(n,l)\eta E_p[P_{nl}], \qquad (5.80)$$

where $E_{HF}[P_{nl}]$ is Slater's expression [Eq. (5.73)], and $E_p[P_{nl}]$ is given by Eq. (5.79). Both are functionals of the radial parts of the PO [Eq. (5.78)].

In order to obtain the absolute minimum of E_T we would have to vary Eq. (5.80) with respect to the orbitals P_{nl}. In this process only the normalization would have to be introduced as a subsidiary condition; there would be no Lagrangian multipliers, since the P_{nl} do not have to obey any orthogonality conditions. The resulting equation would have to be solved by a self-consistent procedure. Substituting the resulting solutions back into Eq. (5.80), we would get the minimum of the energy E_T.

In order to get an idea of the usefulness of the method without first getting into extensive calculations, the author has carried out single Zeta calculations for atoms up to $Z = 36$ (Kr). By "single Zeta" we mean a calculation in which every PO is represented by a single Slater-type orbital (STO) of the form

$$P_{nl}(r) = A_{nl}r^n \exp[-\xi_{nl}r], \qquad (5.81)$$

where A_{nl} is a normalization constant. The term "single Zeta" indicates that, instead of using an expansion of STO's like that used in the Roothaan-type calculations[54] or the one which we used to represent the exact PO's for a valence electron [Eq. (A.15)], we here represent each pseudo-orbital by a *single* STO. We note that the electron orbitals in an atom can be represented by a single STO only if the wave functions are PO's; otherwise the orthogonality conditions would require the use of at least two STO's for most orbitals.

The calculations were carried out in such a way that the PO of Eq. (5.81) was put into Eq. (5.80) and the integrations carried out. The resulting E will be a closed, analytic expression of the Zeta parameters plus the η constant.

$$E_T = E_T(\xi_{n_1 l_1}, \xi_{n_2 l_2}, \dots, \xi_{n_N l_N}, \eta). \qquad (5.82)$$

For the actual calculation a simple computer program was written which determined the parameters by varying them independently. The resulting E_T is the absolute minimum of the energy in the single Zeta approximation. The value of the Zeta's as well as the value of E_T depends on the choice of η, which is not a variational parameter; the energy does not possess a minimum in terms of η. For each choice of η, which must be determined by physical considerations, we obtain a unique set of Zeta's and a unique E_T.

Calculations were carried out for all neutral atoms between Li and Ar and for every second neutral atom between Ar and Kr. The author determined by trial and error that for this range of neutral atoms good results can be obtained by putting

$$\eta = 1.5, \qquad (5.83)$$

for all atoms between $N = 3$ and $N = 36$.

The results are presented in Table 5.1 and in Figures 5.1 through 5.4. In Table 5.1 we have the Zeta parameters along with the computed total energies. For comparison, we have also the total HF energies computed by Froese–Fischer plus the absolute values of the differences between the two energies in percents. In the figures we have the radial densities of the atoms Ne ($Z = 10$), Ca ($Z = 20$), Fe ($Z = 26$), and Kr ($Z = 36$). The APM radial densities (dashed lines) in these diagrams are computed from the formula

$$D = \sum_{n, l} q(n, l) P_{nl}^2(r), \qquad (5.84)$$

which corresponds to Eq. (5.70). For comparison we show also the HF radial densities (full lines) which are plotted from the formula

$$D = \sum_{n, l} q(n, l) \hat{P}_{nl}^2(r). \qquad (5.85)$$

Thus, in accordance with the pseudo-orbital condition, our densities are computed from the nodeless, single-Zeta, pseudo-orbitals [Eq. (5.81)]. The HF densities are computed in the conventional fashion, using the HF orbitals with their $(n - l - 1)$ nodes.

Looking at the total energies we see that the agreement between our values and the HF results is very good. The agreement between our

TABLE 5.1. The Results of the Single Zeta Calculations for the Atoms between Lithium and Krypton

Atom	Z	1s	2s	2p	3s	3p	3d	4s	4p	E	E_{HF}	%
Li	3	2.70	0.65							−7.443	−7.433	0.1
Be	4	3.75	0.97							−14.652	−14.573	0.5
B	5	4.80	1.27	1.20						−24.644	−24.529	0.5
C	6	5.80	1.57	1.55						−37.791	−37.689	0.3
N	7	6.85	1.87	1.90						−54.410	−54.401	0.02
O	8	7.87	2.15	2.22						−74.823	−74.810	0.02
F	9	8.87	2.42	2.57						−99.345	−99.409	0.06
Ne	10	9.92	2.71	2.87						−128.295	−128.547	0.2
Na	11	10.97	3.05	3.42	0.90					−161.689	−161.860	0.1
Mg	12	12.02	3.42	3.92	1.125					−199.520	−199.615	0.05
Al	13	13.07	3.77	4.45	1.32	1.25				−241.856	−241.877	0.01
Si	14	14.12	4.15	4.97	1.52	1.50				−288.910	−288.855	0.02
P	15	15.17	4.50	5.50	1.67	1.72				−340.855	−340.719	0.04
S	16	16.22	4.84	6.04	1.84	1.95				−397.856	−397.506	0.2
Cl	17	17.27	5.20	6.60	2.00	2.19				−460.084	−459.483	0.1
Al	18	18.38	5.54	7.16	2.15	2.40				−527.709	−526.819	0.2
Ca	20	20.50	6.32	8.27	2.62	2.95		1.05		−677.952	−676.760	0.2
Ti	22	22.60	7.00	9.32	2.94	3.42	2.70	1.15		−849.336	−848.409	0.1
Cr	24	24.75	7.70	10.37	3.20	3.82	3.00	1.05		−1043.22	−1043.31	0.01
Fe	26	26.90	8.45	11.42	3.55	4.25	3.80	1.30		−1261.89	−1262.44	0.04
Ni	28	29.10	9.20	12.52	3.85	4.62	4.30	1.35		−1505.16	−1506.88	0.1
Zn	30	31.10	9.85	13.57	4.15	5.07	4.70	1.40		−1774.54	−1777.85	0.2
Ge	32	33.30	10.60	14.62	4.55	5.50	5.50	1.80	1.90	−2071.94	−2075.37	0.2
Se	34	35.40	11.30	15.72	4.85	6.07	6.25	2.00	2.35	−2396.33	−2399.87	0.15
Kr	36	37.52	12.02	16.80	5.27	6.60	7.00	2.20	2.70	−2748.46	−2752.06	0.1

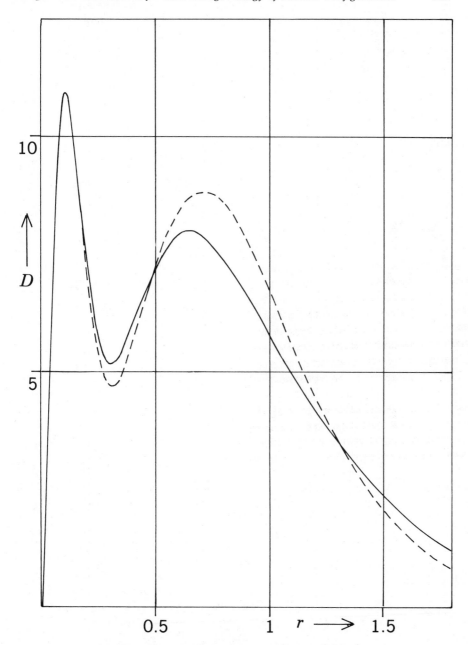

FIGURE 5.1.   The total radial density of the Ne atom.

densities and the HF densities is very good for Ca and Kr and fair for Ne and Fe. It is an important point that the maxima and minima of the densities are at the right place; this shows that the density-dependent pseudopotentials are pushing the electron orbits accurately into the right position; that is,

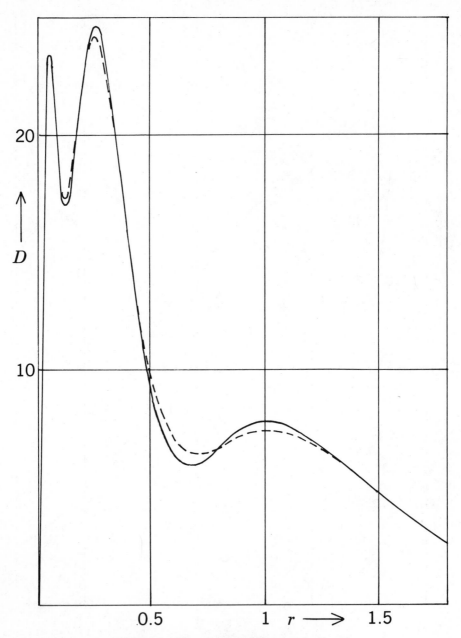

FIGURE 5.2. The total radial density of the Ca atom.

the pseudopotentials are "working" as they should. The simple device of introducing the η parameter is crucial for this, as we found out in trial and error calculations.

Summing up, we may say that for this group of atoms the results for the

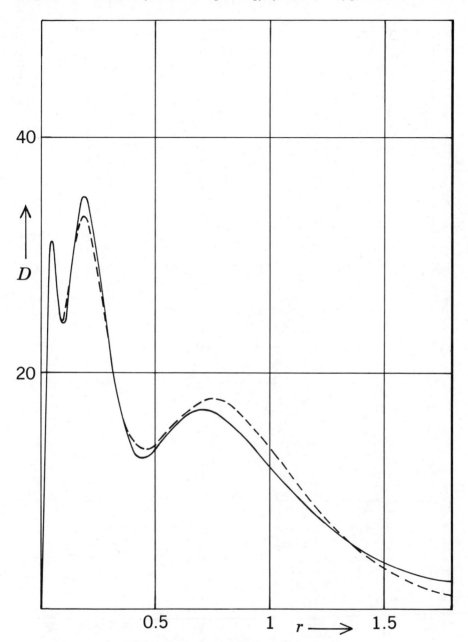

FIGURE 5.3.   The total radial density of the Fe atom.

total energies as well as for the densities are remarkably good considering the extreme mathematical simplicity of this model. As to the applicability of the model to heavier atoms, the same formulas and computer programs can be used for the heaviest atom as for the lightest; whether the single Zeta

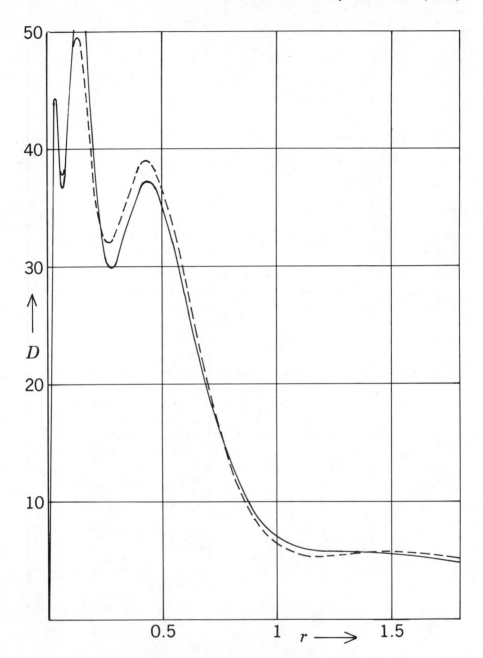

FIGURE 5.4.   The total radial density of the Kr atom.

approximation will be accurate enough for the more complicated orbitals of the heavier atoms remains to be seen. In any case, the results are encouraging for further applications of this model.

5.5. *Xα* MODEL WITH PSEUDOPOTENTIALS

In 1951 Slater suggested[9] that the operator for the exchange interaction in the HF equations could be replaced by a density-dependent local potential derived from the TF model. Later Kohn and Sham have shown[22] that using the exact, universal density functional, a method can be formulated in which the exact energy and density of an atom or molecule can be obtained through the iterative solution of a set of one-electron equations. If, instead of the exact density functional, which is not known at the present time, the TF approximation is used, this method becomes essentially identical with Slater's. The formulation usually referred to as the *Xα* model is the method of Slater and Wood,[55] in which the "strength" of the exchange interaction is regulated by the adjustable factor α.

It was shown by the author[23] that the *Xα* model can be transformed into a pseudopotential formalism. Since the method is completely analogous to the APM model outlined above, here we only refer to the relevant publications. Besides the general formulation, the author has shown that the *Xα* method with pseudopotentials satisfies the Virial theorem just as the original formulation satisfies it.[56] The author has also shown that, in this model, the effective potential in which an electron of an atom is moving depends essentially only on the principal quantum number. It is easy to show that this theorem explains the shell structure of atoms.[23]

chapter 6

The Quantum Theory of Atoms With Two Valence Electrons

6.1. AB INITIO THEORY

A. Introduction

In Section 1.1, where we introduced the concept of pseudopotentials by presenting the ideas of Hellmann, we observed that in many cases the important physical properties of atoms are determined by the valence electrons. For example, the optical spectra of the alkali earth atoms Mg, Ca, Sr, and Ba are typical two-electron spectra although these atoms contain, besides the two valence electrons, many other electrons in their inner shells. We described how Hellmann introduced the method of pseudopotentials for the treatment of atoms of this kind.

At about the same time, the problem was examined from a different angle by Fock. Having decisively contributed to the quantum theory of atoms by supplying the theoretical framework for the Hartree–Fock method for atoms with closed shells,[57] as well as for atoms with one valence electron outside closed shells,[6] Fock turned his attention to atoms with more than one valence electron. It is evident that the HF theory can be extended to such atoms in a straightforward manner; it is also clear, however, that in an atom with two valence electrons the valence electrons are, in many cases, more important than the core electrons. Thus the properties of the valence electrons should be determined with greater accuracy than the properties of the core electrons. The HF method, which treats all electrons on an equal footing, does not satisfy this requirement. hence the idea emerged that, for atoms with two valence electrons, a method should be designed in which the valence electrons are treated as a unit, separately from the core electrons, with the electron correlation between them fully taken into account. For the treatment of the core electrons it is sufficient to apply the HF approximation in which the correlation effects are omitted. These are the ideas embodied in the Theory of Partial Separation of the Variables of Fock, Vaselov, and Petrashen.[58]

The theory of FVP is presented here in detail. This is necessary, since the FVP method serves as the basis for the pseudopotential method; in addition, the discussion will be useful *per se* in view of the fact that, while the major textbooks generally treat the one-electron approximations in great detail, their treatment of the FVP method is either sketchy or it is omitted altogether.

The discussion in this chapter is restricted to two-valence-electron atoms. It will be shown in the next chapter that the theory of atoms with more than two valence electrons is a straightforward generalization of the two-valence-electron theory; thus the ideas presented here are applicable to atoms with any number of valence electrons. Furthermore, we shall show in chapter 8 that the theory of a large class of molecules can be based on the equations developed here. Thus it can be stated without exaggeration that the FVP theory has a pivotal significance for the whole atomic and molecular theory.

B.   The Theory of Fock, Veselov, and Petrashen: Correlated Pair Function for the Valence Electrons

The presentations of Section 6.1 are based on several papers of the author in which the original theory of FVP was simplified as well as extended. The simplification consisted of the introduction of the frozen core approximation.[59] The extensions clarified some mathematical details, creating a more coherent formulation.[60]

Let us consider an atom with N core and two valence electrons. The Hamiltonian will be

$$H = \sum_{i=1}^{N+2} (t_i + g_i) + \frac{1}{2} \sum_{i,j=1}^{N+2} \frac{1}{r_{ij}}, \qquad (6.1)$$

where t_i and g_i are, as before, the kinetic energy operator and the nuclear potential, respectively; r_{ij} is the interelectronic distance. In the absence of the two valence electrons, we have an N-electron system with closed shells, the Hamiltonian of which is

$$H = \sum_{i=1}^{N} (t_i + g_i) + \frac{1}{2} \sum_{i,j=1}^{N} \frac{1}{r_{ij}}. \qquad (6.2)$$

For the wave function of the N-electron atom we put the HF determinant [Eq. (2.1)]:

$$\Psi_F = (N!)^{-1/2} \det[\varphi_1 \cdots \varphi_N], \qquad (6.3)$$

where it is assumed that the spin-orbitals are orthonormal:

$$\langle \varphi_i \mid \varphi_j \rangle = \delta_{ij}, \quad (i, j = 1, 2, \ldots, N). \qquad (6.4)$$

Variation of the total energy (the average value of the Hamiltonian [Eq. (6.2)] formed with the wave function [Eq. (6.3)] with respect to the spin-orbitals gives the HF equations,

$$H_F \varphi_i = \epsilon_i \varphi_i, \quad (i = 1, 2, \ldots, N), \qquad (6.5)$$

where

$$H_F = t + g + U, \qquad (6.6)$$

and

$$U = \sum_{i=1}^{N} U_i. \qquad (6.7)$$

The operator U_i is the same as in Eq. (2.5).

Now let us add the two valence electrons to the atom. In a HF approximation, in which the wave function is a single determinant, we can put the total wave function in the form,

$$\Psi_F = [(N+2)!]^{-1/2} \det[\varphi_1 \cdots \varphi_N \varphi_{N+1} \varphi_{N+2}]$$

$$= [(N+2)!]^{-1/2} \begin{vmatrix} \varphi_1(1) & \varphi_1(2) & \cdots & \varphi_1(N+2) \\ \varphi_2(1) & \varphi_2(2) & \cdots & \varphi_2(N+2) \\ \vdots & & \vdots & \vdots \\ \varphi_N(1) & \varphi_N(2) & \cdots & \varphi_N(N+2) \\ \varphi_{N+1}(1) & \varphi_{N+1}(2) & \cdots & \varphi_{N+1}(N+2) \\ \varphi_{N+2}(1) & \varphi_{N+2}(2) & \cdots & \varphi_{N+2}(N+2) \end{vmatrix}, \qquad (6.8)$$

where we have denoted the spin-orbitals assigned to the valence states by φ_{N+1} and φ_{N+2}. We assume again that the spin-orbitals are orthonormal; therefore, in addition to Eq. (6.4), we have

$$\begin{aligned} \langle \varphi_i \mid \varphi_{N+1} \rangle &= 0 \,, \\ \langle \varphi_i \mid \varphi_{N+2} \rangle &= 0 \,, \end{aligned} \quad (i = 1, 2, \ldots, N) \,. \qquad (6.9)$$

Variation of the total energy (the average value of the Hamiltonian [Eq. (6.1)] formed with the wave function [Eq. (6.8)]) with respect to the spin-orbitals gives the HF equations for the $(N + 2)$-electron atom,

$$H_F \varphi_i = \epsilon_i \varphi_i, \quad (i = 1, \ldots, N, N+1, N+2), \qquad (6.10)$$

where Eq. (6.6) is again valid but now

$$U = \sum_{i=1}^{N+2} U_i \,. \qquad (6.11)$$

Let us form the Laplace expansion of ψ_F in Eq. (6.8) in terms of the last two rows:

$$\Psi_F = [(N+2)!]^{-1/2} \sum_{s=1}^{N+2} \sum_{t=s+1}^{N+2} (-1)^{s+t+(N+1)+(N+2)}$$

$$\times \begin{vmatrix} \varphi_{N+1}(s) & \varphi_{N+1}(t) \\ \varphi_{N+2}(s) & \varphi_{N+2}(t) \end{vmatrix} D^{(N)}(\varphi_{N+1}\varphi_{N+2} \mid s, t) \,. \qquad (6.12)$$

In this expression $D^{(N)}(\varphi_{N+1}, \varphi_{N+2} \mid s, t)$ is an $(N \times N)$ determinant which is the (complementary) minor of the (2×2) determinant standing in front of it, and is obtained from ψ_F by deleting the rows with φ_{N+1} and φ_{N+2} and the columns containing the coordinates s and t. In a simplified notation Eq. (6.12) can be written as

$$\Psi_F = [(N+2)!]^{-1/2} \tilde{A} \left\{ \begin{vmatrix} \varphi_{N+1}(1) & \varphi_{N+1}(2) \\ \varphi_{N+2}(1) & \varphi_{N+2}(2) \end{vmatrix} \det[\varphi_1(3)\varphi_2(4) \cdots \varphi_N(N+2)] \right\},$$

$$(6.13)$$

where $\tilde{A}$ is an antisymmetrizer operator which generates from the curly bracket the double sum shown in Eq. (6.12). Clearly we have

$$\det[\varphi_1(3)\varphi_2(4) \cdots \varphi_N(N+2)] = D^{(N)}(\varphi_{N+1}\varphi_{N+2} \mid 1, 2) \,. \qquad (6.14)$$

In order to introduce the electron correlation between the two valence electrons, we now replace in Eq. (6.12) the (2×2) determinant which represents the valence electrons by a completely arbitrary two-electron function which is subjected, besides the standard requirements of square integrability, single valuedness, and continuity, only to the requirement of antisymmetry; that is, we put as our basic trial function,

$$\Psi_T = [(N+2)!]^{-1/2} \tilde{A}\{\Psi(1,2) \det[\varphi_1(3)\varphi_2(4) \cdots \varphi_N(N+2)]\}. \quad (6.15)$$

This is a wave function in which the electron correlation between the two valence electrons can be fully taken into account, while the core electrons are represented by the one-electron spin-orbitals $\varphi_1 \cdots \varphi_N$. Thus the electron correlation between the core electrons themselves, as well as the core-valence correlation effects, are omitted[*].

The introduction of a two-electron function which appears in the total wave function, Eq. (6.15), together with one-electron orbitals necessitates the introduction of a new type of orthogonality condition. Let us assume that the one-electron orbitals satisfy the condition, Eq. (6.4). We form from these orbitals projection operators according to Eqs. (2.50) and (2.52):

$$\Omega_1 = \sum_{i=1}^{N} \langle 1 | \varphi_i \rangle \langle \varphi_i |, \quad (6.16)$$

and

$$P_1 = 1 - \Omega_1. \quad (6.17)$$

The subscripts indicate that these operators act on functions depending on the coordinate quartet $q_1 \equiv (x_1 y_1 z_1 \sigma_1)$. Using these equations we build up the two-electron operators

$$P \equiv P_1 P_2 = (1 - \Omega_1)(1 - \Omega_2), \quad (6.18)$$

and

$$\Omega = 1 - P. \quad (6.19)$$

In this chapter operators without subscripts will always mean two-electron operators. The properties of P and Ω are summarized in Appendix E.

The function Ψ which appears in Eq. (6.15) does not need to satisfy any orthogonality condition, since the Pauli exclusion principle demands only that the wave function be antisymmetric. For mathematical convenience we introduce now the so-called "strong" orthogonality condition, which is a functional relationship of the following form: the two-electron function $f(1, 2)$ is strong orthogonal to the orbitals $\varphi_1 \cdots \varphi_N$ if it satisfies the relationship

---

[*] A discussion of the additional inclusion of these effects will follow in Chapter 7.

$$\int \varphi_i^*(1)f(1,2)\,dq_1 = 0\,, \quad (i = 1, 2, \ldots, N)\,. \tag{6.20}$$

The relationship must be valid for all values of q_2. Now it is easy to see that if $\Psi(1, 2)$ is an arbitrary function, then we can generate a strong-orthogonal function by putting

$$\Phi(1, 2) = P\Psi(1, 2)\,. \tag{6.21}$$

Indeed, substituting Φ into Eq. (6.20) we obtain

$$\int \varphi_i^*(1)\Phi(1,2)\,dq_1 = \int \varphi_i^*(1)P\Psi(1,2)\,dq_1$$

$$= \int \varphi_i^*(1)P_1P_2\Psi(1,2)\,dq_1 = P_2\int \varphi_i^*(1)(1-\Omega_1)\Psi(1,2)\,dq_1$$

$$= P_2\left\{\int \varphi_i^*(1)\Psi(1,2)\,dq_1 - \sum_{k=1}^{N}\int \varphi_i^*(1)\varphi_k(1)\varphi_k^*(1')\Psi(1'2)\,dq_1\,dq_1'\right\}$$

$$= P_2\left\{\int \varphi_i^*(1)\Psi(1,2)\,dq_1 - \int \varphi_i^*(1')\Psi(1'2)\,dq_1'\right\} = 0\,, \tag{6.22}$$

that is, we see that Φ satisfies Eq. (6.20) regardless of the form of Ψ.

The wave function of Eq. (6.15) has the property that it does not change if Ψ is replaced in it by $\Phi = P\Psi$. In formula:

$$\Psi_T = [(N+2)!]^{-1/2}\tilde{A}\{\Psi(1,2)\det[\varphi_1(3)\cdots\varphi_N(N+2)]\}$$

$$= [(N+2)!]^{-1/2}\tilde{A}\{\Phi(1,2)\det[\varphi_1(3)\cdots\varphi_N(N+2)]\}\,. \tag{6.23}$$

The proof of this statement is given in Appendix F.

This formula means that the total wave function of the atom does not change if we replace in it the two-electron function Ψ by the strong-orthogonalized two-electron function Φ, where the latter is formed according to Eq. (6.21). This theorem is complete analogous to the equivalence of Eqs. (2.6) and (2.7) discussed in Section 2.1. The reason for introducing the Φ, rather than working with the original Ψ, is that the expression for the total energy will be simpler with the former than with the latter. Thus, in the rest of this section our basic wave function will be, instead of Eq. (6.15), the function

$$\Psi_T = [(N+2)!]^{-1/2}\tilde{A}\{\Phi(1,2)\det[\varphi_1(3)\cdots\varphi_N(N+2)]\}\,, \tag{6.24}$$

where we shall assume that Φ is strong-orthogonal to the core orbitals,

$$\langle\varphi_i \,|\, \Phi(1,2)\rangle = \int \varphi_i^*(1)\Phi(1,2)\,dq_1 = 0\,, \quad (i = 1, 2, \ldots, N)\,, \tag{6.25}$$

and that it is normalized,

$$\langle\Phi\,|\Phi\rangle = \int |\varphi(1,2)|^2\, dq_1\, dq_2 = 1\,. \tag{6.26}$$

We now form the expectation value of the Hamiltonian [Eq. (6.1)] with respect to the Ψ_T of Eq. (6.24). Taking into account the subsidiary conditions [Eqs. (6.4), (6.25), and (6.26)], we obtain for the total energy of the atom,

$$E_T = \frac{\langle\Psi_T|H|\Psi_T\rangle}{\langle\Psi_T\,|\,\Psi_T\rangle} = \langle\Phi|H_{12}|\Phi\rangle + E_c\,, \tag{6.27}$$

where

$$H_{12} = H_1' + H_2' + \frac{1}{r_{12}}\,, \tag{6.28}$$

and

$$H_1' = t_1 + g_1 + U(1)\,, \tag{6.29}$$

with

$$U = \sum_{i=1}^{N} U_i\,. \tag{6.30}$$

U_i is again given by Eq. (2.5). The operator H_1' is formally identical with H_F which appears in Eq. (6.5). Despite the identity of the functional forms, however, the two operators are not exactly identical, since in the derivation of Eq. (6.27) it is not assumed that the one-electron orbitals are solutions of Eqs. (6.5). The quantity E_c appearing in Eq. (6.27) is the HF energy of the core electrons. The formulas for the expectation values of the standard many-electron Hamiltonian with respect to correlated functions are given in Appendix G.

In order to obtain the most accurate description of the valence electrons, which is the goal of the FVP theory, we now derive the equation for the best Φ. We vary the energy expression [Eq. (6.27)] with respect to the two-electron function Φ. The variation is carried out in such a way that the subsidiary conditions [Eqs. (6.25) and (6.26)] are taken into account and the core orbitals are kept fixed. The derivation is given in Appendix H. The result is the two-electron equation

$$\left(H_1 + H_2 + P\frac{1}{r_{12}}\right)\Phi = E\Phi\,, \tag{6.31}$$

where

$$H_i = P_i H_i', \quad (i = 1, 2). \tag{6.32}$$

This equation defines the best Φ, and with it the absolute minimum of the total energy of the atom for a given, unspecified set of one-electron orbitals $\varphi_1 \cdots \varphi_N$. While we obtain the absolute minimum of E_T with respect to the two-electron function Φ by solving Eq. (6.31) and substituting the result into Eq. (6.27), the total energy will still depend on the choice of the one-electron orbitals $\varphi_1 \cdots \varphi_N$.

There is one point which must be emphasized here. The generality of Eq. (6.31) is not restricted by the step in which we replaced the function Ψ by its orthogonalized counterpart Φ. We have seen that the total wave function was not changed by this substitution. As the total energy E_T depends only on the total wave function, it has not been changed either. Thus the variation of the total energy has not been in any way restricted by the orthogonalization of the valence-electron wave function to the core orbitals.

The equation for the best Φ, Eq. (6.31), lends itself to a very plausible physical interpretation. In order to see this, we write down here again the operator H_i in detailed form:

$$H_i = P_i H_i' = H_i' - \Omega_i H_i', \tag{6.33}$$

where

$$H_i' = t_i + g_i + \sum_{s=1}^{N} U_s(i), \quad (i = 1, 2). \tag{6.34}$$

The operator H_i consists of the Hamiltonian H_i' and the term $\Omega_i H_i'$, which is equivalent to the Lagrangian multipliers. (For comparison, see Eqs. (2.107)–(2.111). The operator H_i' represents, besides the kinetic energy operator, the complete Coulomb and exchange interaction potentials between a valence electron and the core electrons. The term which corresponds to the Lagrangian multipliers is numerically small, and is the same that we find in the HF equations when these are written down for nonclosed shells. Thus in Eq. (6.31) the term $(H_1 + H_2)$ represents the kinetic energy of the two valence electrons plus their Coulomb and exchange interactions with the core electrons.

We have in Eq. (6.31) the term

$$P \frac{1}{r_{12}} = \frac{1}{r_{12}} - \Omega \frac{1}{r_{12}}. \tag{6.35}$$

Here $1/r_{12}$ is the electrostatic interaction potential between the two valence electrons. The term with the Ω, which is similar to the second term in Eq.

(6.33), is the result of having taken into account the strong orthogonality requirement in the process of variation (i.e., it is the result of the Pauli exclusion principle).

Next we want to demonstrate what form our equation takes for two important special choices of the core orbitals. First let the core be "frozen" in the form of the N-electron atom; that is, let us assume that the one-electron orbitals satisfy the HF equations, Eq. (6.5). In this case the auxiliary operator $\hat{H}$, which we introduced in Appendix E, will be identical with H_F, and P_i will commute with H_F. Thus we obtain[59]

$$\left(H_1 + H_2 + P\frac{1}{r_{12}}\right)\Phi = \left(H_F(1)P_1 + H_F(2)P_2 + P\frac{1}{r_{12}}\right)\Phi$$

$$= \left(H_F(1) + H_F(2) + P\frac{1}{r_{12}}\right)\Phi = E\Phi, \qquad (6.36)$$

where we have used Eq. (E.12). The structure of this equation, along with its physical interpretation, is much simpler than in the case of Eq. (6.31).

Next let the core be frozen in the form of the neutral, $(N+2)$-electron atom; that is, let us assume that the orbitals satisfy Eqs. (6.10). We identify the auxiliary Hamiltonian $\hat{H}$ now again with H_F and put

$$H_i = P_i(H_i' + H_F(i) - H_F(i)). \qquad (6.37)$$

P_i will now commute with H_F. For the rest of the expression we get, using Eqs. (6.34) and (6.11),

$$H_i' - H_F(i) = t_i + g_i + \sum_{s=1}^{N} U_s(i) - t_i - g_i - \sum_{t=1}^{N+2} U_t(i)$$

$$= -U_{N+1}(i) - U_{N+2}(i), \qquad (6.38)$$

and on using Eqs. (6.37) and (6.38) in Eq. (6.31) we obtain

$$\left(H_1 + H_2 + P\frac{1}{r_{12}}\right)\Phi = \left\{P_1(H_F(1) - U_{N+1}(1) - U_{N+2}(1))\right.$$

$$\left. + P_2(H_F(2) - U_{N+1}(2) - U_{N+2}(2)) + P_{12}\frac{1}{r_{12}}\right\}\Phi = E\Phi. \qquad (6.39)$$

Taking into account Eq. (E.12) again and rearranging, we get

$$(H_F(1) + H_F(2) - PS_{12})\Phi = E\Phi, \qquad (6.40)$$

where

$$S_{12} = \frac{1}{r_{12}} - U_{N+1}(1) - U_{N+2}(1) - U_{N+1}(2) - U_{N+2}(2) .$$ (6.41)

Equation (6.40) has a structure similar to Eq. (6.36), except that in the place of the interaction potential, we now have the expression S_{12}.

We summarize now the most important mathematical properties of Eq. (6.31):[60]

1. The solutions of Eq. (6.31) must be strong-orthogonal to the core orbitals. This requirement follows from the derivation, but it also has a plausible physical interpretation. As we have seen, the strong orthogonality is accomplished by multiplying any function by the operator P. Let f be an arbitrary two-electron function, and let us expand it in terms of the complete set of the auxiliary operator $\hat{H}$ which we have introduced in Appendix E. We obtain

$$f(1, 2) = \frac{1}{2} \sum_{s=1}^{\infty} \sum_{t=1}^{\infty} a_{st} \mu_{st} ,$$ (6.42)

where

$$\mu_{st} = [\varphi_s(1)\varphi_t(2) - \varphi_s(2)\varphi_t(1)] ,$$ (6.43)

and the 1/2 is present to compensate for the independent summations in s and t. Operating on the expansion with the operators P and Ω, respectively, we obtain, by simple calculation, the expressions

$$Pf(1, 2) = \frac{1}{2} \sum_{s=N+1}^{\infty} \sum_{t=N+1}^{\infty} a_{st} \mu_{st} ,$$ (6.44)

and

$$\Omega f(1, 2) = \frac{1}{2} \left(\sum_{s=1}^{N} \sum_{t=1}^{N} + \sum_{s=1}^{N} \sum_{t=N+1}^{\infty} + \sum_{s=N+1}^{\infty} \sum_{t=1}^{N} \right) a_{st} \mu_{st} .$$ (6.45)

Thus we see that P has removed all core contributions from the function f; only the valence orbitals are present in the expansion, Eq. (6.44). If f is an arbitrary function which extends into the core area as well as into the valence area of the function space, then Pf will be a function which extends only into the valence area. Thus the strong orthogonalization prevents the two-electron solutions of Eq. (6.31) from "falling into the core."

2. The Hamiltonian operator of Eq. (6.31) is not Hermitian. Nevertheless, it can be shown that this operator is Hermitian with respect to wave functions which are strong-orthogonal to the core orbitals. We will call such operators *partially Hermitian* operators and refer to them as "pH-operators."

3. Exact solutions of Eq. (6.31) exist, and it can be shown easily that the two-electron function Φ_0, which is strong-orthogonal and makes the expectation value of the Hamiltonian of Eq. (6.31) to an absolute minimum, is the exact solution of the equation for the ground state of the atom.

4. Approximate solutions of Eq. (6.31) can be constructed using the conventional Rayleigh–Ritz variation procedure. The only difference between the application of the variation procedure to an ordinary Hermitian operator and its application to Eq. (6.31) is that, in our case, only strong-orthogonal functions can be admitted as trial functions in the variation procedure. The approximate solutions will yield energy values which are upper limits to the exact eigenvalue of Eq. (6.31).

(The mathematical proofs of these statements are presented in Appendix I.)

Reviewing the preceding presentation of the FVP theory, we may say that by introducing the mixed wave functions [Eq. (6.24)] and the strong-orthogonality condition [Eq. (6.25)], Fock and his collaborators succeeded in transforming an $(N + 2)$-electron atomic problem into a two-valence-electron problem. By introducing a two-electron function for the valence electrons, they opened the way for the proper treatment of the electron correlation between these electrons. By introducing the strong orthogonality condition, they achieved a complete separation of the valence electrons from the core; the properties of the valence electrons can be computed from Eq. (6.31) "as if the core did not exist." Indeed, the lowest state of Eq. (6.31) is the ground state of the two valence electrons; core states do not occur in the energy spectrum of Eq. (6.31). The core states enter this theory only in the form of fixed orbitals (like "fixed parameters"), the choice of which will influence the quality of the results to a certain extent.

Thus we may say that, by transforming the $(N + 2)$-electron problem into a two-electron problem, Fock, Veselov, and Petrashen created a theory which accurately reflects the character of certain optical spectra. As we noted in Section 6.1.A, the optical spectra of the alkali earth atoms are typical two-electron spectra although these are not two-electron atoms. In the FVP theory the energy spectrum of such an atom is described by the eigenvalue spectrum of Eq. (6.31). Eq. (6.31) is a two-electron equation with its lowest level corresponding to the ground state of the two valence electrons. Thus the mathematical structure of this equation reflects accurately the physical characteristics of the optical spectra.

C. Some Mathematical Problems Connected with the Calculation of Pair Functions

We have shown in the preceding section that the solutions of Eq. (6.31) can be obtained by the conventional Rayleigh–Ritz variation procedure. While this statement is true *in principle*, there are considerable mathematical difficulties connected with the actual calculations. We want to discuss here some of these difficulties.

The variation method works as follows. We are seeking the exact solution of Eq. (6.31), which we write down here again:

$$\left(H_1 + H_2 + P\,\frac{1}{r_{12}}\right)\Phi = H\Phi = E\Phi. \tag{6.46}$$

Let $f_1, f_2, \ldots, f_k$ be a set of two-electron functions with which we want to describe the electronic structure of the valence electrons. These functions can be chosen in many different ways. They can be part of a complete system, or they can be chosen by physical plausibility while not necessarily being a part of a complete system. Now we have seen that Eq. (6.46) can be changed into the variation principle,

$$E = \text{Min}\{\langle\Phi|H|\Phi\rangle/\langle\Phi\,|\,\Phi\rangle\}. \tag{6.47}$$

The exact solution of Eq. (6.46) will be the absolute minimum of E; approximate solutions will provide upper limits to the exact eigenvalue (for the ground state).

We want to build up our solution from the set of two-electron functions introduced above. In order to do so, we must orthogonalize these functions to the fixed, given, core orbitals. Thus we put

$$\hat{f}_i = Pf_i, \quad (i = 1, 2, \ldots, k), \tag{6.48}$$

and form with these functions the trial function

$$\Phi = \sum_{i=1}^{k} c_i \hat{f}_i. \tag{6.49}$$

On putting Φ into Eq. (6.47) we obtain an expression which is the function of the parameters c_i. Application of the energy-minimum principle leads, in the usual fashion, to a system of linear homogeneous equations for these parameters. From these equations we obtain the secular equation for the determination of the energy value. (We assume that the reader is familiar with these equations, and do not reproduce them here.) If, as additional flexibility, the functions f_i contain some adjustible parameters, those must also be determined from the energy-minimum principle.

In general, the trial function [Eq. (6.49)] will provide an approximate energy value. By skillful choice of the starting set and/or by using very long sets we may be able to obtain energy values which will converge toward the exact eigenvalue. It is a well-known fact that the quality of the approximation will be determined by the choice of the trial function Φ.

A discussion of the merits and demerits of various choices for the trial function is outside the scope of this book. Here we want to show only that the structure of Eq. (6.47) is such that calculations involving certain types of trial functions introduce great mathematical difficulties.

One of the possible choices for the set f_k is to put these functions in the form of (2×2) determinants of one-electron orbitals, or in the form of linear combination of (2×2) determinants. This choice is usually referred to as the configuration interaction (CI) method. Another choice involves the use of Hylleraas type functions. In calculating the energy levels of the He atom, Hylleraas used a trial function which is not separable into a finite set of products of one-electron functions, but depends instead explicitly on the distance between the two electrons of the atom.[61] For the ground state of the He atom, Hylleraas used the trial function,

$$\Phi = \varphi(r_1)\varphi(r_2) \sum_{s,\,t,\,u} c_{stu}(r_1 - r_2)^{2s}(r_1 + r_2)^t r_{12}^u, \qquad (6.50)$$

where $\varphi(r_1)$ and $\varphi(r_2)$ are one-electron functions, r_1 and r_2 are the distances of the two electrons from the nucleus, r_{12} is the distance between the electrons, and c_{stu} is a variational parameter.

Calculations with this method as well as with its improved versions by Hylleraas,[61] Kinoshita,[62] and Pekeris,[63] have shown the usefulness of this kind of trial function. These functions have the property that, with a relatively short expansion (six parameters), a very good approximation to the exact eigenvalue can be obtained. In addition, when the method was used in its most effective form, that is, when long expansions with many parameters were employed, the calculated energy reproduced the ground state with spectroscopic accuracy.

In view of the effectiveness of the Hylleraas method, we would naturally like to have the option of using this kind of trial function for the calculation of approximate solutions of Eq. (6.46). Let us demonstrate what this means mathematically. Choosing for the set $f_1 \cdots f_k$ a Hylleraas-type expansion, we shall obtain, in the variational integral [Eq. (6.47)], matrix elements of the type

$$I_{ij} = \langle Pf_i | H_1 + H_2 + P \frac{1}{r_{12}} | Pf_j \rangle. \qquad (6.51)$$

For the functions f_i and f_j let us put $f_i = f_j = r_{12}$.

As part of a study of electron correlation in atoms, the author has investigated[64] the mathematical structure of matrix components of this type. We have shown that a matrix component of this type will involve not only two-electron integrals which contain the interelectronic coordinate r_{12} but also three- and four-electron integrals containing the combinations $(r_{12}r_{23})$, $(r_{12}r_{23}r_{24})$, $(r_{12}r_{23}r_{14})$, and $(r_{12}r_{23}r_{13})$. The reason for this is that, as we see from

Eq. (6.51), the matrix component is studded with projection operators as well as with exchange operators. (These last are in H_1 and H_2, in the second term of Eq. (2.5).) In order to demonstrate the effect of the projection operator Ω_1 on the Hylleraas-type function r_{12}, let us recall that, according to Eq. (6.16), we have,

$$\Omega_1 r_{12} = \sum_{i=1}^{N} \varphi_i(q_1) \int \varphi_i^*(q_3) r_{32}\, dq_3 . \tag{6.52}$$

Thus the application of Ω_1 to the two-electron function r_{12} made it into a three-electron function; that is, the operator created an additional dimension. As the exchange operator is also an integral operator, its effect is similar. Therefore the repeated application of projection and exchange operators on a Hylleraas-type function will produce integrals with the combinations of interelectronic distances mentioned above.

The calculation of such integrals is very laborious. Thus we realize that the application of the Hylleraas method leads to much greater mathematical difficulties in the case of Eq. (6.46) than in the original case of the He atom. Now when we said that Fock and his collaborators reduced the $(N+2)$-electron problem to a two-valence-electron problem, we used a somewhat imprecise location. Equation (6.46) for the two valence electrons is not a two-electron equation but a two-state equation. When we replace the (2×2) determinant in Eq. (6.13) by an arbitrary two-electron function, we introduced electron correlation between the two valence electron states φ_{N+1} and φ_{N+2}, rather than between the two valence electrons. When we talk about valence electrons we always mean valence *states*, since the electrons are indistinguishable and the distinction core/valence is not possible. The distinction is always between core and valence states, and correlation can be introduced into the wave function by introducing it between electron states.[65]

In summary, we may say that the solutions of Eq. (6.46) are indeed obtainable with the variation procedure; but if, in order to insure fast convergence, we want full flexibility in the choice of the trial function employed in the calculations, then the mathematical difficulties might be very considerable*.

The discussion of the mathematical difficulties encountered in the solving of Eq. (6.46) enables us to elucidate our remarks about the pseudopotential method in Section 1.3. We stated there that the main advantage of the

---

* We have presented this argument in the framework of the variation method for two reasons. First, the existence proof which we presented involved the variation method. Second, until now the most accurate and reliable results have always been obtained with the variation method. (This is a general statement valid for the whole atomic theory.) It remains to be seen whether, among the currently available nonvariational computational techniques, there will be one which will produce results of the same accuracy as the variation method and will achieve this with an appreciable reduction in computational labor.

pseudopotential method is that modelization is much easier in this method than in the ab initio theory. The correctness of this statement will be demonstrated in the case of the FVP theory. In the next section we shall transform the FVP theory exactly into a pseudopotential formalism. In the suceeding section this exact formalism will be transformed into a model, the salient feature of which will be the absence of the mathematical difficulties which we have discussed here in connection with Eq. (6.46). The ab initio equation will be replaced by a model equation which will be a truly two-electron equation, the solution of which will be much easier than the solution of Eq. (6.46).

6.2. PSEUDOPOTENTIAL THEORY: EXACT FORMULATION

In formulating an exact pseudopotential theory for atoms with two valence electrons, we proceed in the same way as we did in the case of one valence electron. Our ab initio equation is Eq. (6.31):

$$H\Phi = E\Phi,\qquad\qquad(6.53)$$

where

$$H = P_1 H_1' + P_2 H_2' + P_1 P_2 \frac{1}{r_{12}},\qquad\qquad(6.54)$$

and H_i' is defined by Eq. (6.29). We introduce the non-orthogonal pseudo-orbital for the two valence electrons. This is the function $\Psi(1, 2)$, which we have already introduced in Eq. (6.15). The connection between the Φ of Eq. (6.53) and the pseudo-orbital Ψ is given by Eq. (6.21):

$$\Phi = P\Psi,\qquad\qquad(6.55)$$

where P is the operator defined by Eq. (6.18),

$$P = P_1 P_2.\qquad\qquad(6.56)$$

Our collection of starting formulas will be complete with Eq. (6.19):

$$\Omega = 1 - P.\qquad\qquad(6.57)$$

We want to transform Eq. (6.53) into a pseudopotential equation; that is, we want to derive the equation for Ψ. Just as in the case of one valence electron, this can be done in several different ways.

We obtain a Phillips–Kleinmann-type transformation by using Eq. (6.55) in Eq. (6.53):

$$H\Phi = HP\Psi = E\Phi = EP\Psi \,, \qquad (6.58)$$

or

$$H(1 - \Omega)\Psi = E(1 - \Omega)\Psi \,, \qquad (6.59)$$

from which we get

$$(H + V_p)\Psi = E\Psi \,, \qquad (6.60)$$

where

$$V_p = -(H - E)\Omega \,. \qquad (6.61)$$

This equation is formally identical with Eq. (2.56), but in this formula we have two-electron operators instead of the one-electron operators of Eq. (2.56).

A Weeks–Rice-type transformation is obtained by applying the formalism presented in Section 2.4. Thus we obtain from Eq. (6.53) the equation

$$(H + V_p)\Psi = E\Psi \,, \qquad (6.62)$$

where V_p is one of the pseudopotentials listed in Eq. (2.115). Taking the first line of that equation, we get

$$V_p = -\Omega(H - E) - (H - E)\Omega + \Omega(H - E)\Omega \,. \qquad (6.63)$$

This expression can be simplified. Because Ω is idempotent, the E-term cancels out from the first and the last term and we obtain,

$$V_p = -(H - E)\Omega - \Omega HP \,. \qquad (6.64)$$

It is easy to show that Eq. (6.64) is equivalent to Eq. (6.61). Let us multiply Eq. (6.60) from the left by P:

$$(PH + PV_p)\Psi = EP\Psi \,. \qquad (6.65)$$

Using Eq. (6.61) we obtain

$$(PH - P(H - E)\Omega)\Psi = E(1 - \Omega)\Psi \,, \qquad (6.66)$$

from which we get

$$(PH - PH\Omega + EP\Omega + E\Omega)\Psi = E\Psi \,. \qquad (6.67)$$

Using Eq. (6.57) and Eq. (E.5) we get

$$(PH - PH(1 - P) + E\Omega)\Psi = E\Psi, \qquad (6.68)$$

and so

$$(PHP + E\Omega)\Psi = E\Psi. \qquad (6.69)$$

This can be written in the form,

$$\{H + [-(H - E)\Omega - \Omega HP]\}\Psi = E\Psi, \qquad (6.70)$$

where the square bracket contains the expression of Eq. (6.64).

Thus we see that the WR-type pseudopotential [Eq. (6.64)] is equivalent to the PK-type pseudopotential [Eq. (6.61)]. This means that if Ψ is an eigenfunction of the PK-type pseudopotential equation [Eq. (6.60)] with the eigenvalue E, then this wave function will also be an eigenfunction of the WR-type pseudopotential equation [Eq. (6.62)] with the same eigenvalue.

Next we show that the PK-type equation [Eq. (6.60)] will reproduce the eigenvalue spectrum of Eq. (6.53). Let us again multiply Eq. (6.60) from the left by P. First we get Eq. (6.65). Taking into account Eq. (6.61), we obtain

$$(PH - PH\Omega + PE\Omega)\Psi = EP\Psi, \qquad (6.71)$$

or

$$PH(1 - \Omega)\Psi = PHP\Psi = EP\Psi, \qquad (6.72)$$

where we have taken into account Eq. (E.5) in eliminating the term with $P\Omega$. Now introduce the orthogonalized orbital Φ, using Eq. (6.55), and get

$$PH\Phi = E\Phi. \qquad (6.73)$$

Taking into account Eq. (6.54), we obtain

$$\begin{aligned}
PH\Phi &= P_1 P_2 \left(P_1 H_1' + P_2 H_2' + P_1 P_2 \frac{1}{r_{12}} \right)\Phi \\
&= \left(P_1 H_1' P_2 + P_2 H_2' P_2 + P^2 \frac{1}{r_{12}} \right)\Phi.
\end{aligned} \qquad (6.74)$$

Using Eq. (E.12) and the idempotent character of P, we get

$$PH\Phi = H\Phi, \qquad (6.75)$$

and putting this into Eq. (6.73) we obtain

$$H\Phi = E\Phi \,. \tag{6.76}$$

Our result is that, if Ψ and E are the solutions of the PK-type equation [Eq. (6.60)], then $\Phi = P\Psi$ will be a solution of Eq. (6.53) with the same eigenvalue. This proves our statement that Eq. (6.60) will reproduce the spectrum of Eq. (6.53). Since we have already shown that the WR-type equation [Eq. (6.62)] will reproduce the spectrum of the PK equation [Eq. (6.60)], we can now make the general statement that both the PK equation and the WR equation are equivalent to Eq. (6.53) in that they reproduce the eigenvalue spectrum of that equation directly and reproduce the eigenfunctions of Eq. (6.53) through the relationship of Eq. (6.55).

In this argument we have used the expression "equivalent" which respect to pseudopotential equations and Eq. (6.53). In order to give a somewhat more precise formulation, we observe that both Eqs. (6.60) and (6.62) will have solutions for which

$$P\Psi = 0 \,. \tag{6.77}$$

In the case of Eq. (6.60) we see this from Eq. (6.72) which was directly obtained from Eq. (6.60). In the case of Eq. (6.62) we substitute Eq. (6.64) and obtain

$$(H + V_p)\Psi = (H - H\Omega + E\Omega - \Omega HP)\Psi$$
$$= (HP + E\Omega - \Omega HP)\Psi = E\Psi \,. \tag{6.78}$$

Rearranging, we get

$$PHP\Psi = EP\Psi \,, \tag{6.79}$$

which is again Eq. (6.72), and it is clear that Eq. (6.77) will give a solution of this equation. Thus we see that both the PK-type equation and the WR-type equation have solutions which satisfy Eq. (6.77). By analogy with the one-valence-electron problem, we shall call these functions core solutions. The fact that these two-electron pseudopotential equations will have core solutions as well as valence solutions is hardly surprising, since their one-electron counterparts also have core solutions, as we saw in Chapter 2. By saying that these equations are equivalent to Eq. (6.53), we mean that the valence solutions of the two-electron pseudopotential equations will reproduce the solutions of Eq. (6.53); but in addition to these valence solutions, the pseudopotential equations will also have core solutions which are absent from the eigenfunctions of Eq. (6.53).

It is easy to establish the meaning of Eq. (6.77). Taking into account that

$P = 1 - \Omega$, we obtain

$$P\Psi = (1 - \Omega)\Psi = 0 , \tag{6.80}$$

and Eq. (6.77) becomes

$$\Psi = \Omega\Psi . \tag{6.81}$$

Now let us expand Ψ in terms of the complete set of the auxiliary Hamiltonian which we defined in Appendix E. Then we obtain an expansion containing all combinations of core and valence orbitals, as in Eq. (6.42). The solutions which satisfy Eq. (6.81) will have the form of Eq. (6.45). In Eq. (6.45) we have an expansion which consists of (2×2) determinants containing two core orbitals and of (2×2) determinants containing one core orbital and one valence orbital. Thus the solutions of the pseudopotential equations which satisfy Eqs. (6.77) or (6.81), and which we call core solutions, will be built from determinants containing two core orbitals and/or from determinants containing one core orbital and one valence orbital.

We have now obtained two exact, two-electron, pseudopotential equations which are the exact equivalents of the ab initio equation, Eq. (6.53). It was shown by Szasz and Brown[66] that there exists a pseudopotential formulation which is simpler than those which we have already presented. We will discuss this formulation here in a form which is a slightly improved version of the original publication.

Instead of relying on the analogy with the one-valence-electron formalism we shall be guided here by physical plausibility. What should a two-electron equation be like in which the Pauli exclusion principle is replaced by potentials? Hellmann has shown that physical plausibility leads to Eq. (1.4) with the Hamiltonian of Eq. (1.3). Besides the kinetic energies, that Hamiltonian contains the Hellmann potentials expressed in both electron coordinates, so that both electrons are kept out of the core. Also, the electrostatic interaction potential of the two valence electrons is present. Let us write down the one-electron operator which will represent the interaction of one of the valence electrons with the core. In a general form, this operator can be written as

$$H_1 + V_1 = P_1 H_1' + V_1 , \tag{6.82}$$

where H_1 is the same as the one-electron operator in Eq. (6.31). V_1 is a pseudopotential, one of the WR-type one-electron pseudopotentials [Eq. (2.115)], which should not be confused with the likewise WR-type two-electron potential [Eq. (6.63)]. In order to emphasize this point, we write out V_1 here using the second line of Eq. (2.115):

$$V_1 = -\Omega_1(H_1 - \epsilon) - P_1(H_1 - \epsilon)\Omega_1 , \tag{6.83}$$

where the Ω_1 and P_1 are the one-electron operators defined by Eqs. (6.16) and (6.17). The parameter ϵ will be left undefined for the time being.

By analogy with Eq. (1.3) our two-electron Hamiltonian should be

$$H = H_1 + V_1 + H_2 + V_2 + \frac{1}{r_{12}}, \tag{6.84}$$

where $H_1 + V_1$ represents the kinetic energy plus the total core/valence interaction potential for the first valence electron, $H_2 + V_2$ represents the same for the second valence electron, and $1/r_{12}$ is the electrostatic interaction. The presence of V_1 and V_2 ensures that the valence electrons are kept out of the core.

It is evident that the operator of Eq. (6.84) follows directly from the principles of physical plausibility, and is the straightforward generalization of Hellmann's Hamiltonian [Eq. (1.3)]. Unfortunately, H is not an exact Hamiltonian. We postulate, however, that the operator in Eq. (6.84) can be made exact by adding the term

$$-\frac{1}{r_{12}}\Omega = -\frac{1}{r_{12}}(1 - P), \tag{6.85}$$

which will be shown to be a small correction to the total energy of the two valence electrons. By adding this term to H we obtain

$$H = H_1 + V_1 + H_2 + V_2 + \frac{1}{r_{12}}P. \tag{6.86}$$

Now consider the wave equation

$$H\Psi = E\Psi, \tag{6.87}$$

where H is given by Eq. (6.86). Let Ψ and E be the solutions of Eq. (6.87). We state that

$$\Phi = P\Psi \tag{6.88}$$

is then a solution of Eq. (6.53) with the same eigenvalue.

In order to prove this theorem let us multiply Eq. (6.87) from the left by P. We calculate first the effect of P on $H_1 + V_1$, which is an argument analogous to the one which follows Eq. (2.126). Taking into account the form of H_1 we obtain

$$PH_1 = P_1P_2H_1 = P_1P_2P_1H_1' = P_2P_1^2H_1' = P_2H_1 = H_1P_2. \tag{6.89}$$

Using Eq. (6.83) we get

$$PV_1 = P_1 P_2 [-\Omega_1 (H_1 - \epsilon) - P_1 (H_1 - \epsilon)\Omega_1]$$
$$= -P_2 P_1^2 H_1 \Omega_1 = (P_1 H_1 \Omega_1) P_2 , \tag{6.90}$$

where we have used twice the relationship $P_1 \Omega_1 = 0$. In view of the form of H_1, we get

$$-P_1 H_1 \Omega_1 P_2 = -P_1 P_1 H_1' \Omega_1 P_2 = -H_1 \Omega_1 P_2 . \tag{6.91}$$

Thus using Eqs. (6.89) and (6.91) we obtain

$$P(H_1 + V_1) = H_1 (1 - \Omega_1) P_2 = H_1 P_1 P_2 = H_1 P . \tag{6.92}$$

Likewise we get

$$P(H_2 + V_2) = H_2 P . \tag{6.93}$$

Now multiply Eq. (6.87) from the left by P and take into account Eqs. (6.86), (6.92), and (6.93). We obtain

$$P\left(H_1 + V_1 + H_2 + V_2 + \frac{1}{r_{12}} P\right)\Psi = \left(H_1 P + H_2 P + P \frac{1}{r_{12}} P\right)\Psi = E P \Psi . \tag{6.94}$$

Using the notation $\Phi = P\Psi$, we obtain

$$\left(H_1 + H_2 + P \frac{1}{r_{12}}\right)\Phi = E\Phi , \tag{6.95}$$

which is identical with Eq. (6.53). Thus the theorem is proved.

It is proved now that Eq. (6.87) with the Hamiltonian of Eq. (6.86) is an exact equation equivalent to Eq. (6.53). The concept of equivalency is used here in the same way as we have used it in connection with the exact PK-type and WR-type equations.

It is easy to show that Eq. (6.87) is not only equivalent to Eq. (6.53) but also to the PK-type equation [Eq. (6.60)] and to the WR-type equation [Eq. (6.62)]. In order to show this let us multiply Eq. (6.87) again from the left by P. Thus we obtain Eq. (6.94). Let us introduce the notation of Eq. (6.54) into Eq. (6.94). We get (6.94) in the form

$$HP\Psi = E P \Psi . \tag{6.96}$$

Using Eq. (6.57) we can write this as

$$H(1 - \Omega)\Psi = E(1 - \Omega)\Psi , \tag{6.97}$$

or

$$(H + V_p)\Psi = E\Psi, \tag{6.98}$$

where V_p is the PK-type pseudopotential [Eq. (6.61)]. Thus the PK-type equation [Eq. (6.60)] is directly obtainable from Eq. (6.87). Since we have already shown that the WR-type equation is obtainable from the PK-type equation, the equivalence of Eq. (6.87) to both equations is thereby demonstrated.

In summary, we have now three different, exact, pseudopotential equations. They are equivalent to each other as well as to the ab initio equation, Eq. (6.53).

In developing the pseudopotential theory of atoms and molecules we now have the option of choosing from these three equations the one that best fits the requirements of a particular problem or calculation. There is one advantage that the Hamiltonian of Eq. (6.86) enjoys over the Hamiltonians of Eqs. (6.60) and (6.62) which should be noted here. Comparing the three Hamiltonians we observe that Eq. (6.86) is the simplest, which is an advantage in most applications. It is also noteworthy that, in contrast to the Hamiltonians of Eqs. (6.60) and (6.62), the operator in Eq. (6.86) depends only on the parameters of the core; it does not depend on the valence-electron quantities E and Ψ.

We shall now identify the energy parameter ϵ in Eq. (6.83). We said that Eq. (6.82) represents the interaction of one of the valence electrons with the core. Let us consider the equation

$$(H_1 + V_1)\psi = \epsilon\psi, \tag{6.99}$$

where ψ is a one-electron PO and ϵ is an energy parameter. In order to show what this equation means, let us write down here again the operators occurring in it,

$$H_1 = P_1 H_1', \tag{6.100}$$

and

$$V_1 = -\Omega_1(H_1 - \epsilon) - P_1(H_1 - \epsilon)\Omega_1. \tag{6.101}$$

H_1' is given by Eq. (6.34) and has the functional form of the HF Hamiltonian operator Eq. (2.3).

At this point the reader is directed to compare these equations with the sequence Eqs. (2.105)–(2.115). It is evident that the two sets of formulas are analogous. Thus the pseudopotential equation [Eq. (6.99)] is equivalent to the HF-type equation

$$H_1\varphi = \epsilon\varphi, \tag{6.102}$$

which is analogous to Eq. (2.111); the HF-type orbital φ is related to the PO solution of Eq. (6.99) by the relationship

$$\varphi = P_1 \psi. \tag{6.103}$$

Transforming Eq. (6.102) into a PO equation with the aid of Eq. (6.103), one obtains Eq. (6.99) with the WR-type operator of Eq. (6.101). Thus ϵ is the energy parameter of the valence electron of a one-valence-electron atom which is obtained from the two-valence-electron atom by removing one of the valence electrons. The remark about Eq. (6.86) depending only on core parameters is still valid in the sense that ϵ can be computed from Eq. (6.99).

The exact pseudopotential equation for the two valence electrons takes an especially simple form for the special choice of the core orbitals which we have discussed in Section 6.1.B. Let the core be frozen in the form of the N-electron atom with the two valence electrons removed. Then our operator H'_i becomes identical with H_F and the core orbitals are solutions of the HF equations [Eq. (6.5)]. The projection operator P will commute with H_F. The ab initio two-electron equation will be Eq. (6.36) in this case. In order to see what Eq. (6.86) will be like, let us first consider $H_1 + V_1$. Using Eqs. (6.100) and (6.101) with $H'_i = H_F(i)$, we obtain

$$H_1 + V_1 = P_1 H_F(1) - \Omega_1(P_1 H_F(1) - \epsilon) - P_1(P_1 H_F(1) - \epsilon)\Omega_1. \tag{6.104}$$

Taking into account that P_1 now commutes with H_F and that is idempotent, we get

$$H_1 + V_1 = H_F(1)P_1 - \Omega_1(P_1 H_F(1) - \epsilon) - H_F(1)P_1\Omega_1 + \epsilon P_1\Omega_1. \tag{6.105}$$

Since $P_1\Omega_1 = \Omega_1 P_1 = 0$, we obtain

$$H_1 + V_1 = H_F(1)[1 - \Omega_1] + \epsilon\Omega_1 = H_F(1) - [H_F(1) - \epsilon]\Omega. \tag{6.106}$$

The second term is the PK-type potential [Eq. (2.56)]. Thus we obtain

$$H_1 + V_1 = H_F + V_p, \tag{6.107}$$

with

$$V_p = \sum_{i=1}^{N} (\epsilon - \epsilon_i)|\varphi_i\rangle\langle\varphi_i|, \tag{6.108}$$

where we have taken into account that the core orbitals are eigenfunctions of H_F. We can now use Eq. (6.107) in Eq. (6.86) and obtain for the exact, two-electron, pseudopotential Hamiltonian

$$H = H_F(1) + V_p(1) + H_F(2) + V_p(2) + \frac{1}{r_{12}} - \frac{1}{r_{12}}\Omega. \tag{6.109}$$

We have seen that the exact equation, [Eq. (6.86)], in its general formulation permitted a very plausible physical interpretation; indeed we have set up the equation guided by that interpretation. A plausible physical meaning is even more evident in the case of Eq. (6.109). In order to demonstrate this let us use Eq. (6.107) in Eq. (6.99):

$$(H_F + V_p)\psi = \epsilon\psi. \tag{6.110}$$

Here we shall anticipate the discussions of Section 6.3 and change the operators in Eq. (6.110) into local potentials. In Section 2.2 we saw how Eq. (6.110) could be changed into an equation with, instead of the exchange and pseudopotential operators, the corresponding local potentials; that is, that Eq. (6.110) could be transformed into the equation,

$$\left(-\frac{1}{2}\Delta + V_M\right)\psi = \epsilon\psi, \tag{6.111}$$

where V_M is the modified potential defined by Eqs. (2.34), (2.40), and (2.41). Replacing $H_F + V_p$ by the Hamiltonian of Eq. (6.111), we obtain for Eq. (6.109):

$$H = -\frac{1}{2}\Delta_1 + V_M(1) - \frac{1}{2}\Delta_2 + V_M(2) + \frac{1}{r_{12}} - \frac{1}{r_{12}}\Omega. \tag{6.112}$$

Apart from the last term this is a Helium-like Hamiltonian with the $-z/r$ potential of the nucleus replaced by the modified potential V_M. It also has the form of Hellmann's Hamiltonian [Eq. (1.3)], where

$$V_M = -\frac{z}{r} + A\frac{e^{-\kappa r}}{r}. \tag{6.113}$$

A comparison of Eq. (6.112) with Eq. (6.109) permits us to express clearly the physical meaning of the latter. We see that $H_F + V_p$ is equivalent to $-1/2 + V_M$. Thus the meaning of the operator $H_F + V_p$ can be understood by analyzing the physical meaning of $-1/2 + V_M$. The latter operator represents the kinetic energy of the valence electron plus the core/valence interaction potential V_M. V_M is the modified potential containing the Coulomb and exchange potentials plus the pseudopotential. (The structure of this function was discussed extensively in Section 2.2.B.) In light of that discussion it is evident that the two terms $V_M(1)$ and $V_M(2)$ which appear in Eq. (6.112) will represent fully the interaction between the core and the two valence electrons. Recalling the structure of V_M we also realize that the two valence electrons are kept out of the core by these potentials. Thus the first four terms of Eq. (6.109) represent the kinetic energies of the two valence electrons plus the total interaction potential with the core. The fifth term of Eq. (6.109) is, of course, the electrostatic interaction between the two valence electrons; and the last term is needed to make the equation exact, but will be shown to be a small correction in Section 6.3.

We summarize now the most important mathematical properties of the exact pseudopotential equation, Eq. (6.86). For the convenience of the reader we write down here again the ab initio, two-electron equation for the orthogonalized function, Eq. (6.31),

$$\left(H_1 + H_2 + P \frac{1}{r_{12}}\right)\Phi = E\Phi , \qquad (6.114)$$

and the equation for the pseudo-orbital, Eq. (6.86),

$$\left(H_1 + V_1 + H_2 + V_2 + \frac{1}{r_{12}} P\right)\Psi = E\Psi , \qquad (6.115)$$

where the solutions of the latter are connected to the solutions of the former by the relationship

$$\Phi = P\Psi . \qquad (6.116)$$

1. The wave function Ψ of the two valence electrons is determined from the wave equation, Eq. (6.115), in which the Hamiltonian operator depends entirely on the core data and does not depend on either Ψ or E. The equation is equivalent to the ab initio equation, Eq. (6.114). The Pauli exclusion principle is taken into account mainly by the one-electron pseudo-potentials V_1 and V_2, and to a small degree also by the presence of the operator P. The solutions of Eq. (6.115) do not need to satisfy any ortho-gonality conditions with respect to the core. By contrast, the Pauli exclusion principle demands that the solutions of Eq. (6.114) be strong-orthogonal to the core; that is, they must be put in the form of Eq. (6.116). The presence of P in the Hamiltonian of Eq. (6.114) is also a consequence of the Pauli principle.

2. The Hamiltonian of Eq. (6.115) is not Hermitian and not partially Hermitian. The basic principles of quantum mechanics are not violated, however, since, as we show in Appendix J, the eigenvalus of Eq. (6.115) are always real.

3. Exact solutions of Eq. (6.115) exist. This follows from the fact that the exact solutions of Eq. (6.114) are related to the exact solutions of Eq. (6.115) by the relationship of Eq. (6.116). We proved that exact solutions of Eq. (6.114) do exist; that proof also proves the existence of the solution of Eq. (6.115). These statements refer to the solutions for which $P\Psi \neq 0$. These are eigenfunctions corresponding to the valence states. In addition to these, Eq. (6.115) will have, in a manner similar to the one-electron pseudopotential equations which we discussed in Chapter 2, solutions for which $P\Psi = 0$. We have called such solutions core solutions, and we have seen that the PK-type

and WR-type equations also had such solutions. The form of these solutions has been mentioned before; Appendix J contains further discussion.

4. Approximate solutions of Eq. (6.115) can be constructed, for the ground state as well as for the excited states, by a method developed by Merzbacher for non-Hermitian operators.[67] This method is presented and discussed in Appendix J.

6.3. PSEUDOPOTENTIAL THEORY: MODEL FORMULATION

A. Introduction

We have seen in the preceding section that the exact pseudopotential Hamiltonians are rather complicated operators. Thus calculations carried out with these operators are likely to be very laborious; in fact, in such calculations we will encounter the same kind of difficulties that we discussed in Section 6.1.C in connection with the ab initio two-electron equation.

In order to overcome these difficulties we shall now introduce a model Hamiltonian H_M which will be formulated in such a way as to replace, for all practical purposes, the exact two-electron pseudopotential Hamiltonians. In setting up this Hamiltonian we shall be guided by physical plausibility. As we have seen in the preceding section, physical plausibility leads to Eq. (6.84). In order to have greater flexibility, we go one step further; as in Section 4.2, we replace the one-electron operator $H_1 + V_1$, which occurs in Eq. (6.84), by the model operator $t + V_m$, which appeared in Eq. (4.5). Thus we postulate that the model Hamiltonian will be

$$H_M = t_1 + V_m(1) + t_2 + V_m(2) + \frac{1}{r_{12}},$$

$$(6.117)$$

and the equation replacing the exact pseudopotential equation will have the form

$$H_M \Psi = E \Psi ,$$

$$(6.118)$$

where the two-electron wave function Ψ does not need to satisfy any orthogonality conditions relative to the core. (In fact Eq. (6.118) can be solved as if the core did not exist.)

Let us write down here the exact pseudopotential equation. Choosing the simplest formulation, Eq. (6.86), we have,

$$H = H_1 + V_1 + H_2 + V_2 + \frac{1}{r_{12}} P .$$

$$(6.119)$$

In this section we want to answer the question: is Eq. (6.117) a good approximation to Eq. (6.119)? Putting it in another way, the question reads: is it possible to derive Eq. (6.117) from Eq. (6.119), thereby giving to the former a sound theoretical justification? The terms "derivation" and "sound theoretical justification" are used as we defined them in Section 3.1.A. We shall answer these questions in the affirmative, and we shall also clarify the conditions which must be fulfilled in order to obtain an affirmative answer.

After the model Hamiltonian of Eq. (1.3), which is a special case of Eq. (6.117), was introduced by Hellmann, the problem of the derivation of a two-electron model Hamiltonian from the exact theory was first tackled by Szasz and McGinn.[5] Starting from the FVP theory they derived exact one-electron pseudopotential equations for the two valence electrons. Analyzing the properties of the valence-electron PO's, it was possible to show that a model Hamiltonian of the structure of Eq. (6.117) can be derived from the FVP theory by reasonable approximations. The problem of modelization was next investigated by Weeks and Rice,[14] who, after formulating the exact pseudopotential, Eq. (6.63), made calculations for the Be and Mg atoms using the exact equations. They also investigated several model formulations, among them Eq. (6.117), and made calculations using this Hamiltonian. The two sets of calculations led to results very close to each other. The modelization problem was again discussed by Szasz and Brown[66] subsequent to the formulation of the exact equation, Eq. (6.86). It was demonstrated that the term given by Eq. (6.85), which is the main difference between the exact and model Hamiltonians, can be viewed as a small perturbation.

In the next section we present the results of Weeks and Rice and the work of Szasz and Brown, which together give the affirmative answer to the question raised above*. In the subsequent section further calculations involving model Hamiltonians are presented.

B. The Derivation of the Model Hamiltonian

We shall derive the model Hamiltonian, Eq. (6.117), in two steps. In the first step, which is based on the work of Weeks and Rice,[14] we shall show that numerical calculations using the two Hamiltonians, in Eqs. (6.117) and (6.119), yield very similar results.

Our starting point is the exact pseudopotential equation, Eq. (6.69),

$$(PHP - EP)\Psi = 0 .\tag{6.120}$$

Weeks and Rice modified the one-electron operators occuring in this equa-

---

* The question of the validity and/or plausibility of Eq. (6.117) is also discussed in other publications, for example, in some of those in which various analytic forms for V_m are suggested. The two investigations cited here are selected for presentation because these are the only ones in which the starting point of the discussion is the exact, two-electron pseudopotential equation, rather than a one-electron approximation to the two-electron equation.

tion by replacing the exact core-valence interaction with the model potential No. 1 in Table 4.3. Then they carried out CI calculations for the Be and Mg atoms using both equations, Eq. (6.120) and Eq. (6.118). The best results obtained were for the Be $E = -27.58$ eV and $E = -27.48$ eV with Eqs. (6.120) and (6.118), respectively; for Mg the corresponding results were $E = -22.66$ eV and $E = -22.51$ eV. The empirical values are $E(\text{Be}) = -27.54$ eV and $E(\text{Mg}) = -22.68$ eV. The agreement with empirical data is very good; what is important here is that the results obtained with the two equations show also a reasonable degree of agreement.

As the second step in the derivation of Eq. (6.117) we turn now to the work of Szasz and Brown.[66] Instead of making numerical comparisons, the goal here will be to derive Eq. (6.117) from Eq. (6.119), and to pinpoint the approximations made in the derivation.

Let us assume that the Hamiltonian of Eq. (6.119) can, for all practical purposes, be replaced by Eq. (6.117). The mathematical formulation of this statement will be that we consider the difference between the two Hamiltonians as a small perturbation. We put, as the perturbation operator,

$$V_{pt} = H - H_M, \tag{6.121}$$

where H is given by Eq. (6.119) and H_M is given by Eq. (6.117). Then we can write the exact Hamiltonian in the form,

$$H = H_M + V_{pt}. \tag{6.122}$$

The exact wave equation becomes

$$H\Psi = (H_M + V_{pt})\Psi = E\Psi. \tag{6.123}$$

Now consider the model equation, Eq. (6.118). Let us denote the exact solutions of that equation by $\hat{\Psi}$ and $\hat{E}$. Then we can write

$$H_M\hat{\Psi} = \hat{E}\hat{\Psi}. \tag{6.124}$$

Formulating the argument for a nondegenerate ground state of an atom, we can solve approximately Eq. (6.123) by considering V_{pt} a perturbation, H_M the "unperturbed" Hamiltonian, and applying perturbation theory. In order to do this, we must assume that H_M is Hermitian. Then we obtain, in the first order of perturbation theory,

$$E = \hat{E} + \int \hat{\Psi}^* V_{pt} \hat{\Psi} \, dq. \tag{6.125}$$

Here E is the approximate solution of the exact equation, Eq. (6.123); $\hat{E}$ and $\hat{\Psi}$ are the exact solutions of the model equation, Eq. (6.124). The model

Hamiltonian H_M will be a suitable replacement for H if the spectrum of H_M will match the spectrum of H in a good approximation; that is, if we are able to show that, in a good approximation, $\langle V_{pt} \rangle$ is negligible,

$$\langle V_{pt} \rangle = \int \hat{\Psi}^* V_{pt} \hat{\Psi} \, dq \approx 0, \qquad (6.126)$$

from which it follows that,

$$\hat{E} \approx E. \qquad (6.127)$$

We evaluate $\langle V_{pt} \rangle$ in the approximation in which the core is frozen in the form of the N-electron atom with the valence electrons removed. Then the operator H_1' in Eq. (6.29) becomes identical with H_F and we obtain, after some simple manipulations,

$$V_{pt} = \tilde{V}_{pt}(1) + \tilde{V}_{pt}(2) + \hat{V}_{pt}(1, 2), \qquad (6.128)$$

where

$$\tilde{V}_{pt} = V_M - V_m, \qquad (6.129)$$

and

$$\hat{V}_{pt} = -\frac{1}{r_{12}} \Omega. \qquad (6.130)$$

The V_M of Eq. (6.129) is introduced according to the relationship $H_F + V_p = t + V_M$ where V_p is the PK expression and V_M is the modified potential.

The perturbation potential consists of two parts. The potentials denoted by $\tilde{V}_{pt}$ are one-electron operators and they are the results of the replacement of the one-electron operator $H_F + V_p$ by the model expression $t + V_m$. The operator $\tilde{V}_{pt}$ is the same one that we introduced in Section 4.2 when we presented the models for one valence electron. On the other hand, $\hat{V}_{pt}$ is a two-electron operator, and this is the term that we have introduced in Eq. (6.85) in order to obtain the exact equation, Eq. (6.86).

We shall try to satisfy Eq. (6.126) by demanding that

$$\langle \tilde{V}_{pt} \rangle \approx 0, \qquad (6.131)$$

and

$$\langle \hat{V}_{pt} \rangle \approx 0. \qquad (6.132)$$

It is evident from the meaning of $\tilde{V}_{pt}$ that the proof of Eq. (6.131) is entirely a one-valence-electron problem. In fact, this problem is identical with the problem of the modelization of the modified potential for the valence electron of a one-valence-electron atom, which we discussed in Chapter 4. Thus, the finding of a model potential V_m for the model Hamiltonian of Eq. (6.117) which will satisfy Eq. (6.131) in a good approximation, is reduced to finding a suitable model potential for the one-valence-electron atom which is obtained from the two-valence-electron atom by removing one valence electron. In fact, Eq. (6.131), which must be satisfied by a good V_m, is equivalent to Eq. (4.19), which we formulated for the one-valence-electron case. Thus any of the model potentials which we have listed in Chapter 4 and which were derived for the one-valence-electron case can be used in the two-electron model Hamiltonian of Eq. (6.117). In the remaining part of this derivation we shall assume that a suitable V_m has been found and that Eq. (6.131) is satisfied.

In $\langle \hat{V}_{pt} \rangle$ we shall carry out the integration in such a way that we approximate the unknown $\hat{\Psi}$ by the product

$$\hat{\Psi} = \psi_0(1)\psi_0(2), \tag{6.133}$$

where ψ_0 is a pseudo-orbital which is the solution of

$$(t + V_m)\psi_0 = \epsilon_0 \psi_0. \tag{6.134}$$

This is a simplified argument formulated for the ^1S ground state of a 2-electron atom. We also assume that ψ_0 is the spatial part of the PO, and in the remaining part of the argument the formulas refer to spinless quantities.

Using Eq. (6.133) for the calculation of $\hat{E}$ as well as for $\langle \hat{V}_{pt} \rangle$ we obtain

$$E = \hat{E} + \langle V_{pt} \rangle = 2\epsilon_0 + \langle \psi_0 | V_0 | \psi_0 \rangle + 2\langle \tilde{V}_{pt} \rangle + 2 \sum_{i=1}^{N} \alpha_i \langle \psi_0 | V_0 | \varphi_i \rangle$$

$$+ \sum_{i,j=1}^{N} \alpha_i \alpha_j \left\langle \psi_0 \psi_0 \left| \frac{1}{r_{12}} \right| \varphi_i \varphi_j \right\rangle, \tag{6.135}$$

where $\alpha_i = \langle \varphi_i | \psi_0 \rangle$ and V_0 is the electrostatic potential

$$V_0(1) = \int \frac{|\psi_0(2)|^2}{r_{12}} dv_2. \tag{6.136}$$

Now we argue as follows. On the basis of the argument presented above we assume that $\langle \tilde{V}_{pt} \rangle \approx 0$. In the integral $\langle \psi_0 | V_0 | \varphi_i \rangle$ the potential V_0 is integrated only over the core area because of the presence of the core orbital φ_i. In the core area the (radius) $\times$ (radial part of ψ_0) will be small, and this is the quantity which will appear in the integral. Combining this fact with the smallness of the α_i, which we see from Appendix A, we conclude that the fourth expression of Eq. (6.135) will be much smaller than the electrostatic interaction integral $\langle \psi_0 | V_0 | \psi_0 \rangle$. The same argument applies a fortiori to the last term of Eq. (6.135). Thus we can write, in a plausible approximation,

$$E = \hat{E},\qquad(6.137)$$

which is the statement we wanted to prove. The argument can easily be extrapolated to the correlated wave function $\hat{\Psi}$, since for any "well-behaving" model potential V_m, $\hat{\Psi}$ will be just as small in the core area as ψ_0.

We consider the model Hamiltonian [Eq. (6.117)] established by the results of Weeks and Rice and by the derivation of Szasz and McGinn. The derivation is, of course, qualitative, as the derivations of most models are. Recalling the main argument of the derivation, we may formulate the following general rule: In order that the model Hamiltonian [Eq. (6.117)] be a good approximation to the exact pseudopotential Hamiltonian, the (pseudo) wave function of the valence electrons should have as small an overlap with the core orbitals as possible.

Observing the fact that the model Hamiltonian [Eq. (6.117)] does not contain any integral operators, we may state that *the main advantage of the modelization is that the model equation* [Eq. (6.118)] *is a true two-electron equation besides being a two-state equation.* It is clear that this property of the model equation permits great simplification in the calculations.

C. Representative Calculations

In this section we review a sample of atomic calculations for two-valence-electron systems which will enable us to form an opinion of the usefulness of the model Hamiltonian, Eq. (6.117). Thus the effective Hamiltonian, in all calculations discussed below, will be

$$H_M = t_1 + t_2 + V_m(1) + V_m(2) + \frac{1}{r_{12}}.\qquad(6.138)$$

Here t is the kinetic energy operator and V_m is the one-electron model potential. (The latter is identified by a reference to Section 4.3.) In identifying the wave function used in the calculations, we shall write down the spatial part of the function.

First, two pioneer calculations must be mentioned, those of Hellmann[2] and those of Gombas and his collaborators.[33] Hellmann computed the wave function for the ground state of the Mg atom using Eq. (4.22) for the model potential V_m. The variational wave function was

$$\Psi(1, 2) = e^{-\alpha(r_1 + r_2)}[1 + cr_{12}],\qquad(6.139)$$

where (α, c) were the variational parameters, r_1 and r_2 the distances of the two electrons from the nucleus, and r_{12} the distance between the two electrons. The result for the total energy of the two valence electrons was $E(\text{Mg}) = -22.21$ eV, while the experimental value is $E_{\exp} = -22.68$ eV.

Gombas and his collaborators used the density-dependent pseudopotential, Eq. (3.108). For an s state their procedure was equivalent to using the modified potential, Eq. (3.13). They made calculations for the Ca atom. For the $(4s)^2\,^1S$ state they have put

$$\Psi(1, 2) = r_1^2 r_2^2\, e^{-\alpha(r_1+r_2)}[1 + cr_{12}]\,, \tag{6.140}$$

where, as before, (c, α) were the variational parameters. The result for the total energy of the two valence electrons was, in eV units, with the experimental energy given in parenthesis, $E(\text{Ca}) = -18.35$, (-17.98). For the $(4s, 5s)\,^3S$ state they have put

$$\Psi(1, 2) = r_1^2 r_2^2\{e^{-\alpha_1 r_1 - \alpha_2 r_2} - e^{-\alpha_1 r_2 - \alpha_2 r_1}\}[1 + cr_{12}]\,, \tag{6.141}$$

where α_1, α_2, and c were variational parameters and the result was $E(\text{Ca}) = -14.55$, (-14.00). In these calculations, for the density of the core which occurs in Eq. (3.13), Gombas used the HF densities.

The calculations of Szasz and McGinn[39] were made with a very simple wave function but for a larger number of atoms. Calculations were made for the ^1S ground states of alkali-earth atoms and for the ^1S ground states of the heavy atoms Zn, Cd, Hg. For the alkali-earth, the modified potential of Eq. (4.22) and for the Zn, Cd, Hg the potential of Eq. (4.27) were used. Two wave function were employed:

$$\Psi_1(1, 2) = e^{-\alpha(r_1+r_2)}[1 + cr_{12}]\,, \tag{6.142}$$

and

$$\Psi_2(1, 2) = e^{-\alpha(r_1+r_2)} + cr_1 r_2\, e^{-\beta(r_1+r_2)} P_1(\cos\theta)\,. \tag{6.143}$$

In Ψ_1 the variational parameters are (α, c), while in Ψ_2 they are (α, β, c). P_1 is the Legendre polynomial with θ being the angle between the two vectors pointing to the two electrons. Ψ_2 is a primitive CI-type wave function with the second term being a $(2p)^2$ configuration.

The calculations have shown that Ψ_2 yields much better results than Ψ_1. The results were as follows: $E(\text{Mg}) = -22.67$, (-22.68); $E(\text{Ca}) = -17.78$, (-17.98); $E(\text{Sr}) = -16.25$, (-16.72); $E(\text{Ba}) = -14.64$, (-15.21); $E(\text{Ra}) = -14.61$, (-15.41); $E(\text{Zn}) = -27.30$, (-27.36); $E(\text{Cd}) = -25.55$, (-25.90); $E(\text{Hg}) = -28.27$, (-29.19).

These calculations are very simple. We turn now to some more sophisticated calculations. The first of these was made by Schwarz.[68] Here the modified potential was given by Eq. (4.28). Two different types of wave functions were employed. First a Roothaan-type[54], analytic HF calculation was made in which the best pseudo-orbitals in the modified potential field of Eq. (6.138) were determined. The second type of wave function was a Hylleraas-type function.[61] These are functions in which the spatial part contains a polynomial of the quantities r_1, r_2, and r_{12}. The wave functions discussed above, Eqs. (6.139)–(6.142), are Hylleraas-type functions. The formula for the actual wave function was not published by Schwarz.

Here are the results obtained for some two-valence-electron systems with the analytic HF procedure, where the numbers in parentheses now mean the results of ab initio all-electron HF calculations: $E(Be) = -26.23$, (-26.16); $E(Mg) = -21.69$, (-21.31); $E(Ca) = -17.77$, (-16.44); $E(Zn) = -25.82$, (-24.18).

Next we list the results obtained with Hylleraas wave function, where the numbers in parentheses are again the experimental values: $E(Be) = -27.30$, (-27.54); $E(Mg) = -22.69$, (-22.68); $E(Ca) = -18.07$, (-17.98); $E(Sr) = -16.83$, (-16.72); $E(Zn) = -26.97$, (-27.36).

We turn next to the calculations carried out by Bardsley, Junker, Nesbet, and Sukumar.[69] Here the model potential employed was No. 6 in Table 4.3. In this model the core polarization is explicitly taken into account, which was not the case in the calculations discussed above. The wave function employed was a superposition of configurations built from Slater-type functions. Calculations were made for the negative alkali ions. Large numbers of configurations were considered; for Rb^- and Cs^- the set included 11 s, 7 p, 4 d, and 3 f orbitals, from which 110 configurations were formed. For Li^- the number of configurations was 86, and for Na^- and K^- there were 104. The results for the electron affinities in eV units, with the empirical values in parentheses, were: $E(Li^-) = 0.62$, (0.62); $E(Na^-) = 0.55$, (0.55); $E(K^-) = 0.53$, (0.50); $E(Rb^-) = 0.51$, (0.49); $E(Cs^-) = 0.49$, (0.47).

Summarizing the results of these calculations, we may say that they have shown the usefulness of the model Hamiltonian, Eq. (6.138). Reviewing the numbers we see that even the simple calculations yielded meaningful results. The sophisticated calculations of Schwarz and Bardsley yielded results which in most cases are comparable to the results of ab initio all-electron calculations. Thus if the one-electron model potentials V_m are chosen carefully and the solution of the wave equation is undertaken with high accuracy, the model Hamiltonian of Eq. (6.138) can be expected to yield accurate results.

chapter 7

The Quantum Theory of Atoms with Arbitrary Number of Valence Electrons

7.1. THE EXACT PSEUDOPOTENTIAL TRANSFORMATION OF THE VALENCE-ELECTRON HARTREE–FOCK EQUATIONS

In this chapter we follow an order of presentation different from the last chapter's. Essentially, the theory of atoms with more than two valence electrons is a generalization of the theory of those with two valence electrons. Nevertheless, the complexity of the formalism increases considerably with the increase in the number of valence electrons. Therefore, it will be better to start here with the treatment of the one-electron HF approximation rather than with the treatment of the approximation in which the valence electrons are represented by a correlated wave function.

Our goal is to formulate an effective pseudopotential Hamiltonian for n valence electrons in an atom. In order to obtain this we must first transform the HF equations of the valence electrons into pseudopotential equations. These tasks were carried out by Kahn, Baybutt, and Truhlar (KBT) whose work will be presented here in detail.[48] The work of KBT is partly based on earlier publications of Kahn and Goddard,[70] Melius, Goddard, and Kahn,[71] and Melius and Goddard.[72]

Let the number of core electrons in an atom be N, the number of valence electrons n, and the corresponding HF orbitals φ_i, $(i = 1 \cdots N)$ and φ_v, $(v = 1 \cdots n)$. The HF equations for the n valence electrons are as follows:

$$H_v \varphi_v = \epsilon_v \varphi_v + \sum_{i=1}^{N} \lambda_{iv} \varphi_i , \quad (v = 1, 2, \ldots, n) , \tag{7.1}$$

where

$$H_v = t + g + U + U_v . \tag{7.2}$$

In these equations t and g are the kinetic energy operator and the nuclear potential, respectively, while U is the complete core potential and U_v is the complete potential of the valence electrons. These last two include the exchange potentials. Since we assume that the core has closed shells, the potential U has the form of Eqs. (2.4) and (2.5); the valence electron potential U_v, which is different for each v, does not need to be specified here, except that it is a functional of all valence orbitals:

$$U_v = U_v[\varphi_v] . \tag{7.3}$$

The λ_{iv} are Lagrangian multipliers which establish orthogonality between the valence and core orbitals. The orthogonality between the valence orbitals themselves is assumed to be established by the angular and spin parts of these orbitals. The Lagrangian multipliers can be incorporated into the Hamiltonian by introducing projection operators.[70]

Using our Ω, we obtain from Eq. (7.1),

$$H_v \varphi_v = \epsilon_v \varphi_v + \Omega H_v \varphi_v , \tag{7.4}$$

or

$$\hat{H}_v \varphi_v = \epsilon_v \varphi_v , \quad (v = 1, 2, \ldots, n) , \tag{7.5}$$

where

$$\hat{H}_v = P H_v . \tag{7.6}$$

We transform now Eq. (7.5) exactly into a pseudopotential equation. The WR-type transformation, which we have described in Section 2.4, can conveniently be used here. Thus we replace Eq. (7.5) by the equation

$$(\hat{H}_v + \hat{V}_p) \psi_v = \epsilon_v \psi_v , \quad (v = 1, 2, \ldots, n) , \tag{7.7}$$

where, using the third line of Eq. (2.115), we have

$$\hat{V}_p = -\Omega\hat{H}_v - \hat{H}_v\Omega + \Omega\hat{H}_v\Omega + \epsilon_v\Omega .   \qquad (7.8)$$

As we have seen in Section 2.4, Eq. (7.7) is exactly equivalent to Eq. (7.5).

Next we want to discuss briefly the spectrum of Eq. (7.7). This is actually the same argument that we have seen in Section 2.4, but, it will be useful to repeat it here because of the slightly changed emphasis. Consider first

$$\hat{V}_p\varphi_v = (-\Omega\hat{H}_v - \hat{H}_v\Omega + \Omega\hat{H}_v\Omega + \epsilon_v\Omega)\varphi_v$$
$$= -\Omega\hat{H}_v\varphi_v = -\Omega(\epsilon_v\varphi_v) = 0 ,   \qquad (7.9)$$

where we have taken into account that Ω annihilates the valence orbitals. Next let φ_k be any core orbital and consider

$$\hat{V}_p\varphi_k = (-\Omega\hat{H}_v - \hat{H}_v\Omega + \Omega\hat{H}_v\Omega + \epsilon_v\Omega)\varphi_k$$
$$= -\Omega\hat{H}_v\varphi_k - \hat{H}_v\varphi_k + \Omega\hat{H}_v\varphi_k + \epsilon_v\varphi_k$$
$$= -(\hat{H}_v - \epsilon_v)\varphi_k ,   \qquad (7.10)$$

where we have taken into account Eq. (B.13). Now using Eqs. (7.9) and (7.10) we obtain,

$$(\hat{H}_v + \hat{V}_p)\varphi_v = \hat{H}_v\varphi_v = \epsilon_v\varphi_v ,   \qquad (7.11)$$

and

$$(\hat{H}_v + \hat{V}_p)\varphi_k = (\hat{H}_v - \hat{H}_v + \epsilon_v)\varphi_k = \epsilon_v\varphi_k .   \qquad (7.12)$$

Thus both φ_v and any of the core orbitals φ_k are the eigenfunctions of $(\hat{H}_v + \hat{V}_p)$ with the eigenvalue ϵ_v. We can form the linear combinations, similar to Eqs. (2.122) and (2.123),

$$\psi = \alpha_v\varphi_v + \sum_{i=1}^{N} \alpha_i\varphi_i ,   \quad (v = 1, 2, \ldots , n) ,   \qquad (7.13)$$

and

$$\psi = \sum_{i=1}^{N} \alpha_i\varphi_i .   \qquad (7.14)$$

The functions given by Eq. (7.13) are pseudo-orbitals corresponding to the valence states; those of the form of Eq. (7.14) are the core solutions.

Our conclusion is that Eq. (7.7) will have, for each valence state $v =$

$1, 2, \ldots, n$, pseudo-orbital solutions of the form of Eq. (7.13). There will also be core solutions, as in the one-valence-electron case. The PO [Eq. (7.13)] will be an exact solution of Eq. (7.7) with an arbitrary set of coefficients. Thus we have the same kind of indeterminacy as in the case of one valence electron.

It was emphasized by KBT that Eq. (7.7) reproduces the valence electron orbital energies only if the ϵ_v in Eq. (7.8) is equal to these orbital energies. Thus we should note that Eq. (7.7) reproduces the orbital energies, but the Hamiltonian $\hat{H}_v$ as well as the operator $\hat{V}_p$ are different for each valence orbital.

It will be convenient now to eliminate the Lagrangian multipliers from $\hat{H}_v$. Using the definitions we obtain,

$$
\begin{aligned}
\hat{H}_v + \hat{V}_p &= \hat{H}_v - \Omega\hat{H}_v - \hat{H}_v\Omega + \Omega\hat{H}_v\Omega + \epsilon_v\Omega \\
&= PH_v - \Omega PH_v - PH_v\Omega + \Omega PH_v\Omega + \epsilon_v\Omega \\
&= PH_v - PH_v\Omega + \epsilon_v\Omega \\
&= H_v - \Omega H_v - H_v\Omega + \Omega H_v\Omega + \epsilon_v\Omega \\
&= H_v + V_p,
\end{aligned}
\tag{7.15}
$$

where V_p indicates again the pseudopotential in the third line Eq. (2.115). In this transformation we have used Eq. (B.20). We obtained the convenient relationship,

$$
\hat{H}_v + \hat{V}_p = H_v + V_p,
\tag{7.16}
$$

and our pseudopotential equation [Eq. (7.7)] becomes

$$
(H_v + V_p)\psi_v = \epsilon_v\psi_v, \quad (v = 1, 2, \ldots, n).
\tag{7.17}
$$

Next we want to introduce the total potential which is felt by a valence electron as a result of the presence of the core but not including the effect of the other valence electrons. In the case of one valence electron this is the quantity which we have called modified potential and defined by Eq. (2.41). Using Eq. (7.2) we can write

$$
H_v + V_p = t + g + U + U_v + V_p.
\tag{7.18}
$$

In this equation U, U_v, and V_p are operators. Let us change these operators into local potentials in the Phillips–Kleinman fashion, and let us define the modified potential of the core as follows:

$$
V_M = \frac{(g + U + V_p)\psi_v}{\psi_v}.
\tag{7.19}
$$

It was pointed out by KBT that by using Eqs. (7.18) and (7.17) this expression can be written in the computationally convenient form

$$V_M = \frac{(\epsilon_v - t - U_v)\psi_v}{\psi_v}. \tag{7.20}$$

In Section 4.2 we have analyzed the structure of V_M for the one-valence-electron case. We have seen that for one valence electron outside closed shells V_M was a strongly l-dependent and weakly n-dependent potential. The properties of Eq. (7.19) were discussed by KBT, who concluded that it is a strongly l-dependent expression. As the next step KBT put V_M in the form of Eq. (4.7); that is, in semilocal form. Thus we replace Eq. (7.17) with the equation

$$(t + V_M + U_v)\psi_v = \epsilon_v \psi_v, \tag{7.21}$$

where V_M is now defined as

$$V_M = \sum_{l=0}^{\infty} V^l(r)\Omega_l(\vartheta, \varphi), \tag{7.22}$$

and Ω_l is the angular momentum projection operator,

$$\Omega_l = \sum_{m=-l}^{+l} |Y_{lm}\rangle\langle Y_{lm}|, \tag{7.23}$$

with Y_{lm} being the normalized spherical harmonics.

It must be emphasized that the transition from Eq. (7.17) to Eq. (7.21) is not an approximation. In order to clarify this point let R_{nl} be the radial part of ψ_v and ϵ_{nl} the corresponding orbital energy. Let us consider the equation

$$[t_R + V^l(r) + U_v(r)]R_{nl} = \epsilon_{nl} R_{nl}. \tag{7.24}$$

In this equation t_R and $U_v(r)$ are radial operators. We obtain Eq. (7.24) from Eq. (7.21) by writing ψ_v, as well as the other valence orbitals which occur in U_v, in central field form and integrating over the spin and over the angles. The function $V^l(r)$ is obtained from Eq. (7.20) by taking ψ_v with the quantum numbers (nl) and eliminating the spin and the angles. Defining the modified potential by Eq. (7.22) and the functions V^l as explained, we obtain a potential which has the property that if we put it into Eq. (7.24), those pseudo-orbitals R_{nl} and eigenvalues ϵ_{nl} which were used in the construction of the potential are exact solutions of Eq. (7.24).

Let us consider now Eq. (7.24). We can solve the equation for V^l and get

$$V^l(r) = \frac{[\epsilon_{nl} - t_R - U_v(r)]R_{nl}}{R_{nl}}. \tag{7.25}$$

We recall now that the pseudo-orbitals which are solutions of Eq. (7.17) have the form of Eq. (7.13). Writing down that equation for the radial parts, we get, similarly to Eq. (7.13),

$$R_{nl} = \sum_{n'=1}^{n} a_{n'l, nl} \hat{R}_{n'l},$$ (7.26)

where we have changed slightly the notation of Eq. (7.13) by incorporating the HF orbital with (nl) into the summation rather than displaying it separately as in Eq. (7.13). $\hat{R}_{n'l}$ is the radial part of the HF orbitals and the $a_{n'l}$ coefficients are arbitrary. The expression above must be normalized after the coefficients are determined.

Examining Eqs. (7.25) and (7.26) we see that we have obtained the potential functions which make up the modified potential of Eq. (7.22) in terms of the HF orbitals, the orbital parameters, and the arbitrary constants $a_{n'l}$. For any set of these constants Eq. (7.26) will be an exact PO and Eq. (7.25) a valid potential function.

Looking at Eq. (7.22) we see that the summation over l runs from zero to infinity. From the derivation it is evident that Eq. (7.25) is valid only for those l values which occur in the core of the atom and only for the n value of the valence shell. In order to remedy this deficiency and define V^l for all l values, KBT proceed as follows. Let l_m be the maximum l value which occurs in the core. Then V^l is given by Eq. (7.25) for $l = 0, 1, \ldots, l_m$. For higher l values the pseudopotential V_p is zero in Eq. (7.17) and the PO's are identical with the HF orbitals. These orbitals will be the solutions of Eq. (7.17) with $V_p = 0$. Let $\hat{R}_{nl}$ be the HF solution of that equation. Repeating the argument which led to Eq. (7.25), we obtain for the case $V_p = 0$,

$$\tilde{V}^l = \frac{[\epsilon_{nl} - t_R - U_v(r)]\hat{R}_{nl}}{\hat{R}_{nl}}, \quad (l = l_m + 1, l_m + 2, \ldots).$$ (7.27)

So we obtain that for $l \leq l_m$ we must use Eq. (7.25); for $l > l_m$ the valid potential is given by Eq. (7.27). In formula,

$$V_M = \sum_{l=0}^{l_m} V^l \Omega_l + \sum_{l=l_m+1}^{\infty} \tilde{V}^l \Omega_l.$$ (7.28)

The potential functions $\tilde{V}^l$ consist of the total electrostatic potential of the core plus the exchange potential. The former is independent of (nl). The exchange potential depends on both n and l but it is small relative to the total electrostatic potential. Thus it is reasonable to assume that $\tilde{V}^l$ will be weakly dependent on both n and l. Thus we put

$$\tilde{V}^l \approx \tilde{V}^{l_m+1}, \quad (l = l_m + 2, l_m + 3, \ldots, \infty).$$ (7.29)

Using this approximation we get for Eq. (7.28),

$$V_M = \sum_{l=0}^{l_m} V^l \Omega_l + \tilde{V}^{l_m+1} \sum_{l=l_m+1}^{\infty} \Omega_l . \tag{7.30}$$

The operator Ω_l satisfies the closure relationship,

$$\sum_{l=0}^{\infty} \Omega_l = 1 , \tag{7.31}$$

and using this relationship we obtain from Eq. (7.30):

$$V_M = \sum_{l=0}^{l_m} V^l \Omega_l + \tilde{V}^{l_m+1} \left[\sum_{l=0}^{\infty} \Omega_l - \sum_{l=0}^{l_m} \Omega_l \right]$$

$$= \tilde{V}^{l_m+1} + \sum_{l=0}^{l_m} (V^l - \tilde{V}^{l_m+1}) \Omega_l . \tag{7.32}$$

This formula lends itself to a plausible physical interpretation. The first term consists of the total electrostatic potential plus the exchange potential computed for $l_m + 1$. The second term consists of a sum of l-dependent potentials where the summation is for those l values which occur in the core. The difference $V^l - \tilde{V}^{l_m+1}$ does not contain any electrostatic terms (those are the same in V^l and $\tilde{V}^{l_m+1}$), but it does contain the exact pseudopotential for all l values for which it is not zero. In addition, V^l contains the correct exchange potential for l and $\tilde{V}^{l_m+1}$ contains the exchange potential for $l_m + 1$. The latter will cancel out with the first term. Thus Eq. (7.32) is exact for $l = 0, \ldots, l_m$ and takes into account the exchange potential in an approximate fashion for $l_m + 1$, $l_m + 2$, and so on.

Up to this point we have transformed the HF equations of the valence electrons into pseudopotential equations, and we have derived the formulas for the modified potential of the core in terms of the solutions of the pseudopotential equations. We now proceed with the description of the actual calculation of the pseudo-orbitals and modified potentials.

We have seen that the PO's are given by the formula

$$R_{nl} = \sum_{n'=1}^{n} a_{n'l, nl} \hat{R}_{n'l} , \tag{7.33}$$

where the $\hat{R}_{n'l}$ are HF orbitals which we assume to be available. Eq. (7.33) is an exact solution of Eq. (7.17) with arbitrary coefficients. In order to define a set of unique coefficients, KBT imposed on Eq. (7.33) the following conditions:

1.   The pseudo-orbitals should satisfy the condition

$$\lim_{r \to 0} \left[\frac{R_{nl}}{r^l} \right] = 0 \, . \tag{7.34}$$

2. The coefficients in Eq. (7.33) should be determined in such a way that a certain functional of the PO's, $I[R_{nl}]$, should be minimized. This functional is

$$I[R_{nl}] = \frac{1}{\lambda + 1} \langle \psi_{nl} | \Omega | \psi_{nl} \rangle + \frac{\lambda}{\lambda + 1} \int_0^\infty \left[\frac{d}{dr} \left(\frac{R_{nl}}{r^l} \right) \right]^2 dr \, , \tag{7.35}$$

where λ is an arbitrary number, ψ_{nl} is the full spin-orbital with the radial part R_{nl}.

Let us first consider the second condition. The minimization of the quantity

$$\langle \Omega \rangle = \langle \psi_{nl} | \Omega | \psi_{nl} \rangle \, , \tag{7.36}$$

means that we minimize the expectation value of the projection operator Ω. As we have seen in Eq. (B.16), Ω projects any function onto the core orbitals. The minimization of this quantity insures that the core components will not dominate in the expansion of Eq. (7.33); that is, the PO will approximate the HF valence orbital as closely as possible in the valence region. The minimization of the second term in Eq. (7.35) will ensure that the PO will have a minimum amount of fluctuations. We would like to have a PO which is nodeless (Slater-type function); that is, one having no zero points other than at $r = 0$ and $r = \infty$. We would also like to minimize those "undulations" which are very much in evidence (e.g., in the PK-type solutions of Szasz and McGinn), and are clearly visible in Figure 2.1. The minimization of the second term of Eq. (7.35), which is like a radial kinetic energy, will act toward the elimination of both the nodes and the undulations. The λ parameter permits the regulation of the relative weights of the first and second terms in Eq. (7.35). For $\lambda = 0$ only the $\langle \Omega \rangle$ term is minimized; one obtains a PO with one less node that $\hat{R}_{nl}$. For $\lambda \to \infty$ only the second term is minimized; this gives a nodeless and very smooth PO with a somewhat larger value for $\langle \Omega \rangle$. The optimal compromise between the two terms is accomplished for $0 < \lambda < \infty$, and it is possible to find a value for which the PO is nodeless, smooth, and has a reasonably small $\langle \Omega \rangle$.

Finally Eq. (7.34) also contributes to minimize undulations. It is easy to show that with Eq. (7.34) imposed, the PO will have the following asymptotic form near the $r = 0$ point:

$$\frac{R_{nl}}{r^l} = a_1 r + a_2 r^2 + \cdots \, , \tag{7.37}$$

and by choosing $a_1 = 0$, the slope of (R_{nl}/r^l) can also be made zero at $r = 0$. This reduces the likelihood of sign changes in the slope of the PO and thus minimizes undulations.

After choosing a definite value for the λ parameter, the two conditions, Eq. (7.34) and the minimization of the functional given by Eq. (7.35), uniquely define the constants in the expansion of Eq. (7.33). Since the HF orbitals are uniquely determined, the pseudo-orbitals obtained with this procedure will also be uniquely determined. These pseudo-orbitals will be referred to as maximum-smoothness pseudo-orbitals, MSPO's.

Having determined the MSPO's, the original goal of the whole procedure, the formation of the modified potential of the core can now be undertaken. The potential functions to be computed are given by Eq. (7.25). Here we introduce an approximation. The valence-electron potential U_v is a functional of the HF valence orbitals $\hat{R}_{nl}$. KBT suggested that we replace, in these potentials, the HF orbitals $\hat{R}_{nl}$ by the pseudo-orbitals R_{nl}; that is, that we make the substitution*

$$U_v[\hat{R}_{nl}] \rightarrow U_v[R_{nl}] \, . \tag{7.38}$$

This is an idea that was, in one form or another, introduced into the pseudopotential theory at several stages of its development. We have seen this approximation used, for example, in Section 3.2, Eq. (3.34). Later it was used by Szasz and McGinn[5] for the justification of the model Hamiltonian, Eq. (6.117). More recently, after the publication of the work of KBT, this approximation was used in the buildup of the APM model (Section 5.3.B).

Calculation of MSPO's and the related modified potentials were carried out by Kahn, Baybutt, and Truhlar for the atoms C, N, O, F, Cl, Fe, Br, and I. For each atom the PO's were built up from HF orbitals which we obtained for the lowest energy multiplet of the ground state configuration. In order to obtain PO's which were not occupied in the ground state configuration, the lowest multiplet of the configuration, arising from a single excitation out of the ground state into the required valence orbital was used. The results of the calculations are presented in their entirety in Appendix K.

In Figure 7.1 we have plotted the $4s$, $4p$, and $4d$ potentials for the Br atom. These potentials are the counterparts of the Cu modified potentials illustrated in Figure 2.3. As we see from comparing the two diagrams, the modified potentials constructed by KBT are of the same basic structure as the modified potentials obtained from the Phillips–Kleinman equation.

We are now able to write down the effective pseudopotential Hamiltonian for the n valence electrons of an atom. Guided by simple physical

---

* In a more recent publication by Hay, Wadt, and Kahn[73] it was pointed out that this approximation might produce unphysical behavior in the modified potential at large r. In such a case it might be necessary to return to the original potential expression which contains the HF orbitals. In connection with this problem, see also the discussion in Section 7.2, of the model potential developed by Christiansen et al.

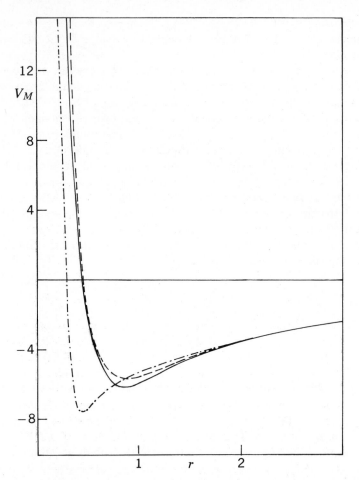

FIGURE 7.1. The modified potentials for a valence electron of the Br atom (————: the s potential; – – – –: the p potential; – · – · – · –: the d potential).

plausibility we obtain[48]

$$H = \sum_{i=1}^{n} (t_i + V_M(i)) + \frac{1}{2} \sum_{i,j=1}^{n} \frac{1}{r_{ij}}, \qquad (7.39)$$

where the modified potential of the core, $V_M(i)$, is defined by Eq. (7.32) and the V^l and $\tilde{V}^l$ functions in the latter expression are given by Eqs. (7.25) and (7.27), respectively. The first sum of Eq. (7.39) is the kinetic energy plus the modified potential of the core, while the second sum is the electrostatic interaction of the valence electrons. The wave equation for the n valence electrons is given by

$$H\Psi = E\Psi, \qquad (7.40)$$

where the n-valence-electron function Ψ does not need to satisfy any orthogonality conditions relative to the core orbitals; that is, Eq. (7.40) can be solved as if the core did not exist.

Summing up, we observe that the exact pseudopotential calculations of Kahn, Baybutt, and Truhlar for many-valence-electrons atoms are the counterparts of the earlier work of Szasz and McGinn for atoms with one valence electron. The method of KBT has several advantages over the work of SMG. In the former the PO's are uniquely determined and the method of calculation ensures very smooth PO's for which the core penetration is small, as demonstrated by the smallness of $\langle \Omega \rangle$. (See Tables K.1 and K.2 in Appendix K.) Also, the PO's and the related potential functions can be determined without the actual construction of the complicated operator $H_v + V_p$ in Eq. (7.17); that is, the exact solutions of Eq. (7.17) are obtained without directly solving this equation. On the other hand, although the $\langle \Omega \rangle$ is very small for all computed PO's, a comparison with the $\langle \Omega \rangle$ computed with the PO's of Szasz and McGinn shows that core penetration is just as small in the latter case as in the case of the KBT calculations. (See Tables A.4 and A.5 in Appendix A.)

Finally, we want to emphasize that, like the work of Szasz and McGinn, the work of Kahn, Baybutt, and Truhlar is an ab initio calculation, which should not be confused with model calculations.

7.2. MODEL POTENTIALS FOR ATOMS WITH MANY VALENCE ELECTRONS

A. Guidelines for the Models and Error Estimation

In Chapter 4 we saw how to set up model potentials for atoms with one valence electron. In this section we generalize those results for many-valence-electron atoms. Here we are talking about one-electron model potentials for an electron in an open or closed valence shell. Most of the presentation will be a straightforward extension of the discussions of Chapter 4; the error estimation will be a new feature here. The formal part of the presentation will be based on the work of the author.[74]

As we saw in the preceding section, the pseudo-orbital of a valence electron in an open shell satisfies Eq. (7.17):

$$(H_v + V_p)\psi_v = \epsilon_v \psi_v, \quad (v = 1, 2, \ldots, n). \qquad (7.41)$$

In this equation H_v is given by

$$H_v = t + g + U + U_v, \qquad (7.42)$$

where U is the potential of the core including exchange and U_v is the potential of the valence electrons. For an open shell we did not specify the form of U_v, except by Eq. (7.3), which said that U_v is a functional of the HF valence orbitals. For a closed valence shell, U_v is defined by Eq. (7.141) below. The pseudopotential V_p is the Weeks–Rice potential [Eq. (2.115)]. For a closed shell, V_p becomes the simpler, Phillips–Kleinman potential.

We introduce now the exact, l-dependent modified potential

$$V_M = \frac{(g + U + V_p)\psi_v}{\psi_v},$$ (7.43)

and then we get Eq. (7.41) in the form

$$(t + V_M + U_v)\psi_v = \epsilon_v \psi_v,$$ (7.44)

where V_M can be written in the semilocal form

$$V_M = \sum_{l=0}^{\infty} V^l \Omega_l,$$ (7.45)

with Ω_l being the angular momentum projection operator [Eq. (7.23)]. The potential functions V^l are n-dependent if Eq. (7.45) is meant as an exact transformation; if the (weak) n-dependence is omitted, Eq. (7.45) is approximate.

Now we imitate the procedures of Section 4.2. Let V_m be a model potential replacing V_M. The equation defining V_m will be

$$(t + V_m + U_v)\psi_v = \epsilon_v \psi_v,$$ (7.46)

and V_m will have the form

$$V_m = \sum_{l=0}^{\infty} V_m^l(r)\Omega_l$$ (7.47)

with V_m^l replacing V^l. As we see from the definition, we define a potential function for each l; if V^l is meant to be n-dependent, we must define a separate V_m^l for each n.

As we see from a comparison with Eq. (4.5), our equation defining the model potential is analogous to the one-valence-electron case. The only difference between the two equations is the presence of the U_v in Eq. (7.46). This term must be present for the following reason. In Eq. (4.5) the model potential V_m represents the core as seen by the single valence electron. In Eq. (7.46) the model potential again represents the core, but now the valence electron moves in the potential of the core and also in the potential of the other valence electrons; the former are represented by V_m, the latter by U_v. Only the core potential is modelized; the potential of the other valence electrons is the same in the model equation as in the original, exact equation.

For the model potential in Eq. (7.46) we recall the same three requirements laid down in Section 4.2: simplicity, reproduction of the exact energy spectrum, and approximation of the pseudo-orbitals. In the case of many-valence electrons we omit the requirement that the model potential should match the empirical spectrum. The eigenvalues of Eq. (7.46) are always required to match the HF energy levels.

As far as the properties of the model potentials are concerned (the (n, l)-dependence, asymptotic form, and so on), the situation is exactly the same as in the case of one valence electron. Thus the discussions of Section 4.2 can be transferred here without changes, except that the remarks made about the V_m^c must be omitted.

Next we shall attempt to estimate the errors committed by the introduction of model potentials.[74] Let V_m be a model potential introduced by some unspecified procedure. We ask the question: under what conditions will V_m yield good results in a *many-electron* calculation? The three conditions laid down above were *one-electron conditions*. Will they be sufficient to ensure accuracy in a many-electron calculation?

The argument presented here follows closely the discussions of Section 6.3. We formulate the argument for two valence electrons; the generalization to many valence electrons is straightforward. Our exact, effective Hamiltonian is given by Eq. (6.119). Let us assume that we can omit the operator P from the last term. Thus we have

$$H = H_1 + V_1 + H_2 + V_2 + \frac{1}{r_{12}}. \tag{7.48}$$

We can rewrite this equation as follows

$$H = t_1 + V_M(1) + t_2 + V_M(2) + \frac{1}{r_{12}}, \tag{7.49}$$

where V_M is the exact modified potential. We now replace V_M by V_m, which is our model potential, and we get Eq. (6.117):

$$H_M = t_1 + V_m(1) + t_2 + V_m(2) + \frac{1}{r_{12}}. \tag{7.50}$$

Let $\hat{\Psi}(1, 2)$ and $\hat{E}$ be the exact solutions of the model equation

$$H_M \hat{\Psi} = \hat{E} \hat{\Psi}, \tag{7.51}$$

and let $\Psi(1, 2)$ and E be the solutions of the exact equation

$$H\Psi = E\Psi, \tag{7.52}$$

where the Hamiltonian is given by Eq. (7.49). Let

$$V_{pt} = H - H_M. \tag{7.53}$$

Then we can write

$$H\Psi = (H + H_M - H_M)\Psi = (H_M + V_{pt})\Psi = E\Psi. \tag{7.54}$$

Viewing V_{pt} as a perturbation and assuming that Eq. (7.51) refers to the ground state of the nondegenerate spectrum of an atom, we obtain, in the first order of perturbation theory

$$E = \hat{E} + \int \hat{\Psi}^* V_{pt} \hat{\Psi} \, dq_{12}. \tag{7.55}$$

Taking into account the form of V_{pt} we obtain

$$\langle V_{pt} \rangle = \int \hat{\Psi}^* \{t_1 + V_M(1) - (t_1 + V_m(1)) + t_2 + V_M(2) - (t_2 + V_m(2))\} \hat{\Psi} \, dq_{12}. \tag{7.56}$$

Thus, as we see from Eq. (7.55), the condition for $\hat{E}$ being a good approximation to E is that $\langle V_{pt} \rangle$ be small or vanishing.

We have seen in Section 4.2 that this kind of argument can be used to show that V_m can be a function very different from V_M and yet yield a good approximation. Here we shall use the argument in a different way. Let us approximate $\hat{\Psi}$ by an antisymmetrized product of one-electron pseudo-orbitals, as we have done in Section 6.3. Thus let

$$\hat{\Psi} = [2!]^{-1/2} \det[\psi_i \psi_j], \tag{7.57}$$

and let us assume that ψ_i and ψ_j are the solutions of the model equation

$$(t + V_m)\psi_i = \epsilon_i \psi_i, \tag{7.58}$$

and similarly for ψ_j. (Here we have switched back, in the spirit of Section 6.3, to the one-valence-electron model equations. The reader will see, however, by a momentary reflection that the argument could be made just as well for many valence electrons. In that case we would have added and subtracted $U_v(1) + U_v(2)$ in the integrand of Eq. (7.56) and used Eq. (7.46) instead of Eq. (7.58).) We have marshalled arguments for the choice of Eq. (7.57) in Section 6.3. On putting $\hat{\Psi}$ into Eq. (7.56) we obtain

$$\langle V_{pt} \rangle = \int \hat{\Psi}^* \{t_1 + V_M(1)\} \hat{\Psi}\, dq_{12} + \int \hat{\Psi}^* \{t_2 + V_M(2)\} \hat{\Psi}\, dq_{12} - (\epsilon_i + \epsilon_j), \qquad (7.59)$$

where we have taken into account Eq. (7.58). Thus $\langle V_{pt} \rangle$ will be small if

$$\int \hat{\Psi}^* (t_1 + V_M(1) + t_2 + V_M(2)) \Psi\, dq_{12} \approx \epsilon_i + \epsilon_j. \qquad (7.60)$$

Using Eq. (7.57) it is easy to see that the above relationship is equivalent to the statement

$$\int \psi_i^* (t + V_M) \psi_i\, dq \approx \epsilon_i, \qquad (7.61)$$

and a similar relationship for ψ_j.

We have not yet specified the model potential V_m. Let us assume that the model potential is constructed to match the HF energy spectrum; that is, the eigenvalues of Eq. (7.58) are equal to the HF energy levels. From that it does not follow that the eigenfunctions of Eq. (7.58) are close to the exact pseudo-orbitals, and thus it does not follow that Eq. (7.61) is satisfied. Equation (7.61) will be satisfied only if ψ_i is close to the exact pseudo-orbital, since the eigenfunctions of $(t + V_M)$ are the exact pseudo-orbitals.

We may state therefore that a one-electron model potential will yield good results in a many-electron calculation if, besides matching the HF energy levels, the eigenfunctions of the model equation will be good approximations to the exact pseudo-orbitals; that is, the model potential will be a good one, if, of the three conditions laid down in Section 4.2, the third is satisfied along with the second*. We emphasize that the analysis presented here does not invalidate the arguments of Section 4.2. Besides constructing good approximations to the pseudo-orbitals, the smallness of $\langle V_{pt} \rangle$ can be accomplished by choosing V_m to be a kind of average of V_M, in the sense that we described it in Section 4.2.

B. A Survey of the Models

In this subsection we look at the main types of model potentials which have been developed for many-valence-electron atoms. Since the description of

---

* It is important to note that the exact, ab initio pseudopotentials, such as the potentials computed by Szasz and McGinn (Section 2.2.B) and those computed by Khan, Baybutt, and Truhlar (Section 7.1), satisfy both the second and third conditions *exactly*. This is true even if the pseudo-orbitals are not uniquely determined, like those computed by Szasz and McGinn. Thus the indeterminacy of the pseudo-orbitals and pseudopotentials does not affect the quality of the results in a many-electron calculation. We note that the importance of the third condition was also emphasized, in a slightly different context, by Melius and Goddard (see Section 7.2.B).

these models requires a more detailed discussion than the description of the one-valence-electron models, we discuss each of them separately instead of tabulating them.

(1)   The Method of Melius and Goddard: Coreless Hartree–Fock Orbitals.[72]
The work of Melius and Goddard does not fit entirely into this section for two reasons. First, the method which we shall describe here is ab initio to a very large extent. Second, the formulation of the method was given first for atoms with one valence electron. We present it here because, while the idea of coreless HF orbitals is ab initio, it has found application mostly in its model form, and while the method was formulated for atoms with one valence electron, its generalization to many electrons is simple.

Our starting point is Eq. (7.41). For one valence electron the equation takes the form

$$(H_v + V_p)\psi_v = \epsilon_v \psi_v, \tag{7.62}$$

where, according to Eq. (7.42),

$$H_v = t + g + U. \tag{7.63}$$

The potential U_v vanishes for one valence electron. The exact solutions of Eq. (7.62) will be of the form

$$\psi_v = \varphi_v + \sum_{i=1}^{N} \alpha_i \varphi_i, \tag{7.64}$$

where φ_v and φ_i are the valence and core HF orbitals, and the α_i's are arbitrary coefficients. Using Eq. (7.64), one can form the exact modified potential according to Eq. (7.43),

$$V_M = \frac{(g + U + V_p)\psi_v}{\psi_v}, \tag{7.65}$$

which, in view of Eq. (7.62), can also be written as

$$V_M = \frac{(\epsilon_v - t)\psi_v}{\psi_v}. \tag{7.66}$$

This expression is an exact modified potential for any set of coefficients α_i.

Melius and Goddard investigated the question of the determination of the coefficients from the point of view of molecular calculations. They concluded that the most advantageous choice for the coefficients requires that the pseudo-orbital ψ_v be as small as possible in the core region. They called such a PO a "coreless Hartree–Fock orbital."

The CHF orbitals were determined as follows. In order to be able to construct ψ_v from as short a basis set as possible, Melius and Goddard demanded that the fluctuations of ψ_v in the core region be as small as possible. This is accomplished, for example, if the coefficients in Eq. (7.64) are subjected to the condition that the pseudo-orbital and its radial derivatives be zero at the center of the atom. (From the higher derivatives we require as many to be zero as there are conditions needed to determine the coefficients.) Another way to determine the coefficients is to require that the PO approximate the Slater orbital with the appropriate (nl). This last procedure leads to pseudo-orbitals which decline smoothly in the core region and thus can be constructed from moderately sized basis sets. Using the CHF orbitals, determined by either procedure, in Eq. (7.66), we obtain an exact, ab initio modified potential.

Having formulated a method for the determination of exact modified potentials, Melius and Goddard investigated the question of how these potentials can best be used in molecular calculations. Such calculations require the potential to be as simple an analytic expression as possible. Let V_m be a semilocal model potential with adjustable parameters. Melius and Goddard suggested that the parameters of V_m be determined in such a way that the condition

$$\langle \psi_\alpha | V_m | \psi_\beta \rangle = \langle \psi_\alpha | V_M | \psi_\beta \rangle , \quad (\alpha, \beta = 1, 2, \ldots) , \qquad (7.67)$$

be satisfied. Here V_M is the exact modified potential determined above, and the ψ_α functions are the members of the basic set used to construct the pseudo-orbital. An alternative formula is

$$\langle \psi_\alpha | V_m | \psi \rangle = \langle \psi_\alpha | V_M | \psi \rangle , \quad (\alpha = 1, 2, \ldots) , \qquad (7.68)$$

where ψ is the exact PO and the ψ_α are chosen to span the space of the valence orbital. Both Eqs. (7.67) and (7.68) are required to be satisfied in a least squares sense. The potential determined this way is a model potential, in contrast to the exact potentials determined by substituting CHF orbitals into Eq. (7.66).

Melius and Goddard have determined exact as well as model potentials for alkali atoms. The final procedure was as follows. CHF orbitals were determined by the STO fitting. Then the model potentials were determined from Eq. (7.68). The pseudo-orbitals used in these equations were the CHF orbitals, and the ψ_α functions were the members of the basic set used in the construction of the pseudo-orbitals.

Melius and Goddard constructed analytic expressions consisting of linear combinations of Gaussian orbitals for the valence orbitals of alkali atoms. These analytic expressions approximated the exact energy levels of the model potentials with high accuracy. It was pointed out that the analytic expressions were much simpler than the similar expressions for HF orbitals.

Thus the conclusion was reached that the CHF orbitals provide a significant reduction in the computational work while at the same time providing ab initio quality in the results.

This last conclusion was further strengthened by the work of Topiol, Moskowitz, and Melius and of Topiol and Osman.[75] In these papers the CHF orbital method was used to construct model potentials for the atoms between Li and Ne and between K and Zn. Thus the method, which was originally developed for one valence electron, was extended to atoms with several valence electrons in open shells. Atomic calculations carried out with these model potentials have shown that the results were in excellent agreement with HF calculations.

(2)   The Method of Durand and Berthelat.[76]   Let us consider an atom with several valence electrons in an open shell. Let the HF orbitals for these slectrons be φ_v, $(v = 1, 2, \ldots, n)$, where these orbitals satisfy Eq. (7.5). Let us consider the model pseudo-orbital. $\tilde{\psi}_v$, which is an STO or a linear combination of STO's. This model PO is, in any case, nodeless. Let us consider the integral

$$I = \int_{R_c}^{\infty} |\tilde{\psi}_v - \varphi_v|^2 \, dq , \qquad (7.69)$$

and let us determine $\tilde{\psi}_v$ in such a way that

$$I = \text{Minimum} . \qquad (7.70)$$

As a subsidiary condition we assume that $\tilde{\psi}_v$ is normalized

$$\langle \tilde{\psi}_v \mid \tilde{\psi}_v \rangle = 1 . \qquad (7.71)$$

The cutoff radius R_c is defined as follows. Let $\hat{R}_{nl}$ be the radial part of φ_v. We define R_c as the point where $\hat{R}_{nl}$ and the core orbital $\hat{R}_{(n-1)l}$ intersect. This quantity can be viewed as a core radius.

It is clear from the description that the model PO determined this way will be smooth and will approximate the HF orbital very closely in the valence region. In the core region it will be small since the bulk of the normalized orbital will be where the bulk of the HF orbital is. Thus the model PO will be like the HF orbital without its inner nodes: it will be a coreless HF orbital.

Now let us consider Eq. (7.46):

$$(t + V_m + U_v)\hat{\psi}_v = \hat{\epsilon}_v \hat{\psi}_v , \qquad (7.72)$$

where V_m is the model potential which we put in the form of Eq. (7.47).

Next we choose a specific analytic form for V_m and determine the parameters of the potential in such a way that the eigenfunctions of Eq. (7.72) will closely approximate the model pseudo-orbitals

$$\hat{\psi}_v \approx \tilde{\psi}_v . \tag{7.73}$$

When this is accomplished, the eigenvalue of Eq. (7.72) will closely approximate the HF levels. This can be seen as follows. The exact PO (which is not identical with the model PO) will have the form of Eq. (7.13),

$$\psi_v = \alpha_v \varphi_v + \sum_{i=1}^{N} \alpha_i \varphi_i \tag{7.74}$$

On putting $\alpha_v = 1$ and normalizing, we get

$$\psi_v = N_v \left(\varphi_v + \sum_{i=1}^{N} \alpha_i \varphi_i \right), \tag{7.75}$$

where the normalization constant will be

$$N_v = \left\{ 1 + \sum_{i=1}^{N} |\alpha_i|^2 \right\}^{-1/2}. \tag{7.76}$$

The quantities $|\alpha_i|^2$ are small, and therefore we can write

$$N_v \approx 1 - \frac{1}{2} \sum_{i=1}^{N} |\alpha_i|^2 + \cdots \tag{7.77}$$

In the valence area, where the core orbitals are practically zero, we get from Eq. (7.75)

$$\psi_v = N_v \varphi_v , \tag{7.78}$$

and thus, to the degree to which the approximation

$$N_v \approx 1 \tag{7.79}$$

is valid, we can write

$$\psi_v \approx \varphi_v . \tag{7.80}$$

We have determined the model PO in such a way as to approximate closely the HF orbital φ_v. This can be expressed by the relationship

$$\tilde{\psi}_v \approx \varphi_v , \tag{7.81}$$

and we have determined the model potential V_m in such a way that the eigenfunctions of Eq. (7.72) will closely approximate the model PO. Thus we shall have

$$\hat{\psi}_v \approx \tilde{\psi}_v, \qquad (7.82)$$

and using Eqs. (7.80) and (7.81) we obtain

$$\hat{\psi}_v \approx \psi_v. \qquad (7.83)$$

Thus the eigenfunctions of the model equation will closely approximate the exact pseudo-orbitals, and thus we can expect that the eigenvalues of Eq. (7.72) will also closely approximate the HF energy levels. We note that a more accurate sequence is

$$\hat{\psi}_v \approx \tilde{\psi}_v \approx \varphi_v \approx \frac{1}{N_v} \psi_v, \qquad (7.84)$$

and so Eq. (7.83) is valid only to the extent to which Eq. (7.79) holds.

The method of Durand and Berthelat produces a model potential which satisfies, in a very good approximation, all three conditions laid down in Section 4.2. In addition, it produces PO's which are very small in the core area and thus will give a small value for the expectation value of the core projection operator Ω. For the actual form of the model potential, Durand and Berthelat used

$$V_m^l = \frac{A_l}{r} + \frac{B_l}{r^2}, \qquad (7.85)$$

where A_l and B_l are adjustible parameters.

(3) The Method of Coffey, Ewig, and Van Wazer.[77] Let us consider a valence electron in a closed valence shell. The HF equation for the orbital φ_v will be Eq. (7.5), which, for a closed shell, takes the form

$$H_v \varphi_v = \epsilon_v \varphi_v, \qquad (7.86)$$

where

$$H_v = t + g + U + U_v. \qquad (7.87)$$

In this equation U_v, the potential of the valence electrons, takes the form of Eq. (7.141) below. In getting Eq. (7.86) from Eq. (7.5) we used the relationships

$$PH_v = H_v P, \qquad (7.88)$$

and

$$P\varphi_v = \varphi_v .$$ (7.89)

The pseudopotential transformation of Eq. (7.86) gives

$$(H_v + V_p)\psi_v = \epsilon_v \psi_v ,$$ (7.90)

where V_p is the Phillips–Kleinman potential.

The goal of Coffey, Ewig, and Van Wazer was to eliminate the difficulties connected with the calculation of the core potential U. They replaced this complicated expression by the model potential

$$g + U = -\frac{Z}{r} + \frac{(1 - e^{-\alpha r})N}{r} .$$ (7.91)

As we see, we have the correct asymptotic relationships

$$\lim_{r \to \infty} (g + U) = -\frac{(Z - N)}{r} ,$$ (7.92)

and

$$\lim_{r \to 0} (g + U) = -\frac{Z}{r} .$$ (7.93)

N is the number of core electrons. Using Eq. (7.91) we get for the model potential

$$V_m = -\frac{Z}{r} + \frac{(1 + e^{-\alpha r})N}{r} + \sum_{i=1}^{N} (\epsilon_v - \epsilon_i)|\varphi_i\rangle\langle\varphi_i| .$$ (7.94)

In this expression, the PK-type pseudopotential is exact for a closed valence shell and approximate for an open shell. The constant α was used to adjust V_m in such a way that the s and p HF levels of an atom are approximated as closely as possible.

(4) The Method of Ewig, Osman, and Van Wazer.[78] Let us consider an atom with several valence electrons in an open or closed shell. The pseudo-orbital of one of the valence electrons is the solution of Eq. (7.17),

$$(H_v + V_p)\psi_v = \epsilon_v \psi_v ,$$ (7.95)

where

$$H_v = t + g + U + U_v .$$ (7.96)

In the preceding equations, U is the nonlocal potential of the core given by Eqs. (2.4) and (2.5), and U_v is the nonlocal potential of the valence electrons. Both of these expressions include exchange. The potential U_v is a functional of the HF valence orbitals as shown by Eq. (7.3). V_p is the Weeks–Rice pseudopotential.

Ewig, Osman, and Van Wazer defined the following wave equation for the pseudo-orbital of the valence electrons,

$$(H_v' + V_p)\psi_v = \epsilon_v \psi_v, \tag{7.97}$$

where

$$H_v' = t + g + \hat{U} + \hat{U}_v, \tag{7.98}$$

and so the valence equation will be

$$(t + g + \hat{U} + \hat{U}_v + V_p)\psi_v = \epsilon_v \psi_v. \tag{7.99}$$

Here the local potential $\hat{U}$ is defined in such a way that it represents the total core-valence interaction. $\hat{U}_v$ is a nonlocal potential representing the valence-valence interactions. This potential has the same functional form as U_v, but with the HF orbitals replaced by the PO's. Thus, $\hat{U}_v$ is a functional of the valence PO's only:

$$\hat{U}_v = \hat{U}_v[\psi_v]. \tag{7.100}$$

What is accomplished in Eq. (7.99) is that the core-valence interaction is written in a form which depends explicitly only on the PO's of the valence electrons. Implicitly, of course, $\hat{U}$ depends on the HF orbitals of the core as well as of the valence electrons.

The function $\hat{U}$ was determined by Ewig, Osman, and Van Wazer as follows. Demanding operator equivalence between Eqs. (7.95) and (7.99), we obtain

$$(U + U_v)\psi_v = (\hat{U} + \hat{U}_v)\psi_v, \tag{7.101}$$

and taking into account that $\hat{U}$ is, by definition, a local potential, we obtain

$$\hat{U} = \frac{(U + U_v - \hat{U}_v)\psi_v}{\psi_v}. \tag{7.102}$$

In order to show the meaning of this expression, we indicate the functional dependence of the operators:

$$\hat{U} = \frac{(U[\varphi_i] + U_v[\varphi_v] - \hat{U}_v[\psi_v])\psi_v}{\psi_v}. \tag{7.103}$$

Here φ_i means the core orbitals, φ_v the HF orbitals for the valence electrons, and ψ_v the pseudo-orbitals for the valence electrons. The difference $(U_v - \hat{U}_v)$ describes the difference between the HF orbitals and the pseudo-orbitals, or, more accurately, the effect of this difference on the core potential. This term is present in the core potential to compensate for the fact that the valence-valence interaction is computed in Eq. (7.97) from pseudo-orbitals rather than from the HF orbitals.

For molecular calculations, Ewig, Osman, and Van Wazer have shown that U can be represented by the simple analytic expression with adjustable parameters,

$$\hat{U} \to \frac{N[1 - \Sigma_i c_i e^{-e_i r^2}]}{r}, \qquad (7.104)$$

which is analogous to the corresponding expression in Eq. (7.91).

(5) The Method of Christiansen, Lee, and Pitzer.[79] This method rests on the Kahn–Baybutt–Truhlar method, and therefore we must look first at Eq. (7.25). We have seen that, in the method of KBT, the modified potential was defined by the formula

$$V^l(r) = \frac{[\epsilon_{nl} - t_R - U_v(r)]R_{nl}}{R_{nl}}, \qquad (7.105)$$

where R_{nl} was the exact PO, ϵ_{nl} the exact HF energy parameter, t_R the kinetic energy operator, and U_v the HF potential of the valence electrons. This last was defined in terms of the HF orbitals, and later was changed by KBT in such a way that the HF orbitals were replaced by the PO's:

$$U_v[\hat{R}_{nl}] \to U_v[R_{nl}]. \qquad (7.106)$$

We have seen that, in some cases, this approximation led to difficulties and the original method, using the HF orbitals, had to be restored. We shall look now at this problem in more detail.

The modified potential, defined by Eq. (7.105), is an exact potential as long as the pseudo-orbital in it is exact. The form of the normalized exact PO is given by

$$\psi_v = N_v \varphi_v + \sum_{i=1}^{N} c_i \varphi_i, \qquad (7.107)$$

where

$$N_v = \left\{ 1 + \sum_{i=1}^{N} |\alpha_i|^2 \right\}^{-1/2}, \qquad (7.108)$$

and

$$c_i = N_v \alpha_i . \tag{7.109}$$

From the form of N_v, we see that $N_v < 1$. Thus in the valence area, where the core orbitals φ_i are practically zero, we obtain

$$\psi_v = N_v \varphi_v , \tag{7.110}$$

and

$$|\psi_v|^2 = N_v^2 |\varphi_v|^2 . \tag{7.111}$$

Since $N_v^2 < 1$, we can write that

$$|\psi_v|^2 < |\varphi_v|^2 , \tag{7.112}$$

in the valence area. From this it follows that making the substitution of Eq. (7.106) means replacing the HF density by the smaller pseudo-density in the electrostatic potential. The electrostatic potential will decline faster and will shield the negative electrostatic potential of the core to a lesser degree when the HF density is replaced by the pseudo-density. Thus the resulting modified potential will become more negative than the exact, after the approximation of Eq. (7.106).

One way of removing this difficulty is to restore the HF orbitals in the potential U_v. The disadvantage of this procedure is that it defeats one of the objectives of the pseudopotential method, which was to avoid the need to work with the orthogonalized HF orbitals. Thus Christiansen, Lee, and Pitzer suggested that the difficulty be removed by the introduction of a model potential. This is done in the following way.

First a model pseudo-orbital is defined by

$$\tilde{\psi}_v = \varphi_v + f_v . \tag{7.113}$$

In this formula φ_v is the HF orbital. We envisage the HF orbital to be cut off at a certain radius r_c. Outside of r_c the model PO is equal to φ_v. Inside of r_c we put $\varphi_v = 0$ and introduce the function f_v, which is zero outside of r_c. Inside of r_c, the auxiliary function f_v is a five-term polynomial in r with a leading power of $(l+2)$. The expansion coefficients are chosen so as to match the amplitude and first three derivatives of the two sections at r_c, and also to make the whole model PO normalized. The cutoff radius r_c is chosen in such a way that it is the innermost point at which the matching results in a nodeless radial pseudo-orbital with no more than two inflections in its entire range. The resulting model PO will be smooth and will be identical with the HF orbital in as large an area as possible. The condition of Eq. (7.34) will be satisfied.

We now generate the model potential by putting the model pseudo-orbital into Eq. (7.105). With the procedure described above, the valence-electron potential U_v generated by the model PO's will be the same, in the valence area, as the potential U_v generated by the HF orbitals. Thus we can write

$$U_v[\tilde{\psi}_v] = U_v[\varphi_v] , \qquad (7.114)$$

in the valence area, and the difficulty with the long-range behavior of the modified potential is removed.

It is clear that the method of Christiansen, Lee, and Pitzer is similar to the method of Melius and Goddard, and that of Durand and Berthelat. The model potential generated by this procedure will satisfy, in a good approximation, all three conditions laid down in Section 4.2. In addition, since the model PO is small in the core area, the expectation value of the core projection operator Ω will be small, which is advantageous in many-electron atomic and molecular calculations.

7.3. AB INITIO AND EXACT PSEUDOPOTENTIAL THEORY WITH CORRELATED WAVE FUNCTION FOR THE VALENCE ELECTRONS

A. Introduction

Having disposed in Section 7.1 of the problem of how the HF approximation can be transformed into a pseudopotential formalism for the n valence electrons of an atom, we now proceed to a theory in which the wave function of the valence electrons is an arbitrary, n-electron function in which the valence-valence correlation effects can fully be taken into account. This theory will have the same goals as the FVP theory presented in Section 6.1; in fact, the theory will be a straightforward generalization of the FVP method. For this reason, we present the basic equations of the theory in a descriptive fashion; there will be no derivations or existence proofs presented. These derivations and proofs are, in every case, the straightforward generalizations of the corresponding arguments for the two-valence-electron case; they were presented in Section 6.1 and in the associated appendixes. The arguments for the n-valence-electron case can easily be constructed by analogy, and do not need to be included in the text.

Some historical background should be provided at this point. The theory of FVP was generalized for more than two valence electrons by Jucys,[80] who developed the formalism with the exception of deriving the equation for the n-valence-electron function analogous to the two-electron equation of the FVP theory. Later, as part of a study of the electron correlation in atoms, the author clarified, starting from the work of Jucys, the most important mathematical properties of correlated wave functions,[65] including the properties of such correlated wave functions as will be used here. The derivation of the *n*-valence-electron equation, which was missing from Jucys' work, was undertaken by the author in a more recent publication, in which the creation of a consistent mathematical background for the whole theory was also attempted. The presentation which follows is based on this work.[60]

B. Ab Initio Theory

Let us consider an atom with N core and n valence electrons, for which the exact Hamiltonian is given by

$$H = \sum_{i=1}^{N+n} (t_i + g_i) + \frac{1}{2} \sum_{i,j=1}^{N+n} \frac{1}{r_{ij}}, \qquad (7.115)$$

where the notation is the same as in the preceding chapters. We represent the atom by the wave function,

$$\Psi_T = [(N+n)!]^{-1/2} \tilde{A}\{\Psi(1, 2, \ldots, n) \\ \times \det[\varphi_1(n+1)\varphi_2(n+2)\cdots\varphi_N(n+N)]\}. \qquad (7.116)$$

In Ψ_T, the valence-electron function Ψ is a completely arbitrary, *n*-electron function, subjected only to the antisymmetry requirement, while the core orbitals are represented by the one-electron spin-orbitals $\varphi_1 \cdots \varphi_N$. $\tilde{A}$ is an antisymmetrizer operator.

The total wave function Ψ_T does not change if we orthogonalize Ψ to the core orbitals. Let

$$P = P_1 P_2 \cdots P_n = \prod_{k=1}^{n} P_k, \qquad (7.117)$$

where the one-electron projection operator P_1 is defined by Eqs. (6.16) and (6.17). Also let

$$\Omega = 1 - P. \qquad (7.118)$$

We orthogonalize Ψ to the core orbitals by forming

$$\Phi(1, \ldots, n) = P\Psi(1, \ldots, n), \qquad (7.119)$$

and now Φ will satisfy the strong orthogonality relationship,

$$\int \varphi_i^*(1)\Phi(1, 2, \ldots, n)\, dq_1 = 0, \quad (i = 1, \ldots, N). \qquad (7.120)$$

The statement that Ψ_T is not changed by the orthogonalization of its correlated part means that

$$\Psi_T = [(N + n)!]^{-1/2} \tilde{A}\{\Psi(1, 2, \ldots, n) \det[\varphi_1(n + 1) \cdots \varphi_N(n + N)]\}$$
$$= [(N + n)!]^{-1/2} \tilde{A}\{\Phi(1, 2, \ldots, n) \det[\varphi_1(n + 1) \cdots \varphi_N(n + N)]\}. \quad (7.121)$$

Thus the subsidiary conditions imposed on our total wave function are as follows. The one-electron orbitals are orthonormal

$$\langle \varphi_i \mid \varphi_k \rangle = \delta_{ik}, \quad (i, k = 1, 2, \ldots, N). \quad (7.122)$$

The n-electron function Φ satisfies the strong orthogonality condition [Eq. (7.120)], and is normalized

$$\langle \Phi \mid \Phi \rangle = 1. \quad (7.123)$$

Next we present the energy expression for the atom; that is, the average value of the Hamiltonian [Eq. (7.115)] with respect to the wave function [Eq. (7.121)]. Using the second line of that equation we obtain (see Appendix G),

$$E_T = \langle \Phi | H' | \Phi \rangle + E_c, \quad (7.124)$$

where

$$H' = \sum_{i=1}^{n} H'_i + \frac{1}{2} \sum_{i,j=1}^{n} \frac{1}{r_{ij}}. \quad (7.125)$$

The operator H'_i, $(i = 1, 2, \ldots, n)$, is an operator expressed in the coordinates of the i-th valence electron. It is the same as in Eq. (6.29); that is, it is given by the formula

$$H'_i = t_i + g_i + U(i), \quad (7.126)$$

where

$$U(i) = \sum_{k=1}^{N} U_k(i). \quad (7.127)$$

In Eq. (7.126) t_i and g_i are the kinetic energy and the nuclear potential, respectively, while U is the total potential, electrostatic and exchange, of the core. Thus H'_i represents the kinetic energy plus the total core-valence interaction of the i-th valence electron. The quantity E_c in Eq. (7.124) is the HF energy of the core, built from the orbitals $\varphi_1 \cdots \varphi_N$.

The equation for the best Φ is obtained by varying the total energy of the atom, Eq. (7.124), with respect to Φ^*. In the process of variation the subsidiary conditions [Eqs. (7.120) and (7.123)] must be taken into account. Carrying out the variation and making use of the orthogonality projection operator [Eq. (7.117)] we obtain

$$H\Phi = E\Phi, \qquad (7.128)$$

where

$$H = PH', \qquad (7.129)$$

and H' is given by Eq. (7.125). E is the total energy of the n valence electrons; that is, this quantity is equal to the first term on the right side of Eq. (7.124). Using E, we can put Eq. (7.124) in the form

$$E_T = E + E_c. \qquad (7.130)$$

The solution of Eq. (7.128) for the ground state of the atom will define the best Φ, and with it the absolute minimum of the total energy of the atom for a given, unspecified set of one-electron core orbitals. As the total wave function [Eq. (7.121)] was not restricted by the orthogonalization of its correlated part, the generality of Eq. (7.128) is not in any way restricted by this orthogonalization either. The exact solutions of Eq. (7.128) will describe the valence electron distribution exactly, including the valence-valence electron correlation.

Equation (7.128) is very general and permits different choices for the core orbitals. We want to show what form the equation takes in two important special cases. The first of these two core cases occurs when the core orbitals $\varphi_1 \cdots \varphi_N$ which occur in Eq. (7.128) are HF orbitals obtained in such a way that, instead of representing the atom by the wave function of Eq. (7.121), the atom is treated in the HF approximation, the core orbitals being our $\varphi_1 \cdots \varphi_N$. In other words, these orbitals are the core solutions of the HF equations for the $(N + n)$ electron atom. It is assumed that the N core electrons are in closed shells, while the n valence electrons are in an open shell with an arbitrary electron configuration.

The second of the two special cases occurs when the core orbitals are HF orbitals for the core electrons of an atom in which both the core electrons and the valence electrons are in closed shells. (More precisely, the core orbitals will be the core solutions of an HF approximation in which the total wave function is a single Slater determinant.)

The HF equations for the n valence electrons of an atom were given by Eq. (7.5). According to that equation we have

$$\hat{H}_v \varphi_v = \epsilon_v \varphi_v, \quad (v = 1, 2, \ldots, n), \qquad (7.131)$$

where

$$\hat{H}_v = PH_v,\qquad (7.132)$$

and according to Eq. (7.2) we have

$$H_v = t + g + U + U_v,\qquad (7.133)$$

where U_v is the potential of the valence electrons. Let us attach the symbol of electron coordinates to the quantities in the last three equations. Doing this and comparing Eq. (7.133) with Eq. (7.126), we obtain

$$H_v(i) = t_i + g_i + U(i) + U_v(i) = H_i' + U_v(i).\qquad (7.134)$$

Now write down Eq. (7.128) in detailed form:

$$H\Phi = PH'\Phi = P\left\{\sum_{i=1}^{n} H_i' + Q\right\}\Phi = E\Phi.\qquad (7.135)$$

In getting this formula we used Eqs. (7.129) and (7.125). The symbol Q is shorthand for the electrostatic interaction of the valence electrons,

$$Q \equiv \frac{1}{2}\sum_{i,j=1}^{n}\frac{1}{r_{ij}}.\qquad (7.136)$$

The core orbitals are in the operator P and in the potential U which is contained in H_i'. We identify these orbitals with the HF orbitals by solving Eq. (7.134) for H_i' and substituting the resulting expression into Eq. (7.135). With this step the formerly unspecified core orbitals become HF orbitals. Carrying out this step, we obtain

$$H\Phi = P\left\{\sum_{i=1}^{n}(H_v(i) - U_v(i)) + Q\right\}\Phi = E\Phi.\qquad (7.137)$$

Using Eq. (7.132) we obtain

$$PH_v(1) = P_1P_2\cdots P_nH_v(1) = [P_1H_v(1)]P_2P_3\cdots P_n = \hat{H}_v(1)[P_2P_3\cdots P_n].\qquad (7.138)$$

Substituting the last equation into Eq. (7.135) and taking into account Eq. (E.15), we obtain the equation for Φ in the form

$$\left\{\sum_{i=1}^{n}\hat{H}_v(i) + PS\right\}\Phi = E\Phi,\qquad (7.139)$$

where the symbol S is defined as

$$S \equiv Q - \sum_{i=1}^{n} U_v(i). \tag{7.140}$$

Eq. (7.139) is the equation for the best Φ for the case when the core orbitals are solutions of the open shell HF equations. It is easy to obtain the best Φ-equation for the closed-shell case. In that case, the HF potential of the valence electrons, $U_v(i)$, is given by the standard, v-independent expression,

$$U_v(i) = \sum_{s=N+1}^{N+n} U_s(i), \tag{7.141}$$

where the operator U_s is the HF potential given by Eq. (2.5). Using Eq. (7.141) in Eqs. (7.140) and (7.134), and putting the resulting expressions into Eq. (7.139), we obtain

$$\left\{ \sum_{i=1}^{n} \hat{H}_F(i) + PS \right\} \Phi = E\Phi, \tag{7.142}$$

where we have denoted the HF Hamiltonian for the whole atom by H_F and

$$\hat{H}_F(i) = P_i H_F(i). \tag{7.143}$$

For an atom with closed shells, the projection operator P_i commutes with H_F. Thus we obtain, by taking into account Eq. (E.15) again, the equation

$$\left\{ \sum_{i=1}^{n} H_F(i) + PS \right\} \Phi = E\Phi, \tag{7.144}$$

which is a straightforward generalization of the similar, two-valence-electron equation, Eq. (6.40).

The most important mathematical properties of Eqs. (7.128), (7.139), and (7.144) are the following:

1.   The exact solutions of the equations will be strong-orthogonal to the core orbitals, and solutions belonging to different eigenvalues will be orthogonal to each other.
2.   The Hamiltonian operators in the equations are partially Hermitian.
3.   Exact solutions of the equations and approximations to the exact solutions can be obtained by the variation method, in which the trial functions are strong-orthogonal to the core orbitals.

Finally we note that the remarks about the FVP theory at the end of

Section 6.1.B are equally valid here for the theory which is formulated for n valence electrons.

C. Exact Pseudopotential Theory

We now transform the ab initio theory presented in the preceding section into an exact pseudopotential formalism. The procedure is closely analogous to the method described in Section 6.2 for the two-valence-electron case. As we have seen, there are three different ways to transform the valence-electron equations into pseudopotential equations. One can use the Phillips–Kleinman method, the Weeks–Rice method, or the Szasz–Brown procedure. We have also seen that the three methods lead to equations which are equivalent, in that their eigenfunctions and eigenvalues are identical for the same atomic states. Since in this section our main goal is to derive an equation not so much for actual calculations but for modelization, we shall work here with the Szasz–Brown formulation, which is the simplest of the three mentioned and as such lends itself most easily to modelization. The presentation which follows is based on a slightly improved version of the author's work[74].

Any one of the three ab initio equations, Eqs. (7.128), (7.139), and (7.144), can be transformed. We shall work with Eq. (7.139) because that equation contains the general HF open-shell Hamiltonian. Let us write down here again the equation,

$$\left\{ \sum_{i=1}^{n} \hat{H}_v(i) + PS \right\} \Phi = E\Phi, \tag{7.145}$$

where S is given by

$$S = Q - \sum_{i=1}^{n} U_v(i), \tag{7.146}$$

and

$$Q = \frac{1}{2} \sum_{i,j=1}^{n} \frac{1}{r_{ij}}. \tag{7.147}$$

The operator $\hat{H}_v$ is defined as

$$\hat{H}_v = PH_v. \tag{7.148}$$

We set up the pseudopotential equation in the same way as in the case of two valence electrons (Section 6.2). We define the pseudo-wave function Ψ by the equation

$$\Phi = P\Psi, \tag{7.149}$$

where Φ is the solution of Eq. (7.145). Ψ is the same function as the n-valence-electron function in Eq. (7.116). The equation for Ψ will be

$$\left\{ \sum_{i=1}^{n} (\hat{H}_v(i) + \hat{V}_p(i)) + SP \right\} \Psi = E\Psi , \qquad (7.150)$$

where $\hat{V}_p$ is the pseudopotential given by the first line of Eq. (2.115):

$$\hat{V}_p = -\Omega \hat{H}_v - \hat{H}_v \Omega + \Omega \hat{H}_v \Omega + \epsilon_v \Omega , \qquad (7.151)$$

and ϵ_v is one of the valence level eigenvalues of the operator $\hat{H}_v$.

It is easy to show that Eq. (7.150) is equivalent to Eq. (7.145). Let Ψ be a solution of Eq. (7.150). Multiply the equation from the left by the operator given by Eq. (7.117):

$$P\left\{ \sum_{i=1}^{n} (\hat{H}_v(i) + \hat{V}_p(i)) + SP \right\} \Psi = EP\Psi . \qquad (7.152)$$

Consider the effect of P on $\hat{H}_v(i) + \hat{V}_p(i)$. Repeating the argument which led to Eq. (6.92) in the two-electron case, we get easily,

$$P(\hat{H}_v(i) + \hat{V}_p(i)) = \hat{H}_v(i)P , \qquad (7.153)$$

and using this relationship we can put Eq. (7.152) in the form

$$\left\{ \sum_{i=1}^{n} \hat{H}_v(i) + PS \right\}(P\Psi) = E(P\Psi) . \qquad (7.154)$$

Using the relationship, Eq. (7.149), we obtain Eq. (7.145). Thus we see that, if Ψ is a solution of Eq. (7.150), then $\Phi = P\Psi$ will be a solution of Eq. (7.145). The equivalence of the two equations is thereby established.

It would be equally easy to transform the closed-shell equation, Eq. (7.144). We can obtain, however, the corresponding pseudopotential equation directly from Eq. (7.150). First let us make use of the relationship, Eq. (7.16):

$$\hat{H}_v + \hat{V}_p = H_v + V_p . \qquad (7.155)$$

Now let us switch to the closed-shell HF equations. Here H_v becomes the v-independent H_F and we obtain

$$H_v + V_p = H_F - \Omega H_F - H_F \Omega + \Omega H_F \Omega + \epsilon_v \Omega . \qquad (7.156)$$

Taking into account that according to Eq. (B.29) the operator P commutes with H_F, and taking into account Eq. (B.20), we obtain

$$H_v + V_p = (H_F - \epsilon_v)\Omega - \Omega H_F P$$
$$= H_F - (H_F - \epsilon_v)\Omega - H_F(\Omega P)$$
$$= H_F - (H_F - \epsilon_v)\Omega .\qquad(7.157)$$

The core orbitals are now eigenfunctions of H_F, so we obtain

$$-(H_F - \epsilon_v)\Omega = \sum_{i=1}^{N} (\epsilon_v - \epsilon_i)|\varphi_i\rangle\langle\varphi_i| .\qquad(7.158)$$

Using Eqs. (7.155), (7.157), and (7.158), we can put Eq. (7.150) in the form,

$$\left\{\sum_{i=1}^{n} (H_F(i) + V_p(i)) + SP\right\}\Psi = E\Psi ,\qquad(7.159)$$

where V_p is now the PK-potential [Eq. (7.158)]. Equation (7.159) is the pseudopotential equation for the n valence electrons of an atom with all electrons in closed shells.

If we want to give a physical interpretation to the pseudopotential equations, Eqs. (7.150) and (7.159), the advantage of having first treated the pseudopotential transformation of the valence-electron HF equations in Section 7.1 will become apparent. Taking a look at the structure of Eq. (7.150), we see that the first term in that equation is the sum of the operators $\hat{H}_v + \hat{V}_p = H_v + V_p$, summed over the coordinates of the n valence electrons. Looking at Eq. (7.17), we see that it contains the same operator, $(H_v + V_p)$. Now in Section 7.1 we saw that the eigenfunctions of this operator were pseudo-orbitals; that is, the valence electron eigenfunctions did not need to be orthogonalized to the core functions. Thus the pseudo-potential incorporated into $(H_v + V_p)$ keeps the valence electrons effectively out of the core. Looking at the second term of Eq. (7.150), we see the many-electron operator S multiplied by the projection operator P. In this term it is the operator P which represents the Pauli exclusion principle with respect to the core electrons. It is clear, therefore, that the Pauli principle is represented in both terms of Eq. (7.150): in the first term by the pseudo-potentials, in the second term by the operator P. Thus the solutions of Eq. (7.150) will not need to be orthogonalized to the core orbitals; one can solve the equation as if the core did not exist. The same is true for the even simpler equation, Eq. (7.159).

The most important mathematical properties of Eqs. (7.150) and (7.159) are analogous to the mathematical properties of the corresponding two-electron equations which we discussed at the end of Section 6.2. Because of the close analogy, these mathematical properties do not need to be discussed here.

Finally we want to put Eqs. (7.150) and (7.159) into a form which will be advantageous for modelization. Using the formulas for S and Q, Eqs. (7.146) and (7.147), and using Eq. (7.155), we obtain,

$$H = \sum_{i=1}^{n} (\hat{H}_v(i) + \hat{V}_p(i)) + SP = \sum_{i=1}^{n} (H_v(i) + V_p(i)) + S(1 - \Omega)$$

$$= \sum_{i=1}^{n} (H_v(i) + V_p(i)) + Q - \sum_{i=1}^{n} U_v(i) - S\Omega . \tag{7.160}$$

Taking into account the form of $H_v + V_p$, Eq. (7.18), we obtain

$$H = \sum_{i=1}^{n} (t_i + g_i + U(i) + V_p(i)) + Q - S\Omega . \tag{7.161}$$

Thus Eq. (7.150) becomes

$$\left\{ \sum_{i=1}^{n} (t_i + g_i + U(i) + V_p(i)) + \frac{1}{2} \sum_{i,j=1}^{n} \frac{1}{r_{ij}} - S\Omega \right\} \Psi = E\Psi . \tag{7.162}$$

The expression in the curly bracket of the first term is the kinetic energy plus the total interaction potential of the i-th valence electron with the core; the summation is over all valence electrons. Thus the first term represents the kinetic energy plus the interaction with the core of the n valence electrons. The second term is the electrostatic interaction of the valence electrons. The third term, $(-S\Omega)$, will be shown to have a definite physical meaning; and it will be shown in Section 7.4 how this term can be treated in modelization.

It can be shown easily that the closed shell equation, Eq. (7.159), leads to an equation of the form of Eq. (7.162), except that, for the closed-shell case, V_p will be the PK-potential, and in the operator S we shall have, instead of U_v, the expression given by Eq. (7.141).

Finally, a note about core-core correlation effects. We have seen in Section 2.5.B, in discussing the work of Öhrn and McWeeny, how the core-core correlation effects can be incorporated into the pseudopotential equation of the valence electron of an atom with one valence electron. In this chapter and the preceding one, we have always assumed that the core electrons are represented by one-electron orbitals. Thus the core-core correlation effects were excluded from the discussions.

A step toward a more general treatment was taken by Kleiner and McWeeny.[81] In this work, which is the extension of the earlier work of Öhrn and McWeeny, the question was investigated how the core-core correlation can be built into a many-valence electron pseudopotential theory. The starting wave function for the atom was put in the form

$$\Phi_T = \tilde{A}\{\Phi_C(1, 2, \ldots, N)\Psi_v(N + 1, \ldots, N + n)\} , \tag{7.163}$$

which is the straightforward extension of Eq. (2.155) for n valence electrons. Both the core function Φ_C and the valence function Ψ_v are arbitrary and may describe fully the core-core and valence-valence correlation effects. Thus, in this treatment, only the core-valence correlation effects were omitted. The Ψ_v is not required to satisfy any orthogonality conditions; therefore, it is a pseudo-wave function.

Kleiner and McWeeny show that if certain core-valence multiple-exchange terms are neglected, Eq. (7.163) leads to the following effective pseudopotential Hamiltonian for the n valence electrons:

$$H_M = \sum_{i=1}^{n} H(i) + \frac{1}{2} \sum_{i,j=1}^{n} \frac{1}{r_{ij}}, \qquad (7.164)$$

where $H(i)$ is the same one-electron pseudopotential Hamiltonian that we obtain in the case of one valence electron. Thus, the inclusion of the core-core correlation effects preserves the additivity of the core-valence interaction potential terms.

7.4. MODEL HAMILTONIANS FOR n-VALENCE-ELECTRON ATOMS

A. The Derivation of the Model Hamiltonian

At the end of Section 7.1 we stated that the effective pseudopotential Hamiltonian for the n valence electrons of an atom is given by the formula

$$H = \sum_{i=1}^{n} (t_i + V_M(i)) + Q, \qquad (7.165)$$

where the semilocal potential V_M, which represents the core-valence interaction, was defined by Eq. (7.32) and by the associated derivation. Q is again the electrostatic interaction of the valence electrons defined by Eq. (7.136). The wave equation for the valence electrons is given by

$$H\Psi = E\Psi, \qquad (7.166)$$

where Ψ is a pseudo-wave function which does not need to satisfy any orthogonality conditions with respect to the core orbitals.

In this section we answer the question: is Eq. (7.166) a good approximation to the exact pseudopotential equations given in Section 7.3? In answering this question we shall present an argument based on some recent work of the author[112] which is an improved version of an earlier study of the problem[82]. We shall argue that, in a good approximation, the exact pseudopotential Hamiltonian can be written in the form

$$H = \sum_{i=1}^{n} (t_i + V_M(i)) + \hat{\eta} Q, \tag{7.167}$$

where $\hat{\eta}$ is a constant for which an explicit expression will be given. It will be shown that in some cases the value $\hat{\eta} = 1$, that is, the Hamiltonian of Eq. (7.165), is a good approximation; it will also be shown, however, that in some other cases Eq. (7.167) with an $\hat{\eta} \neq 1$ will have to be used.

We proceed now to derive Eq. (7.167). The derivation will be carried out for the exact equation, Eq. (7.159), which is for atoms with closed shells. After the derivation is presented, it will be easy to show that the results are equally valid for the more general open-shell equation, Eq. (7.150).

Let us write down here Eq. (7.159) again:

$$\left\{ \sum_{i=1}^{n} (H_F(i) + V_p(i)) + SP \right\} \Psi = E\Psi. \tag{7.168}$$

Here H_F is the HF Hamiltonian, V_p is the PK potential, and S is given by Eq. (7.140). For closed shells, the potential of the valence electrons takes the form of Eq. (7.141). Changing the notation slightly, we put

$$S = Q - W, \tag{7.169}$$

where

$$W = \sum_{i=1}^{n} \sum_{j=1}^{n} U_i(j), \tag{7.170}$$

and U_i is the HF potential generated by the HF valence orbital φ_i, and is given by Eq. (2.5). Taking into account the form of S, we obtain from Eq. (7.168),

$$\left\{ \sum_{i=1}^{n} (t_i + g_i + U(i) + V_p(i)) + Q - S\Omega \right\} \Psi = E\Psi. \tag{7.171}$$

This equation, which is also identical with Eq. (7.162), is our starting point.

Let us consider first the operator in the bracket which appears in the first term of Eq. (7.171). That operator is the same as the operator in Eq. (7.18) with the U_v removed from the latter. As we have seen in the argument following Eq. (7.18), this operator can be transformed exactly into the semilocal potential V_M, which was defined by Eq. (7.19) and for which the final form of Eq. (7.32) was derived. As we see from Eq. (7.19), the potential V_M depends on the core orbitals as well as on the valence orbitals of the neutral atom. Thus the introduction of V_M rests on a rigorous application of the frozen-core approximation: V_M represents the core-valence interaction for that case in which the core is frozen into the same state in which it was in the neutral atom in the HF approximation.

We now assume that the operator in Eq. (7.171) can be transformed this way, and replace that equation by

$$\left\{ \sum_{i=1}^{n} (t_i + V_M(i)) + Q - S\Omega \right\} \Psi = E\Psi .$$ (7.172)

We emphasize again that the core-valence interaction V_M now represents the core frozen into the same state as in the neutral atom. Thus whenever Eq. (7.172) or equations derived from it are used in an atomic or molecular calculation, this condition must be kept in mind. (We recall that the derivation following Eq. (7.19) was done for open shells. The formulas for closed shells are special cases of that derivation, with U_v everywhere replaced by the expression of Eq. (7.141).)

Now comparing Eq. (7.172) with Eq. (7.167), we see that the only difference between the two equations is the presence of the term $(-S\Omega)$. Consider the quantities

$$\langle S\Omega \rangle = \int \Psi^* S\Omega\Psi \, dq ,$$ (7.173)

and

$$\langle S \rangle = \int \Psi^* S\Psi \, dq ,$$ (7.174)

where Ψ is the eigenfunction of Eq. (7.172). Here $\langle S \rangle$ is the expectation value of the operator S and $\langle S\Omega \rangle$ is the expectation value of the core part of S. As we indicated at the end of Section 7.3, the operator P in the last term of Eq. (7.168) represents the many-electron part of the Pauli exclusion principle. The operator $S\Omega$ in Eq. (7.172) comes from this term. Since $P = 1 - \Omega$, we get

$$\langle SP \rangle = \langle S \rangle - \langle S\Omega \rangle ,$$ (7.175)

which means that $S\Omega$ must be subtracted from S in order to satisfy the Pauli exclusion principle.

In order to estimate the size of this term, we have calculated approximately the ratio

$$\eta = \frac{\langle S\Omega \rangle}{\langle S \rangle} ,$$ (7.176)

and obtained the result,[112]

$$\eta = 1 - \prod_{k=1}^{n} (1 - \langle \Omega \rangle_k), \tag{7.177}$$

where $\langle \Omega \rangle_k$ is the expectation value of the one-electron projection operator [Eq. (6.16)] with respect to the pseudo-orbital ψ_k:

$$\langle \Omega \rangle_k = \int \psi_k^* \Omega \psi_k \, dq. \tag{7.178}$$

The one-electron pseudo-orbitals $\psi_1 \cdots \psi_n$ are the solutions of the valence-electron pseudopotential equations

$$(H_F + V_p)\psi_k = \epsilon_k \psi_k, \quad (k = 1, 2, \ldots, n), \tag{7.179}$$

where $(H_F + V_p)$ is the same operator as in Eq. (7.168).

We have examined further Eq. (7.172), and in order to compare the relative size of Q and $S\Omega$ we have computed the ratio

$$\hat{\eta} = \frac{\langle Q \rangle - \langle S\Omega \rangle}{\langle Q \rangle}. \tag{7.180}$$

and obtained[112] that in a good approximation

$$\hat{\eta} = 1 + \eta. \tag{7.181}$$

Assuming that the relationship of Eq. (7.180) can be transferred to the operators themselves, we obtain

$$Q - S\Omega \approx \hat{\eta} Q, \tag{7.182}$$

and using this approximation, we obtain from Eq. (7.172),

$$\left\{ \sum_{i=1}^{n} (t_i + V_M(i)) + \hat{\eta} Q \right\} \Psi = E\Psi, \tag{7.183}$$

with which the derivation of Eq. (7.167) is completed.

In order to compare these results with those of Section 6.3 we have recalculated the quantity which we have called there V_{pt}. It is easy to show[112] that

$$-\langle \hat{V}_{pt} \rangle = +\left\langle \frac{1}{r_{12}} \Omega \right\rangle = \eta \left\langle \frac{1}{r_{12}} \right\rangle, \tag{7.184}$$

where η is again given by Eq. (7.177). Using this result, we can write Eq. (6.118) in the form

$$\left(t_1 + V_m(1) + t_2 + V_m(2) + (1 - \eta)\frac{1}{r_{12}}\right)\Psi = E\Psi , \qquad (7.185)$$

which has now, apart from the different structure of the constant in front of the electrostatic interaction term, the same general form as Eq. (7.183). We note that the difference in the constant is not due to the difference in the number of valence electrons, but is a result of the core orbitals being different in the two equations. In Eq. (7.183) the core orbitals are the same as in the neutral atom; in Eq. (7.185) they are the same as in the doubly positive ion.

The results contained in Eqs. (7.183) and (7.185) can be summarized as follows: the many-electron part of the Pauli exclusion principle can be taken into account, in a good approximation, by multiplying the electrostatic interaction term by a constant which depends only on the overlap between the core orbitals and the valence electron pseudo-orbitals.

It will be shown in the work of the author[112] that the result given by Eq. (7.177) rests exclusively on the assumption that the pseudo-wave function, representing the valence electrons, has a reasonably small core penetration, and that this penetration is accurately represented by the expectation values $\langle\Omega\rangle_k$.

Next we want to take a close look at η and $\hat{\eta}$. From Eqs. (7.177) and (7.181) we see immediately that

$$\lim_{\langle\Omega\rangle_k \to 0} \eta = 0 , \qquad (7.186)$$

and

$$\lim_{\langle\Omega\rangle_k \to 0} \hat{\eta} = 1 . \qquad (7.187)$$

Eq. (7.186) shows that $\eta \to 0$ when the overlap between the core orbitals and the pseudo-orbitals is negligible. We see from the formulas that, generally,

$$\eta \leqq 1 , \qquad (7.188)$$

and

$$\hat{\eta} \geqq 1 . \qquad (7.189)$$

The result $\eta \leqq 1$ is physically plausible: in the exact pseudopotential equation, the electrostatic interaction is reduced by the presence of the operator P, thus it must also be reduced when that operator is replaced by $(1 - \eta)$ as in Eq. (7.185).

The result $\hat{\eta} > 1$ is surprising at first sight. It becomes plausible if we look at Eq. (7.182). We must keep in mind that $\hat{\eta}$ is not the relationship between $S\Omega$ and S but between $S\Omega$ and Q. Thus $\hat{\eta}$ does not directly represent the effect of the Pauli principle on an operator; only $(1 - \eta)$ has that physical meaning. It is easy to show that

$$Q - S\Omega = Q - Q\Omega + W\Omega \approx Q - \eta Q + \eta W = (1 - \eta)Q + \eta W. \tag{7.190}$$

From this relationship we see that Q is indeed reduced by the factor $(1 - \eta)$ which replaces the operator P and is always less than one. The second term, ηW, is the result of having $(W\Omega)$ as well as $(-Q\Omega)$ in Eq. (7.172). The relationship between Q and W is such that

$$\langle W \rangle / \langle Q \rangle = 2. \tag{7.191}$$

Transferring this relationship to the operators themselves gives the factor $\hat{\eta} = 1 + \eta$ on the right side of Eq. (7.190), and likewise in Eq. (7.183).

We proceed now to the discussion of Eqs. (7.183) and (7.185). First a remark about the structure of η and $\hat{\eta}$. It is interesting that the effect of the Pauli principle on the many-electron terms can be taken into account by multiplying the electrostatic interaction by a constant; it is equally interesting that this constant does not depend appreciably on the details of the pseudo-wave function of the valence electrons. Although the pseudo-orbitals are assumed to be the exact solutions of Eq. (7.179), this assumption is not explicitly used in the derivation of Eq. (7.177). In fact, the derivation is valid for any set of PO's which have a reasonably small overlap with the core; that is, for any set of "well-behaving" pseudo-orbitals. Also, Eq. (7.177) is a good approximation even if the valence electrons are represented by a correlated, n-electron function.

Next we look at Eq. (7.185) and ask whether the conclusions in Section 6.3 were correct. There we used perturbation theory to estimate the size of the term $((1/r_{12})\Omega)$. We concluded there that this term is negligible. Our result here is that

$$\frac{1}{r_{12}}\Omega \approx \eta \frac{1}{r_{12}}. \tag{7.192}$$

For a *purely qualitative judgment,* it is enough to look at Eqs. (7.186) and (7.188). From these formulas, as well as from Eq. (7.177), we see that if the overlap between the valence wave function and the core is reasonably small, then $\eta \to 0$ and the $((1/r_{12})\Omega)$ term is indeed negligible. Thus Eq. (7.185) strengthens the conclusions of Section 6.3. This is, by the way, true also for our requirement that the PO's have as small an overlap with the core as possible.

Turning to Eq. (7.183), we can reach there a similar qualitative judgment.

By the same arguments we have just used, we can say that if the valence-core overlap is small, then $\hat{\eta} \approx 1$ and the Hamiltonian, Eq. (7.165), will be a good approximation. Thus the results of this section establish qualitatively the validity of the model Hamiltonians which we have formulated earlier.

The fact that we have an explicit formula for η enables us to go one step further. We can now calculate the value of η for any set of pseudo-orbitals. In Appendix A we have listed the numerical values of $\langle \Omega \rangle_k$ for atoms with one valence electron. The PO's in that case are the exact solutions of the PK equations. Using these numbers we get from Eq. (7.177) the following values for the two-valence-electron atoms: $\eta = 0.10$ (Be); $\eta = 0.11$ (Al$^+$); $\eta = 0.36$ (Ca); $\eta = 0.26$ (Zn). Appendix K contains the values of $\langle \Omega \rangle_k$ for atoms with more than two valence electrons. The PO's there are the Maximum-Smoothness Pseudo-Orbitals, which are the exact solutions of the pseudo-potential equation, Eq. (7.17). We obtain from Eq. (7.177) the following η values: $\eta = 0.10$ (F); $\eta = 0.44$ (Cl); $\eta = 0.12$ (Fe); $\eta = 0.63$ (I).

When looking at these numbers we must keep in mind that although the results of this section are physically plausible, numerical values obtained from them are approximate. We must also be aware of the fact that this list of numbers is a very small random selection from the very large number of atoms to which pseudopotential theory can be applied. Nevertheless, treating these numbers with due caution, we are able to draw some conclusions.

For the atoms with two valence electrons, the results for Be, Al$^+$, and Zn bear out the accuracy of the remark by Weeks and Rice to the effect that the core-valence overlap is "about 10%" of the valence orbitals.[14] Thus for these atoms the approximate conclusion $\eta \approx 0$ is justified. On the other hand, the result for Ca does not fit very well into the picture. At this point we must recall that in the calculations which are summarized in Appendix A, there was no effort made to minimize the overlap integrals $\langle \Omega \rangle_k$. Thus the large value of this atom may be fortuitous.

In the case of atoms with more than two valence electrons, the results for η present a mixed picture. We recall that the PO's used in these cases are MSPO's which are determined by trying to make $\langle \Omega \rangle_k$ small. Presumably these results for η represent the smallest or nearly the smallest η values. Now the η values are small for F and Fe, but this follows mainly from the fact that the F core does not contain p electrons, so we have only the contributions from the $2s$ electrons in Eq. (7.177). Likewise in the Fe atom, the $3d$ electrons do not contribute to the η, leaving only the ($4s$) contributions. For these atoms $\eta \approx 0$ and $\hat{\eta} \approx 1$.

On the other hand, for the Cl and especially for the I atom, η is not very small and $\hat{\eta}$ is not very close to 1. From these results, as well as from the Ca results, we must conclude: There may be cases, especially when the number of valence electrons is large, in which it is not a very good approximation to put $\eta = 0$ and $\hat{\eta} = 1$.

For the choice of η and $\hat{\eta}$, a very simple procedure presents itself. The prescription is as follows: calculate η and $\hat{\eta}$ from the PO's and core orbitals if these orbitals are available. If $\hat{\eta} \approx 1$, then Eq. (7.165) is a good starting point for the calculations. If $\hat{\eta}$ is appreciably different from 1, then there are two options. The first is to use the calculated value of $\hat{\eta}$ in Eq. (7.165). The second is to treat $\hat{\eta}$ as an adjustable parameter similar to the α of the $X\alpha$ model or the η of the APM model (Chapter 5). In this case, the calculated $\hat{\eta}$ value can serve as an initial approximation.

In the whole discussion presented in this section we have worked with the modified potential V_M. In other words, the core-valence interaction was represented by the semilocal potential defined by Eq. (7.32). This potential represented the core-valence interaction exactly in the framework of the HF approximation. As the last step of the modelization of the exact n-electron pseudopotential equation, we shall now assume that V_M can be replaced by a model potential V_m; that is, we make the substitution

$$V_M \rightarrow V_m . \tag{7.193}$$

This step is actually the same that we made in Section 7.2. The modified potential V_M is a one-electron potential, and we replace it by another one-electron potential V_m. Thus the procedure outlined in Section 7.2 can be transferred here directly, regardless of the fact that here we use these potentials in the many-electron equation, Eq. (7.166).

Summing up, the model Hamiltonian for the n valence electrons of an atom is given by

$$H_M = \sum_{i=1}^{n} (t_i + V_m(i)) + \hat{\eta}Q , \tag{7.194}$$

where V_m is a model potential representing the core-valence interaction. For V_m, any of the model potentials listed in Section 7.2 can be used, or V_m can be identified with the exact modified potential V_M. The $\hat{\eta}$ is a numerical constant for which a theoretical value is given by Eqs. (7.177) and (7.181). As we have seen, in many cases it is a good approximation to put $\hat{\eta} = 1$. This constant can also be considered adjustable, like the parameters in the $X\alpha$ and APM models.

chapter 8

The Pseudopotential Theory of Molecules

8.1. INTRODUCTION

In this book we divide all molecules into two groups, called Group I and Group II. We put into Group I all molecules which have atomic cores, or at least one atomic core. An atomic core is defined as a group of electrons which have, in the molecule, essentially the same electron distribution as in the free atom from which the molecule is built. In other words, an atomic core is a group of electrons which move, in the molecule, in the same atomic orbitals in which they were moving in the free atom. We put into Group II all other molecules; that is, Group II contains the molecules which do not even partially exhibit an atomic structure. By the *pseudopotential theory of molecules* we shall mean the theory of molecules in Group I. We shall apply pseudopotential theory only to molecules with (one or more) atomic cores; there will be no attempt made to formulate a pseudopotential theory for the second group.

The application of the pseudopotential theory to the first group only reflects the situation at the time of writing, and should not imply the final exclusion of the second group from the pseudopotential theory. In this connection we recall that a similar situation existed for atoms until quite recently: the theory was applied to the valence electrons but never to the core electrons. The development of the all-electron pseudopotential model (APM), presented in Chapter 5, demonstrated that pseudopotential theory can be applied to the inner electrons as well. In Section 3.3.C a small step was already taken toward the application of the theory to some of the molecules in Group II. In that section we formulated a density-dependent pseudopotential for cylindrical symmetry. Such a pseudopotential would, of course, not be useful for molecules with atomic cores, but it would be well-suited to an all-electron pseudopotential treatment of molecules with cylindrical symmetry. Thus the extension of the theory, at some future time, to molecules of all types cannot, and should not, be ruled out.

Our goal in this chapter is to show that pseudopotential theory is a reliable tool for the calculation of the properties of molecules which possess atomic cores. The presentation will be made in two main steps. First we shall show how to set up the effective pseudopotential Hamiltonian for a molecule. The second step will consist of the description of a few representative calculations.

At this point let us ask the question: how do we determine whether a molecule contains, or does not contain, atomic cores? Before starting a set of molecular calculations, how does one decide whether pseudopotential theory can be applied to a particular molecule with a reasonable change of success?

The procedure to be followed can be described in the case of diatomic molecules. Analyzing the spectra of diatomic molecules, Herzberg concluded that there are molecules in which some of the electrons must be in atomic cores.[83] Looking at some special cases, for example, the N_2 molecule, Herzberg concluded that all electrons of this molecule will move in molecular orbitals with the exception of the electrons in the K shells of the constituent atoms. On the other hand, experimental evidence has indicated that in the HI molecule the electrons in the K, L, M, and N shells of the I atom remain in their atomic core in the molecule; only the $(5s)$ and $(5p)$ electrons of the I and the $(1s)$ electrons of the H atom, move into molecular orbitals when the molecule is formed. Thus, in the N_2 molecule we can hardly talk about atomic cores, since the presence of the atomic K shells is not a significant deviation from a strictly molecular structure. In the HI molecule, which has 54 electrons, we can say that 46 electrons remain in an atomic core and only 8 electrons move into molecular orbitals. Thus, in comtemplating molecular structure calculations for these two molecules, we must proceed with an all-electron calculation for N_2; for HI the pseudopotential method can be applied with a reasonable change of success (see Section 8.5).

Another way of determining the presence of atomic cores is to look at the HF densities of the constituent atoms.[5] HF calculations have been carried out for many atoms and ions; the HF electron densities are available for all neutral atoms.[3] Let us consider a diatomic molecule of AB type. Let R_0 be the equilibrium internuclear distance. We can draw a diagram in which the

HF densities of the individual electrons are shown in such a way that the centers of the two atoms are placed at the distance R_0 from each other. From the diagram it will be evident that certain orbitals of atom A will show a strong overlap with certain orbitals of atom B at this internuclear distance. There will be some orbitals which do not show appreciable overlap. Those orbitals, which show only negligible or no overlap, can be viewed as belonging to atomic cores. Thus, there will be atomic cores in a diatomic molecule if there are atomic orbitals which do not show an appreciable overlap at the equilibrium nuclear distance R_0.

The most reliable conclusion can, of course, be reached if the presence of atomic cores is deduced from the molecular spectra as well as from the atomic densities. For example: the for Cl_2 molecule the internuclear distance is $R_0 = 1.988$ a.u. At this distance, the $(3s)$ and $(3p)$ electrons overlap strongly.[3] Thus, we must assume that these electrons will move into molecular orbitals; the inner electrons in the $(1s)^2(2s)^2(2p)^6$ configuration will remain in atomic cores. Indeed, this description agrees with Herzberg's conclusion, reached on the basis of spectroscopic evidence.[83]

Summing up, we determine the assumed presence of atomic cores in a molecule either by looking at the spectroscopic and other experimental data or by looking at the HF atomic densities. If there are atomic cores (at least one) in a molecule, then pseudopotential theory can be applied; if this is not the case, an all-electron calculation must be carried out.

8.2. SUMMARY OF EARLIER DEVELOPMENTS

As we pointed out in Section 1.1, the idea of using pseudopotentials in a molecular calculation was first put forward by Hellmann,[2] who suggested for the treatment of the molecules Na_2, K_2, Cs_2, and so on, the Schroedinger equation

$$H\Psi = E\Psi, \tag{8.1}$$

with the effective, two-electron, pseudopotential Hamiltonian,

$$H = \sum_{i=1}^{2} \left\{ -\frac{1}{2}\Delta_i + V_H(r_{ai}) + V_H(r_{bi}) \right\} + \frac{1}{r_{12}}, \tag{8.2}$$

where V_H is the Hellmann potential [Eq. (1.1)], and r_{ai} and r_{bi} are the distances of the i-th electron from nuclei a and b, respectively. As we indicated in Section 1.1, the Hamiltonian, Eq. (8.2), embodies the idea that the cores of the Na, K, Cs, and so on, atoms remain atomic in the molecule; that is, the molecular binding is entirely due to the valence electrons of these atoms. The pseudopotentials are incorporated into V_H. This pioneering idea, along with some calculations which will be presented below, has set the stage for further developments.

The first attempt to derive from first principles an effective pseudopotential Hamiltonian for diatomic molecules was carried out by Szasz and McGinn.[5] Starting from the quantum theory of atoms with two valence electrons, presented in Section 6.1, Szasz and McGinn derived the model pseudopotential Hamiltonian, Eq. (6.117). Extrapolating these results to molecules, it was postulated that the effective pseudopotential Hamiltonian for a diatomic molecule is given by the formula,

$$H = \sum_{i=1}^{n} \left\{ -\frac{1}{2}\Delta_i + V_{HF}(r_{ai}) + V_{HF}(r_{bi}) + V_R(r_{ai}) + V_R(r_{bi}) \right\} + \frac{1}{2}\sum_{i,j=1}^{n} \frac{1}{r_{ij}}. \tag{8.3}$$

In this formula we have the Hamiltonian for an AB-type diatomic molecule with n binding electrons. The interaction of the binding electrons with the atomic cores centered on a and b are given by the local potentials V_{HF} and V_R. V_{HF} is the HF potential consisting of the nuclear attraction g plus the potential in Eq. (A.9), which is the sum of the electrostatic and exchange potentials. The V_R is the PK-type pseudopotential given by Eq. (A.11).

The Hamiltonian of Eq. (8.3) is an extrapolation of atomic results to molecules. The merit of this formula is that here for the first time an exact one-electron pseudopotential was incorporated into molecular theory.

In order to derive a theoretically-better-established Hamiltonian, Schwarz started with the molecular open-shell HF equations of Roothaan.[54] By a set of plausible approximations Schwarz has shown[40] that the effective Hamiltonian of the n binding electrons of a molecular can be written in the form,

$$H = \sum_{i=1}^{n} \left\{ -\frac{1}{2}\Delta_i + \sum_{\substack{a \\ (\text{centers})}} W^a(i) \right\} + \frac{1}{2}\sum_{i,j=1}^{n} \frac{1}{r_{ij}}. \tag{8.4}$$

Here $W^a(i)$ is the complete interaction potential of electron i with the atomic core centered on nucleus a. Schwarz has shown that $W^a(i)$ must have the form,

$$W^a(i) = \sum_{l=0}^{\infty} W_l^a(i)\Omega_l, \tag{8.5}$$

where $W^a(i)$ is a local potential and Ω_l is an angular momentum projection operator,

$$\Omega_l = \sum_{m=-l}^{+l} |Y_{lm}\rangle\langle Y_{lm}|. \tag{8.6}$$

Thus $W^a(i)$ is a semilocal potential like the one in Eq. (4.7). The potential $W^a(i)$ contains the nuclear attraction of nucleus a plus the electrostatic, exchange, and pseudopotentials of the atomic core centered on nucleus a.

For practical calculations Schwarz suggested that $W^a(i)$ should be put in the form of analytic expressions with adjustable parameters which should be determined by matching the optical spectrum of those neutral atoms or positive ions from which the molecule is built.

According to Schwarz's analysis, the three main approximations inherent in Eq. (8.4) are as follows. First, it is assumed that in the presence of atomic cores in the molecule, the interaction potential of each binding electron with the cores is the sum of additive terms each of which depends only on the parameters of one atomic core. Second, it is assumed that the one-electron potential of the cores is a semilocal potential which depends on the azimuthal quantum number of the valence electron but does not depend on its orbital energy parameter. Third, it is assumed that the electrostatic interaction between the valence electrons can be computed by using pseudo-orbitals instead of HF orbitals; in other words, in the calculation of the valence-valence interaction the Pauli exclusion principle need not be taken into account. Schwarz has pointed out that the errors introduced by these approximations compensate each other to a considerable degree.

The next important step in the development of molecular pseudopotential theory was made by Kahn and Goddard.[70] We discussed in Section 2.5.C the G1 method of Goddard; and we have seen that the effective potential introduced in that theory for the valence electron of an atom contained a special form of the exact pseudopotential. We have also seen that Goddard's method can be formulated for atoms as well as for molecules. In order to check the reliability of the molecular pseudopotential Hamiltonian, Kahn and Goddard made two sets of calculations for the molecules LiH, Li_2, BH, and LiH_2. In one set of calculations the all-electron G1 method was applied, in its molecular formulation, to the calculation of some properties of these molecules. In the other set of calculations it was assumed that the atomic cores of the Li and B atoms remain atomic in the molecules, and that the binding of the molecules is entirely due to the valence electrons. For this case Kahn and Goddard employed an effective valence electron Hamiltonian with the G1 pseudopotential taking care of the Pauli exclusion principle. For the two binding electrons of the LiH molecule the effective Hamiltonian was written in the form

$$H = h_1 + h_2 + \frac{1}{r_{12}}, \tag{8.7}$$

where

$$h_i = -\frac{1}{2}\Delta_i + \hat{U}_{Li}(r_{ai}) + g_{Li}(r_{ai}) + g_H(r_{bi}). \tag{8.8}$$

In this formula the Li core is centered on nucleus a, and the H core is the nucleus b. The g's are the appropriate nuclear potentials and $\hat{U}_{Li}$ is the interaction potential of the atomic core of the Li with the binding electrons.

The pseudopotential in this quantity is the G1 pseudopotential mentioned in Section 2.5.C. Kahn and Goddard especially emphasized that $\hat{U}_{Li}$ is a semilocal potential; that is, it has the form of Eq. (4.7).

Having carried out the two sets of calculations, Kahn and Goddard observed that the agreement between the two sets of results was very good. For example, they obtained for the equilibrium internuclear distance and for the dissociation energy of the Li_2 molecule the following results: $R_0 = 5.05$ a.u., $D_0 = 0.33$ eV (pseudopotential calculations); $R_0 = 5.05$ a.u., $D_0 = 0.40$ eV (all-electron calculations). The results for LiH were: $R_0 = 3.01$ a.u., $D_0 = 1.85$ eV (pseudopotential); $R_0 = 3.01$ a.u., $D_0 = 1.92$ eV (all-electron). In judging these results the reader should remember that the dissociation energy is the difference between two much larger numbers; it is the difference between the total energy of the two binding electrons and the ionization potentials. A very small error in the calculation of the total energy appears magnified in the dissociation energy. Based on the good agreement achieved in the two sets of calculations, Kahn and Goddard reached the conclusion that the pseudopotential Hamiltonian in Eq. (8.7) represents accurately the effective Hamiltonian of the binding electrons.

Summing up, we may say that the derivations of Szasz and McGinn, and of Schwarz, and the work of Kahn and Goddard, have shown the usefulness of the pseudopotential method in molecular calculations, and the basic soundness of Hellmann's model Hamiltonian in the improved form given by Eqs. (8.3), (8.4), and (8.7).

Despite the successes achieved in calculations which were based on the effective Hamiltonians listed above, it cannot be overlooked that all four approaches mentioned in this section were handicapped by some weakness. The idea of Hellmann, ingenious as it was at an early stage of the development of molecular quantum mechanics, was nevertheless purely intuitive. The derivation of Szasz and McGinn was an extrapolation of atomic work to molecules. The work of Schwarz was restricted by being based on the HF approximation, omitting thereby the important valence-valence correlation effects. The calculations of Kahn and Goddard were made for rather small molecules. Thus even if the Hamiltonians derived in these publications led to good numerical results (in these papers and in other publications in which various forms of model, one-electron pseudopotentials were used), it is evident that there is a need for a more accurate derivation of the molecular pseudopotential Hamiltonian, a derivation which would not suffer the shortcomings mentioned. Such a derivation is presented in the next section.

8.3. THE DERIVATION OF THE EFFECTIVE PSEUDOPOTENTIAL HAMILTONIAN FOR A DIATOMIC MOLECULE

Motivated by the observations made at the end of the preceding section the author has undertaken to construct a derivation of the effective molecular

Hamiltonian which does not suffer from the shortcomings of previous efforts. The contents of this section are based on the author's paper.[74]

Let us consider a diatomic molecule of AB type. The constituent atoms A and B should have K_A and K_B core electrons and n_A and n_B valence electrons, respectively. Let us assume that we have reason to believe that the diatomic molecule which is formed will retain the atomic cores. Thus in the molecule we shall have the atomic cores with the K_A and K_B electrons; the number of binding electrons will be $n = n_A + n_B$.

In Figure 8.1 we have schematically illustrated the configurations of the system before and after the formation of the molecule. Configuration 1 at the top of the figure shows the separated atoms. Configurations 2 and 3 show the molecule in the vicinity of the equilibrium internuclear distance; in configuration 2 the cores do not overlap while in configuration 3 they overlap slightly but are still assumed to be atomic.

We start by writing down the pseudopotential equations for the separated atoms. The formalism of Section 7.3 will be used. We put for atom A the total wave function [Eq. (7.116)],

$$\Psi_T^A = [(K_A + n_A)!]^{-1/2} \tilde{A}\{\Psi_A(1, 2, \ldots, n_A) \det[\varphi_1^A(n_A + 1) \cdots \varphi_{K_A}^A(n_A + K_A)]\}.$$

(8.9)

Here the valence electrons are represented by the n_A-electron function Ψ_A and the core electrons by the spin-orbitals $\varphi_1^A, \ldots, \varphi_{K_A}^A$. As we have seen in Section 7.3, the rigorous application of the variation principle leads to Eq. (7.162) for the best Ψ_A,

$$\left\{\sum_{i=1}^{n_A} (t_i + g^A(i) + U^A(i) + V_p^A(i)) + Q_A - S_A \Omega_A\right\}\Psi_A = E_A \Psi_A.$$

(8.10)

In this equation, t_i is the kinetic energy operator, $g^A(i)$ is the Coulomb potential of nucleus A, $U^A(i)$ is the HF potential of core A,

$$U^A(1)f(1) = \sum_{l=1}^{K_A} \int \frac{\varphi_l(2)\varphi_l^*(2)f(1)\,dq_2}{r_{12}} - \sum_{l=1}^{K_A} \int \frac{\varphi_l(1)\varphi_l^*(2)f(2)\,dq_2}{r_{12}},$$

(8.11)

and V_p^A is the pseudopotential of core A.

$$V_p^A = -\Omega_A H_v^A - H_v^A \Omega_A + \Omega_A H_v^A \Omega_A + \epsilon_v^A \Omega_A,$$

(8.12)

with Ω_A being the projection operator onto the core A, and H_v^A the HF valence Hamiltonian:

$$H_v^A = t + g^A + U^A + U_v^A,$$

(8.13)

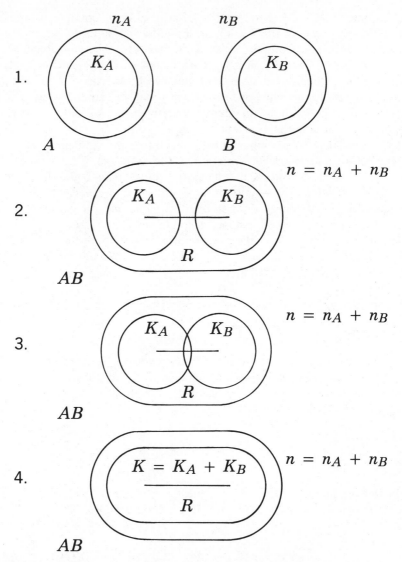

FIGURE 8.1. Schematic representation of the molecular binding in the AB-type diatomic molecule with atomic cores.

where U_v^A is the HF potential of the valence electrons. The symbol Q_A means the electrostatic interaction of the valence electrons,

$$Q_A = \frac{1}{2} \sum_{i=1}^{n_A} \sum_{j=1}^{n_A} \frac{1}{r_{ij}}, \qquad (8.14)$$

and the operator S_A is defined by Eq. (7.140),

$$S_A = Q_A - \sum_{i=1}^{n_A} U_v^A(i). \tag{8.15}$$

Equation (8.10) is the exact equation for the best Ψ_A with the core orbitals being the solutions of the open-shell HF equations. We obtain similarly for atom B,

$$\left\{ \sum_{i=1}^{n_A} (t_i + g^B(i) + U^B(i) + V_p^B(i)) + Q_B - S_B\Omega_B \right\} \Psi_B = E_B\Psi_B, \tag{8.16}$$

where this equation defines the best Ψ_B under the same conditions as Eq. (8.10) defined the best Ψ_A for atom A.

We have now the exact pseudopotential equations for the separated atoms. Our goal is to derive the equation for configuration 2. (The same equation will serve for configuration 3 with the core-core interaction properly taken into account.) The trouble with configuration 2 is that the valence electrons are represented by a molecular wave function, while the cores have atomic symmetry.

Now if the reader reviews Section 7.3 it will become evident that that part of the section which deals with the derivation of the equation for the best valence-electron function can be easily generalized to a diatomic molecule, provided that all quantities, the valence-electron function as well as the core orbitals, become molecular quantities. We shall utilize this property of the exact theory. Thus let us consider now configuration 4, which is a diatomic molecule in which the core does not have atomic structure. We divide the electrons of the molecule into two groups. We shall designate the K inner electrons arbitrarily as core electrons and the outer n electrons as "valence" electrons. We put $K = K_A + K_B$ and $n = n_A + n_B$. Configuration 4 is *constructed* in such a way that configuration 2 is a special case of configuration 4. We obtain configuration 2 from configuration 4 by increasing the internuclear distance in such a way that the molecular orbitals of the inner electrons in configuration 4 become the atomic orbitals of the core electrons in configuration 2. So our next step will be to derive the equation for the best valence-electron function in configuration 4; and using that we shall construct the equation for configuration 2.

The ab initio Hamiltonian of configuration 4 is

$$H = \sum_{i=1}^{N} (t_i + g_i) + \sum_{i,j=1}^{N} \frac{1}{r_{ij}}, \tag{8.17}$$

where the total number of electrons is $N = K + n$. Let the total wave function of the molecule be

$$\Psi_T = (N!)^{-1/2} \tilde{A}\{\Psi(1, 2, \ldots, n) \det[\varphi_1(n+1)\varphi_2(n+2)\cdots\varphi_K(n+K)]\}, \tag{8.18}$$

where $\varphi_1 \cdots \varphi_K$ are spin-orbitals representing the inner electrons, and Ψ is an n–electron function representing the "valence" electrons. The spin-orbitals are MO's and the Ψ is also a molecular function in the Hund–Mulliken sense; that is, it extends over the whole molecule.

We want to derive the equation for the best Ψ with fixed one-electron orbitals $\varphi_1 \cdots \varphi_K$. The derivation leading to the equation is a straightforward, one-to-one generalization of the atomic case, presented in Section 7.3. We present here only the main steps and the result. First we form the molecular projection operator,

$$P = P_1 P_2 \cdots P_n = \prod_{k=1}^{n} P_k, \tag{8.19}$$

where

$$P_i = 1 - \Omega_i, \tag{8.20}$$

and

$$\Omega_i = \sum_{l=1}^{K} \langle i \mid \varphi_l \rangle \langle \varphi_l \mid. \tag{8.21}$$

The spin-orbitals in Ω_i are the inner orbitals appearing in Eq. (8.18). Also let

$$\Omega = 1 - P. \tag{8.22}$$

Quantities without index will be n-electron operators. Introducing the nonrestrictive, strong orthogonality condition, we put

$$\Phi = P\Psi, \tag{8.23}$$

where Φ is now strong-orthogonal to the inner orbitals. We obtain the total energy of the molecule in the form,

$$E_T = \langle \Phi | H' | \Phi \rangle + E_c, \tag{8.24}$$

where E_c is the HF-type energy of the inner electrons and H' is

$$H' = \sum_{i=1}^{n} H'_i + \frac{1}{2} \sum_{i,j=1}^{n} \frac{1}{r_{ij}}, \tag{8.25}$$

with

$$H'_i = t_i + g_i + U(i), \tag{8.26}$$

where

$$U(1)f(1) = \sum_{l=1}^{K} \int \frac{\varphi_l(2)\varphi_l^*(2)f(1) \, dq_2}{r_{12}} - \sum_{l=1}^{K} \int \frac{\varphi_l(1)\varphi_l^*(2)f(2) \, dq_2}{r_{12}} . \quad (8.27)$$

Variation of E_T with respect to Φ^* under the subsidiary conditions of strong orthogonality and normalization gives

$$H\Phi = E\Phi , \quad (8.28)$$

where

$$H = PH' , \quad (8.29)$$

and

$$E_T = E + E_c . \quad (8.30)$$

Eq. (8.28) is the equation for the best strong-orthogonal valence-electron function. In order to get the equation for the best pseudo-wave function Ψ, we apply the transformation introduced by the author.[60] We state that the equation for the best Ψ is

$$\left\{ \sum_{i=1}^{n} (H_i' + V_p(i)) + \left(\frac{1}{2} \sum_{i,j=1}^{n} \frac{1}{r_{ij}} \right) P \right\} \Psi = E\Psi , \quad (8.31)$$

where H_i' is given by Eq. (8.26) and V_p is a one-electron, molecular, WR-type pseudopotential,

$$V_A(i) = -\Omega_i H_i' - H_i'\Omega_i + \Omega_i H_i'\Omega_i + \epsilon\Omega_i . \quad (8.32)$$

Let us prove that Eq. (8.31) is indeed the exact equation equivalent to Eq. (8.28). Let us write out the latter in detail:

$$P\left\{ \sum_{i=1}^{n} H_i' + \frac{1}{2} \sum_{i,j=1}^{n} \frac{1}{r_{ij}} \right\} \Phi = E\Phi . \quad (8.33)$$

Operate on Eq. (8.31) from the left by P:

$$\left\{ P \sum_{i=1}^{n} (H_i' + V_p(i)) + P\left(\frac{1}{2} \sum_{i,j=1}^{n} \frac{1}{r_{ij}} \right) P \right\} \Psi = EP\Psi . \quad (8.34)$$

Now consider the effect of P on the first term. Using Eq. (B.19), which is equally valid for atomic and molecular projection operators, we obtain

$$P \sum_{i=1}^{n} (H_i' + V_p(i)) = P \left(\sum_{i=1}^{n} H_i' \right) P. \qquad (8.35)$$

Using Eq. (8.35) in Eq. (8.34) and using the notation of Eq. (8.23), we obtain immediately Eq. (8.33). The equivalence of Eqs. (8.31) and (8.33) is thereby proved.

In order to see better the similarity between the atomic and molecular equations, let us write Eq. (8.31) in the form

$$\left\{ \sum_{i=1}^{n} (t_i + g_i + U(i) + V_p(i)) + Q - Q\Omega \right\} \Psi = E\Psi, \qquad (8.36)$$

where we have used Eq. (8.26), and have introduced the symbol

$$Q = \frac{1}{2} \sum_{i,j=1}^{n} \frac{1}{r_{ij}}, \qquad (8.37)$$

and the operator Ω is defined by Eq. (8.22).

We have now obtained the exact pseudopotential equations for the configurations 1 and 4. The equations for the separated atoms of configuration 1 are Eqs. (8.10) and (8.16). The molecular equation for configuration 4 is Eq. (8.36). All three equations are exact in the sense that they give the best valence-electron pseudo-wave function for a fixed set of core orbitals.

Inspecting the structure of these equations, we see that they are very similar. The first term, which is a sum over the valence electrons, contains the kinetic energies, the nuclear potentials, plus the electrostatic and exchange potentials of the core. The last term in this sum, designated by V_p, is the pseudopotential keeping the valence electrons out of the core. The term Q is the Coulomb interaction of the valence electrons. The last term of the effective Hamiltonian, which has the form $(-S\Omega)$ in the atomic equations and $(-Q\Omega)$ in the molecular equation, is the many-electron part of the Pauli exclusion principle; that is, the core projection of the Coulomb interaction Q in the molecular equation, and the core projection of the operator S in the atomic equation. The effect of the Pauli exclusion principle on these many-electron operators is that the core projections must be subtracted out. (The difference arises from the fact that in the atomic equations the core orbitals are the solutions of the open shell HF equations, while in the molecular equation no special assumption was made about the inner orbitals.)

We now introduce an approximation. Based on the discussions of Section 7.4 and using Eq. (7.182), we change the operators in the atomic equations as follows:

$$Q_A - S_A \Omega_A \approx \hat{\eta}_A Q_A, \qquad (8.38)$$

and

$$Q_B - S_B \Omega_B \approx \hat{\eta}_B Q_B, \tag{8.39}$$

where $\hat{\eta}_A$ and $\hat{\eta}_B$ are the constants that we introduced in Section 7.4. Next we *extrapolate* the results of Section 7.4 by assuming that they can also be used in the molecular equation and we put

$$Q - Q\Omega \approx (1 - \eta_M)Q, \tag{8.40}$$

where η_M is a constant similar to the η introduced in Section 7.4. Again, on the basis of the results of Section 7.4, we put

$$\hat{\eta}_A \approx 1, \quad \hat{\eta}_B \approx 1, \tag{8.41}$$

and

$$\eta_M \approx 0, \quad (1 - \eta_M) \approx 1, \tag{8.42}$$

where the first two approximations for the separated atoms are frequently justified, as we have seen in Section 7.4, and the second line, which is also frequently justified in the atomic case, is extrapolated to the molecular situation. Further discussion of these approximations will follow in the next section.

Introducing the approximations given by [(8.38), (8.39), (8.41)] into the atomic equations, and the approximations given by Eqs. (8.40) and (8.42) into the molecular equations, we obtain these three equations in the following form:

$$\left\{ \sum_{i=1}^{n_A} (t_i + g^A(i) + U^A(i) + V_p^A(i)) + Q_A \right\} \Psi_A = E_A \Psi_A, \tag{8.43}$$

$$\left\{ \sum_{i=1}^{n_B} (t_i + g^B(i) + U^B(i) + V_p^B(i)) + Q_B \right\} \Psi_B = E_B \Psi_B, \tag{8.44}$$

and finally,

$$\left\{ \sum_{i=1}^{n} (t_i + g_i + U(i) + V_p(i)) + Q \right\} \Psi = E \Psi. \tag{8.45}$$

Let us recall now that Eqs. (8.43) and (8.44) are the equations for the separated atoms of configuration 1, while Eq. (8.45) is the equation for configuration 4. What we want is the equation for configuration 2. We have seen that configuration 2 is a special case of configuration 4; we have seen that the former can be obtained from the latter by increasing the internuclear distance in such a way that the molecular orbitals of the inner electrons of configuration 4 become the atomic orbitals of the cores in configuration 2. The structure of Eq. (8.45) shows that in this process only the operators $U(i) + V_p(i)$ would change, since only these operators depend on the inner orbitals [see Eqs. (8.27) and (8.32)]. The operator Q depends only on the valence electron coordinates; thus it will be the same in configuration 2 as it is in configuration 4. Therefore, on the basis of Eq. (8.45) we conclude that the effective Hamiltonian of configuration 2 will have the form

$$H = \sum_{i=1}^{n} (t_i + V(i)) + Q, \tag{8.46}$$

where the $V(i)$ is the operator which will represent the complete interaction between the atomic cores and the valence electron.

The next step of the derivation is to change the core-valence interaction operators of the atomic equations into local (or semilocal) potentials. We have seen in the discussion following Eq. (7.17) how this can be done. In the atomic equations, Eq. (8.43) and (8.44), the core orbitals are the solutions of the open-shell HF equations. The pseudopotential equation for a valence electron of atom A is given, in the HF approximation, according to Eq. (7.17), by the equation,

$$(H_v^A + V_p^A)\psi_v^A = \epsilon_v^A \psi_v^A, \tag{8.47}$$

where

$$H_v^A = t + g^A + U^A + U_v^A. \tag{8.48}$$

Selecting a particular eigenfunction of Eq. (8.47), we can define the modified potential according to Eq. (7.19),

$$V_M^A = \frac{(g^A + U^A + V_p^A)\psi_v^A}{\psi_v^A}, \tag{8.49}$$

which can also be written in the form

$$V_M^A = \frac{(\epsilon_v^A - t - U_v^A)\psi_v^A}{\psi_v^A}, \tag{8.50}$$

where U_v^A is the potential of the valence electrons. We have seen in Section 7.1 that the modified potential is strongly l-dependent; thus we must introduce the semilocal potential, Eq. (7.22), which, in our case, will become,

$$V_M^A = \sum_{l=0}^{\infty} V_l^A(r)\Omega_l, \tag{8.51}$$

where the V_l^A is a local potential and Ω_l is given by Eq. (7.23).

We introduce now V_M^A into the atomic equation, Eq. (8.43), which then takes the form

$$\left\{ \sum_{i=1}^{n_A} (t_i + V_M^A(i)) + Q_A \right\} \Psi_A = E_A \Psi_A. \tag{8.52}$$

Let us define a semilocal modified potential for atom B in a similar fashion. We obtain,

$$V_M^B = \sum_{l=0}^{\infty} V_l^B(r)\Omega_l, \tag{8.53}$$

where V_l^B is a local potential. Introducing this into Eq. (8.44), we obtain

$$\left\{ \sum_{i=1}^{n_B} (t_i + V_M^B(i)) + Q_B \right\} \Psi_B = E_B \Psi_B. \tag{8.54}$$

We are now ready to write down the equation for configuration 2. We state that the equation will be

$$\left\{ \sum_{i=1}^{n} (t_i + V_M^A(i) + V_M^B(i)) + Q \right\} \Psi = E\Psi, \tag{8.55}$$

where the summation is over all binding electrons in configuration 2, the semilocal potentials V_M^A and V_M^B are given by Eqs. (8.51) and (8.53), respectively, and the symbol Q, the Coulomb interaction of the binding electrons, is given by Eq. (8.37). As we see, the equation has the form of Eq. (8.46).

In proving that Eq. (8.55) is the correct equation for configuration 2, we shall not attempt to derive this equation from Eq. (8.45), which is valid for configuration 4, despite the fact that, as we have pointed out, configuration 2 is a special case of configuration 4. We needed Eq. (8.45) only to establish that the correct equation will have the form of Eq. (8.46); that is, we needed Eq. (8.45) only to establish the core-valence separability. For proving the correctness of Eq. (8.55), a much simpler procedure presents itself. Inspecting Figure 8.1 we see that if the operators representing the effects of the core on the valence electrons can be separated from the operators (potentials) representing the valence-valence interactions, then the former must be the same in configuration 2 as they are in the separated atoms of configuration 1. This follows simply from looking at the diagrams of Figure 8.1, if we assume that the "atomic core approximation" is strictly valid; that is, if we assume that the atomic cores remain intact in the transition from configuration 1 to configuration 2. Therefore, we shall prove the correctness of Eq. (8.55) by showing that the effective Hamiltonians of the separated atoms are the limiting case of the effective Hamiltonian of Eq. (8.55) for infinite internuclear distance; that is, for $R \to \infty$.

Let us now assume that the diatomic molecule which is in the state described by configuration 2 dissociates into the separated atoms of configuration 1. From the n binding electrons of the molecule, the electrons $(1, 2, \ldots, n_A)$ move to atom A and electrons $(n_A + 1, \ldots, n_A + n_B)$ move to atom B. The cores remain intact in the dissociation. We shall indicate this process by the symbol $R \to \infty$.

Let us consider the term with $V_M^A(i)$ in Eq. (8.55). Writing it out, we get

$$\sum_{i=1}^{n} V_M^A(i) = V_M^A(1) + \cdots + V_M^A(n_A) + V_M^A(n_A + 1) + \cdots + V_M^A(n_A + n_B).$$

$$(8.56)$$

The modified potentials in this formula are given by Eq. (8.51), which was written down on the basis of Eqs. (8.49) and (7.19). We have seen in Appendix K that V^A can be written in the form

$$V_l^A = -\frac{Z_A - K_A}{r} + \hat{V}_l^A,$$

$$(8.57)$$

where $\hat{V}_l^A$ is a short-range, exponentially declining, potential.

Let us write down now the potential Q:

$$Q = \frac{1}{2}\sum_{i,j=1}^{n}\frac{1}{r_{ij}} = \frac{1}{2}\sum_{i,j=1}^{n_A}\frac{1}{r_{ij}} + \frac{1}{2}\sum_{ij=n_A+1}^{n_A+n_B}\frac{1}{r_{ij}} + \sum_{i=1}^{n_A}\sum_{j=n_A+1}^{n_A+n_B}\frac{1}{r_{ij}}.$$

$$(8.58)$$

In this formula, the first term is the Coulomb interaction of the electrons $(1, \ldots, n_A)$, the second term is the interaction of the electrons $(n_A + 1, \ldots, n_A + n_B)$, and the last term is the Coulomb interaction between the two groups of electrons. We can write

$$Q = Q_A + Q_B + Q_{AB}.$$

$$(8.59)$$

Let us consider now the electrons $(n_A + 1, \ldots, n_A + n_B)$. In the dissociation process these are moving to atom B. As the internuclear distance increases, these electrons will move out of the range of the short-range potentials $\hat{V}_l^A$. They will still be in the range of the potential $-(Z_A - K_A)/r$, which is the Coulomb potential of core A. Let us select the electron j in such a way that $j = n_A + 1, \ldots, n_A + n_B$. Then, for large R, we can write in good approximation,

$$V_l^A(r_j) \approx -\frac{Z_A - K_A}{r_j} \approx \sum_{i=1}^{n_A} \frac{1}{r_{ij}}, \quad (R \to \infty), \tag{8.60}$$

where we have used the fact that $n_A = Z_A - K_A$. Thus for $R \to \infty$, the long-range term of the potential of Eq. (8.57) will be compensated by the long-range Coulomb interaction between the two groups of electrons.

Let us write down now the term with $V_M^B(i)$ in Eq. (8.55):

$$\sum_{i=1}^{n} V_M^B(i) = V_M^B(1) + \cdots + V_M^B(n_A) + V_M^B(n_A + 1) + \cdots + V_M^B(n_A + n_B). \tag{8.61}$$

The local potentials which occur in this expression can be written in the form

$$V_l^B = -\frac{Z_B - K_B}{r} + \hat{V}_l^B. \tag{8.62}$$

Consider now the electron i in the first group; that is, let $i = 1, \ldots, n_A$. As the internuclear distance increases, this electron will move to atom A, and thus out of the range of the short-range potential $\hat{V}_l^B$. It will be still in the range of the potential $-(Z_B - K_B)/r$, which is the long-range Coulomb potential of core B. But this potential will be compensated again by the long-range Coulomb interaction between the two groups. We have shown that, seen from the position of the electron j, on the atom B, the interaction term compensated the Coulomb potential of the core A, according to Eq. (8.60). Seen from the position of electron i, on the atom A, this same interaction term will compensate the long-range potential of core B, since, for $R \to \infty$, we can write, in good approximation,

$$V_l^B(r_i) \approx -\frac{Z_B - K_B}{r_i} \approx \sum_{j=n_A+1}^{n_A+n_B} \frac{1}{r_{ij}}, \quad (R \to \infty), \tag{8.63}$$

where $(Z_B - K_B) = n_B$. Thus the long-range Coulomb potential in the modified potentials V_M^A and V_M^B will be compensated by the long-range potentials of Q_{AB}. This is the same as to say that, for $R \to \infty$, the long-range potentials of the cores are shielded by the valence electrons of both atoms.

Our conclusion is that, as $R \to \infty$, the electrons $(1, 2, \ldots, n_A)$ will feel the core potentials $V_M^A(1) + \cdots V_M^A(n_A)$ only; the potentials $V_M^B(1) + \cdots + V_M^B(n_A)$ will not be felt by these electrons. Likewise, the electrons $(n_A + 1, \ldots, n_A + n_B)$ will feel the core potentials arising from core B only.

Naturally, both groups will generate the Coulomb interactions Q_A and Q_B among themselves. Denoting the effective Hamiltonian of Eq. (8.55) by $H(AB)$ and the Hamiltonians of Eqs. (8.52) and (8.54) by $H(A)$ and $H(B)$, our conclusions can be summarized in the statement,

$$\lim_{R \to \infty} H(AB) = H(A) + H(B) . \tag{8.64}$$

It is therefore proved that the correct effective Hamiltonian for the AB-type diatomic molecule is given by Eq. (8.55). A detailed discussion of this derivation follows in the next section.

8.4. DISCUSSION OF THE DERIVATION OF THE MOLECULAR EFFECTIVE HAMILTONIAN

Since the effective molecular Hamiltonian has probably the widest ranging applicability in the whole pseudopotential theory, it is worthwhile to analyze in detail the meaning and the derivation of this operator and of the associated wave equation. We start by writing down our equations in detail.

The wave equation for an AB-type diatomic molecular is given by Eq. (8.55),

$$\left\{ \sum_{i=1}^{n} \left(-\frac{1}{2}\Delta_i + V_M^A(i) + V_M^B(i) \right) + \frac{1}{2} \sum_{i,j=1}^{n} \frac{1}{r_{ij}} \right\} \Psi(1, 2, \ldots, n) = E\Psi(1, 2, \ldots, n) . \tag{8.65}$$

In this equation Ψ is an n-electron function for the n binding electrons of the molecule. The semilocal potentials V_M^A and V_M^B have the form

$$V_M^A = \sum_{l=0}^{\infty} V_l^A(r)\Omega_l , \tag{8.66}$$

and

$$V_M^B = \sum_{l=0}^{\infty} V_l^B(r)\Omega_l . \tag{8.67}$$

The local potential V_l^A is given by

$$V_l^A = \frac{(\epsilon_v^A - t - U_v^A)\psi_v^A}{\psi_v^A} , \tag{8.68}$$

where ψ_v^A is an exact pseudo-orbital for one of the valence electrons in the open shell of atom A; U_v^A is the HF potential of the valence electrons; t is the kinetic energy operator; and ϵ_v^A is the orbital parameter associated with ψ_v^A. The fact that V_l^A is an l-dependent radial function can be seen by writing the PO in the central field form,

$$\psi_v^A = R_{nl}^A(r) Y_{lm_l}(\vartheta, \varphi) \eta_{ms}(\sigma), \tag{8.69}$$

where η is the spin-function, Y_{lm_l} is the spherical harmonics, and $R_{nl}^A(r)$ is the radial part of the PO. On putting Eq. (8.69), into Eq. (8.68) we obtain

$$V_l^A(r) = \frac{[\epsilon_{nl}^A - t_R - U_v^A(r)] R_{nl}^A}{R_{nl}^A}, \tag{8.70}$$

where t_R and $U_v^A(r)$ are radial operators obtained from t and U_v^A. The pseudo-orbital ψ_v^A is the exact solution of the equation

$$(H_v^A + V_p^A)\psi_v^A = \epsilon_v^A \psi_v^A, \tag{8.71}$$

where

$$H_v^A = t + g^A + U^A + U_v^A. \tag{8.72}$$

In the last two equations, g^A is the nuclear potential, U^A is the HF potential, and V_p^A is the WR-type pseudopotential of core A.

A set of similar formulas can be written down for atom B. The local potential V^B in Eq. (8.67) is given by

$$V_l^B = \frac{(\epsilon_v^B - t - U_v^B)\psi_v^B}{\psi_v^B}, \tag{8.73}$$

and the rest of the formulas are analogous to Eqs. (8.69)–(8.72). We note that in Eq. (8.65) the index i, in the $V_M^A(i)$, means the coordinates of the i-th electron relative to the nucleus A; and the same index in $V_M^B(i)$ means the coordinates relative to nucleus B.

Next we want to show that if the core orbitals of atom A or atom B or both are the orbitals of a positive ion without any valence electrons, the formulas become much simpler. Let core A be the core of the positive ion of atom A. Then U_v^A vanishes, since the ion does not have any valence electrons. H_v^A becomes the HF Hamiltonian for the core electrons

$$H_v^A \to H_F^A = t + g^A + U^A. \tag{8.74}$$

The pseudopotential V_p^A becomes the PK potential. Adding one valence electron to the core, the equation for this electron will be

$$(H_F + V_p^A)\psi_v^A = \epsilon_v^A \psi_v^A, \tag{8.75}$$

and for the potential V_l^A we get

$$V_l^A = \frac{(\epsilon_v^A - t)\psi_v^A}{\psi_v^A}. \qquad (8.76)$$

We obtain the modified potential by putting V_l^A into Eq. (8.66).

To begin with, we must mention some obvious omissions. No attempt has been made here to include in the derivation the correlation effects between the electrons in the atomic cores; nor have the correlation effects between the electrons of the cores and the binding electrons been included. Another omission is the core-core interaction, which has not been mentioned.

As we pointed out in the derivation, the equation which we have derived, Eq. (8.65), is valid for configuration 2, but its validity can be extended to configuration 3 as long as the overlapping cores remain atomic. We can take into account the core-core interaction as follows. For strictly non-overlapping cores, the Coulomb-type interaction energy

$$E_{cc} = \frac{(Z_A - K_A)(Z_B - K_B)}{R}, \qquad (8.77)$$

must be added to the molecular energy expression.[5,48] If there is a small overlap between the cores, then Eq. (8.77) can be replaced, for example, by an expression derived from Thomas–Fermi theory.[34] The TF model provides a simple expression for the core-core interaction; the formula depends only on the total densities of the cores [Gombas I, p. 143].

It should also be mentioned that although the derivation of the effective molecular Hamiltonian was made for a diatomic molecule, the results can be meaningfully generalized to any molecule in which there is at least one atomic core. This statement is valid, regardless of how many nuclei the molecule contains and regardless of the symmetry of the molecule.

The effective Hamiltonians discussed in Section 8.2 are special cases of Eq. (8.65). Comparing Hellmann's expression with Eq. (8.65), we see that in Hellmann's expression $n = 2$ and $V_M = V_H$. The Hamiltonian derived by Szasz and McGinn, Eq. (8.3), is obtained from Eq. (8.65) by putting for V_l^A, which occurs in V_M^A,

$$V_l^A = V_{HF} + V_R, \qquad (8.78)$$

where V_{HF} and V_R are the l-dependent expressions in Eqs. (A.9) and (A.11). The Hamiltonian of Schwarz, Eq. (8.4), becomes identical with Eq. (8.65) if the local potential V_l^A in Eq. (8.65) has the form of Eq. (8.76). The Hamiltonian of Kahn and Goddard is obtained from Eq. (8.65) by putting

$$V_l^A = g_A + \hat{U}^A, \qquad (8.79)$$

where $\hat{U}^A$ is the potential in Eq. (8.8).

Next we shall analyze certain special features of the effective molecular Hamiltonian. As we have seen in the derivation, the "atomic core approximation" is rigorously built into the Hamiltonian. The modified potentials V_M^A and V_M^B, which represent the interaction between the cores and the binding electrons of the molecule, are the same in the molecular Hamiltonian as in the Hamiltonians of the separated atoms. This means that the potentials of the cores are supposed to be exactly the same in the molecule as in the separated atoms; that is, they are not supposed to be influenced by the fact that the valence electrons of the separated atoms have moved into molecular orbitals as the system moved from configuration 1 to configuration 2. This assumption is an extension of the "frozen core" idea first introduced in Section 2.1 for atoms with one valence electron. We assumed there that the orbitals of the core will be the same in the neutral atoms as they are in the positive ion; that is, the core orbitals do not "feel" the presence of the valence electron. We shall call this assumption the "positive ion" approximation. It is obvious that in those cases in which this is a good approximation, the assumption about the core being the same in the molecule is also a good approximation. If the core does not feel the presence of the valence electron in the atom, it will not feel the presence of the binding electrons in the molecule either.

A somewhat different assumption underlines the approximation which we have also called "frozen core," and in which the core orbitals were the same as in the neutral atom. In the molecular situation, we assume that the core potentials of the molecule are the same as they are in the neutral atom. Here we assume that the core will not feel the difference between the valence electrons of the neutral atom and the binding electrons of the molecule. We call this assumption the "neutral atom" approximation. It is evident that the "atomic core" approximation includes both the "positive ion" and the "neutral atom" approximations. In fact, we have seen that the modified potential will have different forms for the two cases; for the former it will be given by Eq. (8.76), and for the latter by Eq. (8.68). The difference is the absence of the valence electron potential U_v^A in the modified potential for the "positive ion" case.

It is also evident that the quality of the "atomic core" approximation will be different for the two cases. There is no way of establishing how good the atomic core approximation will be in a particular case. This is a *basic assumption* of the pseudopotential model, one built into the theory *a priori*. In cases in which this approximation is not valid, the pseudopotential theory cannot be used. We can say, however, something about the difference between the "positive ion" and the "neutral atom" approximations. From the preceding discussion it is evident that the positive ion approximation will be accurate only for those molecules which are built from atoms with one or two (or at most three) valence electrons. This follows simply from the fact that the positive ion approximation is usable only for atoms with (approximately) not more than three valence electrons. If there are more than three valence electrons, it is extremely unlikely that the core will not feel whether these valence electrons are there or not. On the other hand, the neutral atom approximation can be applied to molecules with many binding electrons. Thus we can describe the situation as follows. The neutral atom approximation can be applied to a much larger group of molecules than the positive ion approximation. At the same time, for those cases for which the positive ion approximation can be applied, it is a better approximation than the "neutral atom." This is clear since, if the core does not feel the presence of the valence electron in the atom, it will not feel the transition from configuration 1 to configuration 2 either. In such a case the atomic core approximation that we have built into Eq. (8.65) will be very good.

Another special feature of the effective Hamiltonian is that it is valid for any internuclear distance from $R = \infty$ to about the equilibrium internuclear distance, or more precisely, to an internuclear distance at which the cores are still atomic; that is, they overlap only slightly. The validity of this statement is evident from the derivation; H was constructed this way. Note also that, as is evident from the discussion, there exists an exact pseudopotential Hamiltonian for every R, in the above-mentioned range, which does not depend on either the binding-electron wave function Ψ or on the binding-electron energy E. For configuration 1 the exact Hamiltonians were in Eqs. (8.10) and (8.16); for configurations 2 and 3, which are special cases of configuration 4, we had the exact Hamiltonian in Eq. (8.36). Thus the derivation presented in Section 8.3, which is based on the exact pseudopotential transformation introduced by the author,[60] is the first in which it is shown that there are exact Hamiltonians underlying the models which are usually used in the calculations. The reason why this can easily be shown from the author's formulation is that, while this formulation does not go beyond the PK and WR formulations conceptually, it is simpler mathematically. Simplicity of formulation is always an advantage in modelization.

Next we want to point out that in Eq. (8.65) the valence-valence correlation can be fully taken into account. This is evident from the fact that Eq. (8.65) is an equation for the n-electron function $\Psi(1, 2, \ldots, n)$. In solving Eq. (8.65) the correlation between the valence electrons can be taken into account by writing Ψ in the appropriate correlated form. The form of Ψ is not restricted to an antisymmetrized product of one-electron pseudo-orbitals. In fact, the exact solutions of Eq. (8.65) will include exactly the valence-valence correlation energy.

In order to elucidate this point a little more, we must ask ourselves whether the derivation of the modified potentials V_M^A and V_M^B will affect valence-valence correlation. We have seen that the forms of these potentials were determined by the atomic core approximation. The modified potentials were determined to be the same as in free atoms treated in the HF approximation; that is, neglecting correlation. Thus the question is: do the core potentials depend on whether the valence electron wave function is correlated or not?

The answer is simple in the positive core approximation. In that case the core potentials will not depend on the valence-valence correlation simply because the core does not depend on whether any valence electrons are present, correlated or not. In the neutral atom approximation there is no clear-cut answer, but we can argue as follows. In that approximation we have assumed that the core potentials do not feel when the valence-electron orbitals change into molecular orbitals. If that is an acceptable assumption, then the presence or absence of correlation in the valence electron wave function will not have an appreciable effect on the core potentials since the $AO \rightarrow MO$ transition is certainly a larger effect than the transition from no correlation to a correlated wave function.

The atomic core approximation which we have just discussed is, strictly speaking, not an approximation but a built-in *assumption*. The derivation of Eq. (8.65) clearly shows that there are only two "real" approximations involved in obtaining the effective Hamiltonian. The first was the replacement of the core-valence interaction operators by the semilocal modified potentials. The second was the omission of the operator $(-S\Omega)$ from the atomic equations and $(-Q\Omega)$ from the molecular equation. We shall now discuss these two approximations.

The first of these approximations consists of the replacement of the core-valence interaction operators in Eq. (8.47) by the modified potential given by Eq. (8.49). What we have done is to replace the equation,

$$(H_v + V_p)\psi_v = \epsilon_v\psi_v, \tag{8.80}$$

by another which reads,

$$(t + V_M)\psi_v = \epsilon_v\psi_v. \tag{8.81}$$

Here we have omitted the A index, since the argument is the same for both atoms. Also, the replacement of the core-valence operators in the molecular equation, Eq. (8.55), amounts to this step.

In these two equations the symbols are defined as follows:

$$H_v = t + g + U + U_v, \tag{8.82}$$

and

$$V_p = -\Omega H_v - H_v\Omega + \Omega H_v\Omega + \epsilon_v\Omega, \tag{8.83}$$

where U is the HF potential of the core, U_v is the HF potential of the valence electrons in the open shell of the atom, and V_p is a WR-type pseudopotential. The modified potential V_M was defined as

$$V_M = \frac{(g + U + V_p)\psi_v}{\psi_v}, \tag{8.84}$$

where ψ_v is an exact PO solution of Eq. (8.80).

First we want to show that the introduction of the modified potential is not really an approximation. We have pointed this out before, but, since this step is crucial in the derivation of Eq. (8.65), an even more detailed discussion will be useful. Let ψ_v and ϵ_v be the solutions of Eq. (8.80). With V_p being the operator in Eq. (8.83), these solutions will not be unique. Let us assume that we have made the solution unique be imposing some mathematical condition, for example, the conditions given by Eqs. (7.34) and (7.35) of KBT. Then ψ_v and ϵ_v are unique solutions of Eq. (8.80). Let us consider now the equation,

$$(t + V_M)\psi = \left(t + \frac{(g + U + V_p)\psi_v}{\psi_v}\right)\psi = \epsilon\psi, \tag{8.85}$$

and let us substitute into the left side $\psi = \psi_v$; then we get

$$(t + V_M)\psi_v = (t + g + U + V_p)\psi_v = (H_v + V_p)\psi_v. \tag{8.86}$$

Since ψ_v is the solution of Eq. (8.80), we obtain

$$(t + V_M)\psi_v = (H_v + V_p)\psi_v = \epsilon_v\psi_v. \tag{8.87}$$

Thus we see that ψ_v and ϵ_v are the exact solutions of Eq. (8.85). Moreover, since these are unique solutions, they will also be unique solutions of Eq. (8.85). Thus the modified potentials, which is constructed with a unique, exact solution of Eq. (8.80), will be such that Eq. (8.81) will have the same eigenfunction and eigenvalue as Eq. (8.80).

Now let us suppose that such a procedure is possible for all valence levels. Let us denote these levels by their (nl) values and let us switch to the formulas from which the spin and angular parts are eliminated. Then we can construct a potential function for each (nl); and we shall have, according to Eq. (8.70),

$$V_l^{(n)}(r) = \frac{[\epsilon_{nl} - t_R - U_v(r)]R_{nl}}{R_{nl}}, \quad (n = n_v, n_v + 1, \ldots, \infty). \tag{8.88}$$

We can combine all these potentials into the semilocal potential

$$V_M^{(n)} = \sum_{l=0}^{\infty} V_l^{(n)}(r)\Omega_l, \quad (n = n_v, n_v + 1, \ldots, \infty), \tag{8.89}$$

where $V_l^{(n)}(r)$ is given by Eq. (8.88). In order to extend the discussion, we have attached the n index to V_l as well as to V_M. This index indicates that these quantities depend on the principal quantum number n as well as on l.

Let us now place $V_M^{(n)}$ into Eq. (8.81):

$$(t + V_M^{(n)})\psi = \left(t + \sum_{l=0}^{\infty} V_l^{(n)}(r)\Omega_l \right)\psi = \epsilon\psi, \quad (n = n_v, n_v + 1, \ldots). \quad (8.90)$$

If, as we have assumed, a potential function like $V_l^{(n)}$ in Eq. (8.88) can be constructed for all (nl), then Eq. (8.90) will be the exact equivalent of Eq. (8.80), since Eq. (8.90) will now exactly reproduce the unique eigenfunctions and eigenvalues of Eq. (8.80). Thus the replacement of the core-valence interaction operators by a modified potential constructed this way is not an approximation. This statement is valid regardless of whether this substitution is taking place in an atomic or in a molecular effective Hamiltonian.

The question now arises whether a potential function can be constructed for all (nl). The difficulty lies in the fact that, if the R_{nl} has nodes, the potential will have points where $V_l \to \infty$. Let the azimuthal quantum numbers which occur in the core be $l = 0, 1, \ldots, l_m$. Let the principal quantum number of the electrons in the valence shell be n_v, and the azimuthal quantum numbers $l = 0, 1, \ldots, l_m, l_m + 1, \ldots, (n_v - 1)$. Then, for $l \leq l_m$, the solutions of Eq. (8.80) will be nodeless pseudo-orbitals; that is, the construction of Eq. (8.88) will be straightforward. For $l = l_m + 1, \ldots, n_v - 1$, the pseudopotential will be zero and the solutions of Eq. (8.80) will be HF orbitals. Since these azimuthal quantum numbers do not occur in the core, the solutions of Eq. (8.80) will still be nodeless, and the construction of the potentials [Eq. (8.88)] will still be straightforward.

Let us consider now the next higher shell with $n = n_v + 1$. For an electron in any of the excited states with $l \leq l_m$, Eq. (8.80) will be valid. In this argument we shall assume that one electron is moved from the n_v-shell into the next higher shell, while the orbitals of the other valence electrons are frozen in the n_v-shell (along with the core orbitals which are also frozen). Since the potential U_v may depend on ψ_v, Eq. (8.80) must be solved in a self-consistent fashion, despite the other orbitals being frozen. The result will be a pseudo-orbital, but it will not be nodeless, since the pseudopotential V_p replaces the orthogonality requirement with respect to the core orbitals but not with respect to the orbitals of the valence shell. Thus the PO with $n = n_v + 1$ and $l = 0 \cdots l_m$ will have one node. It is easy to see that this statement is valid also for the states with $n = n_v + 2, n_v + 3, \ldots$, and so on, with the difference that the number of nodes will be $2, 3, \ldots$, and so on.

Now consider those states for which $n = n_v + 1$ and $l > l_m$. Among these states there might be some with azimuthal quantum numbers which occur in the lower valence shell. These orbitals will have one node. On the other hand, there might be some orbitals with an azimuthal quantum number which does not occur in the valence shell. Such orbitals will again be nodeless.

The point of this argument is that as we go to excited states with $n = n_v + 1, n_v + 2, \ldots$, and so on, more and more of the orbitals will have nodes. For these states, the potential functions [Eq. (8.88)] will have points where $V_l \to \infty$. It is not *a priori* certain that for such states a meaningful modified potential can be constructed.

In order to see how this problem is treated in actual calculations, we shall look at the two major efforts in which exact modified potentials have been constructed. For the lowest states of the valence electron of atoms with one valence electron, PK-type, exact, modified potentials have been calculated by Szasz and McGinn;[11] later these calculations were extended to the excited states by McGinn.[84] In the terminology of this section, these calculations were done in the positive ion approximation. For atoms with many valence electrons, MSPO's and the associated modified potentials have been calculated by Kahn, Baybutt, and Truhlar.[48] In our terminology, these calculations were done in the neutral atom approximation. The calculations of SMG were presented in Section 2.2.B and in Appendix A; the calculations of KBT can be found in Section 7.1 and in Appendix K.

Taking a look first at the more general case, the many-valence-electron atoms, we see from Eq. (7.32) that KBT have constructed a modified potential which is exact for $n = n_v$ and for $l = 0, 1, \ldots, l_m$. For $l = l_m + 1, l_m + 2$, and so on, the pseudopotential is zero; for these states Eq. (7.32) is approximate, in that the exchange potential in it is the potential for l_m and not the correct potential for $l_m + 1, l_m + 2$, and so on. It is obvious that Eq. (7.32) is approximate for all n values for which $n > n_v$. Thus we see that in the construction of Eq. (7.32) the tacit assumption has been made that the modified potentials are not strongly n-dependent; also it was assumed that the l-dependence is also weak for those states for which the pseudopotential is zero; that is, those for which the l-dependence is only in the exchange potential.

Turning to atoms with one valence electron, we have seen that SMG have calculated exact, PK-type pseudo-orbitals and modified potentials for the lowest valence shell where $n = n_v$, and for the azimuthal quantum numbers which occur in the core; that is, for $l = 0, 1, \ldots, l_m$. For these states the PO's are nodeless, so there is no difficulty in constructing a modified potential from Eq. (8.88). The calculations were extended by McGinn to a considerable number of excited states. These are presented below in Section 9.2. It was shown by McGinn that although the pseudo-orbitals for many excited states do have the nodes discussed above, these nodes do not prevent the construction of meaningful modified potentials. This can be seen if we go back to the original definition of the modified potential, Eq. (8.84). In the positive ion approximation, the V_p becomes the PK potential, so we get from, Eq. (8.84),

$$V_M = g + U + \sum_{i=1}^{K} (\epsilon_v - \epsilon_i)\langle \varphi_i \,|\, \psi_v \rangle \frac{\varphi_i}{\psi_v}. \tag{8.91}$$

The singularity of this expression arises from (φ_i/ψ_v) where the ψ_v will now have nodes. It turns out, however, that since the ψ_v represents excited states, the node will be in that far out region where φ_i and V_M are very small. By spacing properly the points at which V_M is computed (numerically), one finds that the node of ψ_v causes only a small "bump" in the V_M.

Thus the calculations carried out for atoms with one valence electron have shown that exact, modified potentials can be constructed for the states of the lowest valence shell as well as for excited states. A further important result is that the exact modified potentials turned out to be largely n-independent. This is the same conclusion that we drew from the properties of the density-dependent pseudopotentials, and it is the same assumption which underlies Eq. (7.32).

We sum up now the discussion of the replacement of the interaction operators by semilocal potentials. If exact potential functions can be constructed either from Eq. (8.88) or from Eq. (8.84), then this step is not an approximation. For atoms with one valence electron, potentials have been constructed for the ground state as well as for excited states. It may be stated that, for such atoms, potential functions can be constructed for all (nl). For atoms with many valence electrons, modified potentials have been constructed for those states for which the pseudo-orbitals are nodeless; that is, for those states whose azimuthal quantum number occurs in the core. For these states there is no problem with the construction of the modified potentials. It is not known whether modified potentials could be constructed for the excited states in which the pseudo-orbital has nodes. The calculations for atoms with one valence electron have shown, however, that the modified potentials are not strongly n-dependent. Thus, it is probably a good approximation to consider the modified potentials which were computed for $n = n_v$ to be valid also for the excited states.

We turn now to that step in the derivation of the effective Hamiltonian in which we omitted the operators $(-S\Omega)$ from the atomic equations, and constructed the molecular Hamiltonian of Eq. (8.65) in such a way that it does not contain an operator of this structure. We have discussed the meaning of such operators in Section 7.4. We have seen that although the mathematical structure of these operators is complex, their physical meaning is very simple. In the separated atoms the function $(\Omega\Psi)$ meant the core projection of the valence electron wave function. The expectation value of the operator $S\Omega$, with respect to the valence wave function Ψ, is roughly equivalent to the core projection of the operator S; this is the term which must be subtracted from the expectation value of S because of the many-electron part of the Pauli exclusion principle. In Section 7.4 it was shown that, instead of subtracting this term, one can take this effect into account by introducing the constants η and $\hat{\eta}$, for which expressions were derived giving these constants in terms of the one-electron pseudo-orbitals of the valence electrons. We have also seen that, in many instances, especially for atoms with only a few valence electrons, these effects are small and can be omitted.

We have also seen that, for atoms with many valence electrons, the effect of these terms is not necessarily negligible, and it might be necessary to take them into account by using the constants η and $\hat{\eta}$.

In the derivation of the molecular Hamiltonian, we have omitted such a term; specifically, we replaced Eq. (8.36) by Eq. (8.45); and we postulated, on the basis of Eq. (8.45), that the form of the correct molecular Hamiltonian is given by Eq. (8.46), in which the interaction term Q was not multiplied by any projection operator. The physical meaning of this step is evident. The expectation value of the operator $(Q\Omega)$ is roughly the core projection of the operator Q; omitting this term means that we do not subtract from the expectation value of Q the core part. It is plausible to assume that, just as in the atomic case, the subtraction of $(Q\Omega)$ can be taken into account by introducing a factor $(1 - \eta_M)$ in front of Q, with $\eta_M < 1$. In writing down Eq. (8.46) we have assumed that it is a good approximation to put $\eta_M = 0$; this step is an extrapolation of the atomic case with a few valence electrons. As we have seen in Section 7.4, in the case of many valence electrons it might be necessary to take this effect into account. It is plausible to assume that this last statement is also true for the molecular case. For molecules with a larger number of binding electrons, it might be necessary to modify the interaction term Q by a multiplicative constant, the value of which might be determined by theory; or the constant might be treated as adjustable. This is one of the points of pseudopotential theory at which further research might be useful.

Finally, we would like to synchronize our results with the conclusions reached by Schwarz.[40] As we have seen in Section 8.2, Schwarz pointed out three approximations in the molecular effective Hamiltonian. The second approximation was that the model potential is energy-independent; that is, it does not depend on the HF orbital parameter of the valence electron, but only on its azimuthal quantum number. What is under discussion here is not the n-independendence of the modified potential; that is, not the assumption that the modified potential is approximately the same for the (n, l) and $(n + 1, l)$ states. Rather, in this case, the energy-independence refers to the assumption that the modified potential is approximately the same in the (n, l) valence state of an atom with one valence electron as it is in the (n, l) valence state of the same atomic core with more than one valence electrons. Thus the second approximation of Schwarz is actually the difference between the positive ion approximation and the neutral atom approximation. In our treatment, these two approximations entered the discussion as initial assumptions; the effective molecular Hamiltonian in Eq. (8.65) is valid for both cases. There is a difference between the two cases in the determination of the modified potential; in the positive ion approximation the modified potential must match the spectrum of the atom with one valence electron, while in the neutral atom approximation it must match the energy levels of the valence electron in the presence of other valence electrons.

We sum up now the discussion of the effective molecular Hamiltonian which is the Hamiltonian in Eq. (8.65):

1. The Hamiltonian contains only one "real" approximation, namely the omission of the projection operator from the many-valence electron interaction term. This statement is correct if the one-electron interaction potentials representing the atomic cores are the exact, (n, l)-dependent modified potentials. The additivity of the potentials for different cores is a built-in consequence of the frozen core approximation.

2. If the one-electron interaction potentials are assumed to be different from the exact modified potentials—that is, if they are n-independent and/or model potentials—then this assumption is the second real approximation introduced into the effective Hamiltonian. As we have pointed out, the exact modified potentials as well as the approximate model potentials obey different rules in the "positive core" and "neutral atom" approximations; these are treated not as "real" approximations, but rather, as built-in assumptions.

We conclude that the effective molecular Hamiltonian which stands in Eq. (8.65) is reasonably well established, although further research into the finer details of the derivation, especially into the size of the errors introduced by the approximations, might be useful.

8.5. REPRESENTATIVE CALCULATIONS

In this section we review a sample of representative molecular pseudopotential calculations. From the large number of calculations which have been already carried out, and to which new efforts are being added at a high rate at the time of writing, we have selected a few which will demonstrate the scope of the applications of the pseudopotential method to molecules. Let us consider an arbitrary molecule of Class I, and let us denote the atomic cores by $A = 1, 2, \ldots, M$. Let the number of binding electrons be $i = 1, 2, \ldots, n$. The effective molecular Hamiltonian is given, on the basis of Eq. (8.55), by the formula

$$H = \sum_{i=1}^{n} \left(-\frac{1}{2} \Delta_i + \sum_{A=1}^{M} V_m^A(i) \right) + \frac{1}{2} \sum_{i,j=1}^{n} \frac{1}{r_{ij}} + \frac{1}{2} \sum_{A,B=1}^{M} E_{AB}. \qquad (8.92)$$

In this formula, $V_m^A(i)$ is a model potential representing the core A, expressed in the coordinates of the i-th binding electron; that is, the index i means the coordinates of the i-th electron relative to the center of core A. E_{AB} is the interaction energy of the atomic cores A and B.

The wave equation of the binding electrons is

$$H\Psi(1, 2, \ldots, n) = E\Psi(1, 2, \ldots, n), \qquad (8.93)$$

where Ψ is an antisymmetric, n-electron wave function which does not need to satisfy any orthogonality conditions; that is, we can solve Eq. (8.93) as if the cores did not exist.

We shall write down now the basic types of expressions which can be used for V_m^A. In this discussion we shall treat the exact, ab initio modified potentials as one of the models.

First, the atomic core (or cores) may be treated in the positive ion approximation. Then the relevant model potentials are those discussed in Sections 4.2 and 4.3. The model potential is defined by the atomic equation

$$(t + V_m^A)\psi = \epsilon\psi ,\tag{8.94}$$

and we determine V_m^A by matching the empirical spectrum of the atom which consists of the core A plus one valence electron. In this case the model potential is generally one of the following two types,

$$V_m^A = \sum_{l=0}^{\infty} V_m^l(r)\Omega_l ,\tag{8.95}$$

or

$$V_m^A = \sum_{l=0}^{\infty} V_m^l(r)\Omega_l + V_m^c .\tag{8.96}$$

In these equations the semilocal potential is the modified potential. In Eq. (8.96), V_m^c represents the core-valence correlation (polarization).

If the model potential is identical with the exact, ab initio modified potential, then

$$V_m^A = V_M ,\tag{8.97}$$

and V_M is defined by

$$V_M = (g + U + V_p)\psi/\psi ,\tag{8.98}$$

where U is the electrostatic and exchange potential of the core and V_p is a Phillips–Kleinman-type pseudopotential.

These are the basic formulas for the positive ion approximation. If the core (or cores) are treated in the neutral atom approximation, the relevant model potentials are those discussed in Section 7.2. In this case, the model potential will be defined by Eq. (7.46):

$$(t + V_m^A + U_v)\psi = \epsilon\psi ,\tag{8.99}$$

where U_v is the HF potential (including exchange) of the valence electrons. We determine V_m^A by matching the HF orbital parameters of the valence electrons in the neutral atoms. The model potential will be of the form of Eq. (8.95).

If the model potential is identical with the exact modified potential, then

$$V_m^A = V_M, \qquad (8.100)$$

and V_M is defined by Eq. (8.98), with V_p now being the Weeks–Rice potential. From the point of view of the model potential, the difference between the positive ion and neutral atom approximations is that in the former, V_m^A is defined by Eq. (8.94), while in the latter it is defined by Eq. (8.99). In the neutral atom approximation, the equation defining the model potential must contain the potential of the valence electrons, U_v.

We proceed now with the discussion of the calculations. The serial number is for reference and reflects, approximately, the chronological order of publication. The numbers placed in parentheses following the results of the calculations are the experimental values unless otherwise specified. The energies will be in eV units and the distances in atomic units.

1. First, two pioneer calculations must be mentioned. After introducing the basic idea of pseudopotentials with the potential given by Eq. (1.1), Hellmann used this potential to calculate the ground state energies of the K_2 and KH molecules.[2] Treating the molecules as two-electron problems, and applying the Heitler–London method [Slater III, p. 41], Hellmann obtained for the dissociation energy and equilibrium internuclear distance the following results: $D = 0.19$, (0.52); $R = 7.69$, (7.40). For the KH molecule, a Heitler–London wave function augmented by an ionic term gave $D = 0.79$, (1.92) and $R = 3.96$, (4.24).

Considering the date (1935) when these calculations were made and the simplicity of the Hellmann potential, we must judge these results very respectable. Hellmann analyzed the results and pointed out that the absence of electron correlation in the trial wave function is responsible for the poor results for the dissociation energy.

The second pioneer calculation which we want to mention is the work of Preuss.[85] Preuss carried out calculations for the ground states of the positive ions of alkali-hydrides. These calculations were made much later than Hellmann's but still before the pseudopotential method was discovered by the majority of scientists. For V_m^A Preuss used Eq. (4.22). The wave function employed in the variational calculation was

$$\Psi = e^{-\alpha(r_a + r_b)}\{1 + \gamma(r_a - r_b)^2\}, \qquad (8.101)$$

where α and γ were variational parameters, and r_a and r_b the distances of the single electron from the centers. The results for the binding energies were, Na_2^+: $D = 1.06$, (1.03); K_2^+: $D = 0.93$, (0.76); Rb_2^+: $D = 0.80$, (0.73); Cs_2^+: $D = 0.80$, (0.66).

2. The calculations of Szasz and McGinn[18] were the first ab initio molecular pseudopotential calculations; that is, these were the first calculations in which the one-electron modified potentials were the exact Phillips–Kleinman types rather than a model potential. The modified potential was given by Eq. (A.20), for which the analytic expressions were given by Eqs. (A.22) and (A.23). A polarization potential was added to the modified potential, and the expression used was Eq. (A.24). Thus the V_m^A of Eq. (8.92) was of the form

$$V_m^A = V_M + V_m^c,\qquad(8.102)$$

which shows that the potential was of the form of Eq. (8.96), with the l-dependence neglected in the modified potential. Only the s potential was used.

For the molecules Li_2, Na_2, and K_2 a Heitler–London-type wave function, built from Slater functions, was used. The results were,

$$
\begin{array}{lll}
Li_2 & D = 0.24,\ (1.05); & R = 6.13,\ (5.05)\,.\\
Na_2 & D = 0.25,\ (0.74); & R = 6.37,\ (5.80)\,.\\
K_2 & D = 0.24,\ (0.52); & R = 6.60,\ (7.40)\,.
\end{array}
$$

For the molecules LiH, NaH, and KH, the wave function employed was a Heitler–London type, augmented by an ionic term which was multiplied by a Hylleraas-type correlation factor. The one-electron functions were STO's. The results were,

$$
\begin{array}{lll}
LiH & D = 1.46,\ (2.52); & R = 3.60,\ (3.01)\,.\\
NaH & D = 1.46,\ (2.27); & R = 3.78,\ (3.57)\,.\\
KH & D = 1.23,\ (1.92); & R = 4.21,\ (4.24)\,.
\end{array}
$$

Analyzing the results, Szasz and McGinn pointed out that for the A_2 types, about one-third of the binding energy was obtained, while for the hydrides the result was about two-thirds. It was pointed out that the better results for the hydrides were due to the fact that in the trial functions of these molecules some of the electron correlation was taken into account. Comparing the results for the A_2 types with the results for the hydrides, Szasz and McGinn advanced the conclusion that the gap between the calculated and observed binding energies was not due to some general deficiency of the pseudopotential method, but was the result of electron correlation not being fully taken into account in the variational trial function.

Another observation made from the calculations was that the presence of the polarization potential is not very important for the binding energy, but is important for the internuclear distance. In order to obtain good results for R, the polarization of the cores must be properly taken into account.

Finally, we note that, as we have seen at various points above, the Phillips–Kleinman pseudopotentials, which were employed in the calculations of Szasz and McGinn, are not uniquely determined. Here we refer to the discussion following Eq. (7.61), where it was pointed out that the indeterminacy does not affect the quality of the results in a many-electron calculation as long as an ab initio potential is employed.

3. The calculations of Simons[86] are characterized by an extremely simple model potential. The potential is No. 5 in Table 4.3. Simons treated the H_2O and HF molecules in the positive ion approximation; that is, the parameters of the model potential were adjusted to the spectrum of the O^{5+} and F^{6+} ions. The method of calculation was a single center SCF procedure. An interesting feature was the introduction of a screening constant, $(1 - \eta)$, in front of the interaction terms, for which the value $\eta = 0.1$, $(1 - \eta) = 0.9$ was chosen (see Section 7.4). The results obtained for H_2O were as follows: distance O–H was $R_{O-H} = 1.83$, (1.81); the angle $\theta_{H-O-H} = 105.2$, (104.52); total energy of the molecule $E(H_2O) = -75.96$ atomic units, (-75.92). For the HF molecule the results were, $R = 1.72$, (1.73); total energy $E(HF) = -100.002$ atomic units, (-100.005).

4. Under this number we collected some model potential calculations. The model potential used is No. 6 in Table 4.3. As we see, this model potential, which was introduced by Bardsley, is still quite simple, although much more sophisticated than the original Hellmann formula. The core polarization is included here with a separate term. Using this method, Bardsley and Junker calculated[69] the ground state energies of the positive ions of the diatomic alkali molecules and obtained, for the dissociation energies the following results: $D(Li_2^+) = 1.31$, (1.57); $D(Na_2^+) = 1.02$, (1.03); $D(K_2^+) = 0.88$, (0.76); $D(Rb_2^+) = 0.84$, (0.73); $D(Cs_2^+) = 0.88$, (0.66).

Using the same model potential, calculculations were made by Bardsley, Junker, and Norcross[87] for the molecules Na_2^+, Na_2, and Na_2^-. Here the calculations were extended to some excited states. The trial wave functions were of valence-bond type built from Slater functions. We quote the results for the dissociation energies and internuclear distances of some states of the Na_2 molecule:

$$X\,^1\Sigma_g \qquad D = 0.71,\ (0.74); \qquad R = 5.75,\ (5.67)\,.$$
$$A\,^1\Sigma_u \qquad D = 0.94,\ (1.02); \qquad R = 6.79,\ (6.87)\,.$$
$$B\,^1\Pi_u \qquad D = 0.24,\ (0.33); \qquad R = 6.30,\ (6.43)\,.$$
$$a\,^3\Pi_u \qquad D = 1.16,\ (1.04); \qquad R = 5.77,\ (5.96)\,.$$

In the calculations of Stevens, Karo, and Hiskes,[88] the model potential of Bardsley was again used (No. 6 in Table 4.3). Multiconfiguration SCF was carried out for the $X\,^1\Sigma^+$ and $a\,^3\Sigma^+$ states of the molecules LiH, NaH, KH, RbH, and CsH, and for the $X\,^2\Sigma^+$ states of their negative ions. The results for the dissociation energies (in cm^{-1} units) and for the internuclear distances (in Å units) are as follows. ($X\ ^1\Sigma^+$ state):

LiH	$D = 18{,}861,\ (20{,}286);$	$R = 1.58,\ (1.59)$.
NaH	$D = 15{,}204,\ (15{,}760);$	$R = 1.88,\ (1.89)$.
KH	$D = 14{,}490,\ (14{,}550);$	$R = 2.24,\ (2.24)$.
RbH	$D = 14{,}294,\ (14{,}240);$	$R = 2.30,\ (2.37)$.
CsH	$D = 15{,}384,\ (15{,}000);$	$R = 2.42,\ (2.49)$.

5. In Section 7.1 we saw how the open-shell HF equations were transformed into pseudopotential equations and how the MSPO's were constructed by Kahn, Baybutt, and Truhlar.[48] We turn now to the discussion of some calculations in which the ab initio modified potentials of KBT were used.

The modified potential was given by Eq. (7.32). The functions in that formula were defined by Eqs. (7.25) and (7.27). The analytic fits to these functions were given by Eqs. (K.4) and (K.5).

Kahn, Baybutt, and Truhlar carried out calculations for obtaining the potential energy curves of hydrogen halides and diatomic halogens. The molecules studied were of the HX and X_2 types, with X being F, Cl, Br, and I. The binding electrons in the halogens were the $(ns)^2(np)^5$ electrons. The approximation applied here is, in contrast to the calculations discussed above, the neutral atom approximation.

The calculations were carried out in two steps. First, a multi-configurational self-consistent-field (MCSCF) calculation was carried out. In order to ensure the proper behavior of the wave function in dissociation, for the $X\ ^1\Sigma^+$ state of the HX molecules the two-configuration function

$$\Psi = \tilde{A}\{1\sigma^2(C_1 2\sigma^2 - C_2 3\sigma^2)1\pi_x^2 1\pi_y^2\}, \tag{8.103}$$

was used ($\tilde{A}$ = antisymmetrizer). For the $X\ ^1\Sigma_g^+$ state of the X_2 molecules, the wave function used was

$$\Psi = \tilde{A}\{1\sigma_g^2 1\sigma_u^2(C_1 2\sigma_g^2 - C_2 2\sigma_u^2)1\pi_{xu}^2 1\pi_{xg}^2 1\pi_{yu}^2 1\pi_{yg}^2\} \tag{8.104}$$

First self-consistent calculations were carried out with these functions (which are also referred to as generalized valence bond (GVB) functions). Then a configuration interaction (CI) calculation was made, with Eqs. (8.103) and (8.104) as reference configurations. These calculations were carried out for HBr and Br_2. In the CI calculations all single and double substitutions were included. The number of configurations employed for HBr were 1637 and for Br_2, 3396.

The results for the dissociation energies and equilibrium nuclear distances were, with the MCSCF wave function:

HF	$D = 4.19$, (6.12);	$R = 1.79$, (1.73).
HCl	$D = 3.86$, (4.62);	$R = 2.53$, (2.41).
HBr	$D = 3.19$, (3.91);	$R = 2.66$, (2.67).
HI	$D = 2.68$, (3.21);	$R = 3.12$, (3.04).
F_2	$D = 0.99$, (1.68);	$R = 2.80$, (2.68).
Cl_2	$D = 1.92$, (2.51);	$R = 3.81$, (3.76).
Br_2	$D = 0.96$, (1.99);	$R = 4.50$, (4.31).
I_2	$D = 0.8$, $\ (1.56)$;	$R = 5.32$, (5.04).

The results obtained with CI were, for the total energies and equilibrium internuclear distances: $E(HBr) = -14.1387$ a.u. at $R = 2.66$ and $E(Br_2) = -26.9714$ a.u. at $R = 4.50$. It was pointed out by KBT that the equilibrium internuclear distances obtained with configuration interaction did not differ from those obtained with the generalized valence bond (GVB) functions. The results for the dissociation energies improved with CI in such a way that for HBr and GVB approach gave 82% of the dissociation energy, while the CI approach gave 92%. The corresponding values for Br_2 were 49% and 63%, respectively.

Kahn, Baybutt, and Truhlar also compared their calculations to all-electron studies and observed good agreement between the two approaches. In view of our discussion in Section 8.4, the good agreement between a carefully executed ab initio pseudopotential calculation and an equally carefully executed all-electron calculation means two things:

1. The basic assumption of the pseudopotential effective Hamiltonian, Eq. (8.92), which is the frozen-core, or more precisely in this case, the neutral atom approximation, appears to be valid.

2. The approximation of omitting the projection operators or equivalent constants from the valence-valence interaction term is justified.

The calculations of KBT are important for two reasons. First, like the calculations of Szasz and McGinn, these were ab initio studies. Second, the actual numerical calculations for obtaining the molecular energies were carried out very carefully and an effort was made to include electron correlation in the valence electron wave function.

6. The next calculations which we want to discuss are the work of Schwarz, Chang, and Habitz.[89] In this work calculations were performed for a number of diatomic and triatomic molecules from the first row of the periodic table. The molecules were H_2O, N_2, BeO, LiF, LiH, BeH_2, LiH^+, Li_2^+, and HeH. These calculations are important, not for the results *per se*, but because Schwarz *et al.* used these calculations to test the molecular effective Hamiltonian given by Eq. (8.92).

As we indicated in Sections 8.2 and 8.4, Schwarz pointed out that if the underlying assumption for the molecular effective Hamiltonian was the positive ion approximation, the the Hamiltonian contained three "real" approximations, the additivity of the cores, the energy parameter independence of the one-electron model potentials, and the omission of the projection operators from the interaction term. From the derivation of Section 8.3, based on the author's work, we have seen that the first of these approximations was the consequence of the rigorous application of the frozen core approach, and thus we have considered it an initial assumption. The second approximation did not appear in that derivation, since we treated the positive ion and the neutral atom approximations as separate cases for which Eq. (8.92) is equally valid, although with different rules for V_m^A. The derivation identified Schwarz's third approximation as the only "real" approximation.

Now let us take a look at the molecules investigated. The number of binding electrons in the sequence above are 8, 10, 8, 8, 2, 4, 1, 1, 1. For all these molecules SCH used the positive ion approximation; that is, that form of the frozen core approximation in which the core orbitals are the same in the molecule as in the positive ion. SCH used model potential No. 8 in Table 4.3, with the parameters adjusted to the empirical spectrum of the $(1s)^2$ core plus one valence electron. According to our discussions in Section 8.4, this is a good approximation for atoms with about one to three valence electrons, but not, for example, for the O atom where there are six valence electrons.

Schwarz pointed out earlier[40] that the error introduced by this approximation is, to a large extent, compensated by the omission of the projection operators from the valence-valence interaction term. The calculations discussed here, along with some earlier atomic calculations[90], where carried out in order to test whether the errors introduced by these two approximations really cancel each other.

Taking a look at the sequence of molecules again, we see that the number of binding electrons is fairly large at the beginning of the set and only one at the end. Thus as we pass along the sequence, the positive ion approximation is poorly satisfied at the first few molecules and very well satisfied for the molecules with one binding electron. The omission of the projection operators from the valence-valence interaction term can also be very well investigated by this set. The omission means a significant approximation for eight or ten binding electrons, but it is much less significant for two or four. For one binding electron there is no interaction term, and thus the approximation is absent in this case.

The method used by SCH for checking the effects of these two approximations was to compare the calculation step-by-step with the corresponding ab initio all-electron calculations. Using this procedure for the above sequence of molecules, Schwarz, Chang, and Habitz were able to show that the errors introduced by the two approximations—the use of the positive ion approximation for a larger number of valence electrons and the omission of the projection operators from the valence-valence interaction term—cancel each other to a considerable degree.

7.   Psuedopotential calculations were carried out for sodium clusters by

Flad, Stoll, and Preuss.[91] The model potential used was No. 10 in Table 4.3. A special feature of these calculations was the use of the density-functional model for the calculation of the correlation energies. Studied were the equilibrium geometries of clusters Na_M for $M = 2 - 8$. For each of the clusters, Na_3 to Na_8, total energies of two isomers were computed. Ionization energies were obtained by calculating the total energies of the neutral and ionized clusters. The computed ionization energies were compared with recent measurements and it was shown that for Na_4, Na_6, and Na_8, the calculated ionization energies of those structures which contained the maximum number of Na_2 units had a better agreement with experiment than the ionization energies of the isomer structures. The maximum deviation from experiment in the ionization energies was 3%.

Here we have reported the use of density-functional theory for the calculation of correlation energies. Looking at our list of calculations, we see that the correlation energy has been a problem in molecular pseudopotential calculation from the very beginning. Hellmann had pointed out that the poor results of his calculations were due to the lack of electron correlation in the valence electron wave function. The same conclusion was reached by Szasz and McGinn. Even in the elaborate CI calculations of Kahn, Baybutt, and Truhlar, which were carried out for HBr and Br_2, there was still a sizeable gap between the calculated and empirical values which was clearly caused by electron correlation. The reason for this situation is that in many molecules the binding is almost entirely a correlation effect; that is, calculations carried out in the framework of the HF approximation either do not give binding or provide only a fraction of the binding energy.

Recognizing the importance of this problem, Preuss *et al.* suggested the use of a density-functional formalism for the calculation of the correlation energy in the framework of the pseudopotential model. This method was used in work on sodium clusters. More generally, the method was tested in atomic calculations.[92] Molecular calculations were carried out by Preuss, Stoll, Wedig, and Krüger[93] in which the correlation energy was computed from a density functional formula. In these calculations the positive ion approximation was used in conjunction with the model potential No. 11 in Table 4.3. Calculations were made for a wide range of molecules. For the molecules CH_4, C_2H_4, C_2H_6 and B_2H_6, the calculated orbital energies were compared with all-electron calculations, and reasonable agreement was found. For the first and second row monohydrides, the dissociation energies were computed at the experimental internuclear distances, and good agreement was found with the all-electron results as well as with the experimental values. Here we quote the results for the diatomic alkali molecules:

$$Li_2 \quad D = 0.87, (1.05); \quad R = 5.11, (5.05).$$
$$Na_2 \quad D = 0.68, (0.74); \quad R = 5.71, (5.80).$$
$$K_2 \quad D = 0.41, (0.52); \quad R = 7.27, (7.40).$$

Thus, using the density-functional formula for the correlation energy, Preuss *et al.* obtained 78%, 92%, and 79% of the dissociation energies for the molecules Li_2, Na_2, and K_2.

8.   In this group we put some more recent calculations which were carried out for molecules with a larger number of binding electrons. Calculations were carried out for the low-lying states of NaAr and NaXe by Laskowski, Langhoff, and Stallcop.[94] The method used was the ab initio procedure of Kahn, Baybutt, and Truhlar (see above). The molecules are treated as nine-electron systems and the method of calculation for the valence electron wave function was a SCF procedure plus configuration interaction. Potential curves were computed for $x\,^2\Sigma^+$, $A\,^2\Pi$, $B\,^2\Sigma^+$, $C\,^2\Sigma^+$, $(4)\,^2\Sigma^+$ and $(1)\,^2\Delta$. The computed potential curves agreed well with those deduced from high-energy scattering data.

For the molecule Cu_2, model calculations were carried out by Pelissier.[95] The method used was the model potential of Berthelat and Durand (see Section 7.2). The $(3d)$ and $(4s)$ electrons of the Cu atom were treated as valence electrons; that is, the Cu_2 molecule was treated as a 22-electron problem. Here again the importance of electron correlation·between the binding electrons became evident; it was pointed out that SCF calculations were not sufficient to get good results. Only after the inclusion of electron correlation in the wave function, in the form of CI calculations, was it possible to get good results. The calculated values for the dissociation energy and the internuclear distance were: $D(Cu_2) = 2.00$, (2.05); $R = 4.24$, (4.17).

For the homonuclear diatomic molecules of Ar, Kr, and Xe, model potential calculations were carried out by Christiansen, Pitzer, Lee, Yates, Ermler, and Winter.[96] The method used was the model potential procedure of Christiansen, Lee, and Pitzer (see Section 7.2). Only the outer two *s* electrons and the *p* electrons were considered to be particitating in the molecular binding. The calculated spectroscopic constants have shown excellent agreement with the all-electron results. We quote some figures here. (In contrast to our previously used notation, the numbers in the parentheses mean the all-electron results and not the experimental values.)

$$Ar_2^+ \quad D = 1.28,\ (1.27); \quad R = 4.64,\ (4.68)\,.$$
$$Kr_2^+ \quad D = 1.13,\ (1.30); \quad R = 5.28,\ (5.23)\,.$$
$$Xe_2^+ \quad D = 1.05,\ (1.10); \quad R = 6.09,\ (6.07)\,.$$

9.   In the last group of calculations we have placed publications dealing with compounds of considerable complexity. The electronic structure of transition-metal complexes $Ni(CO)_4$, $Cd(CO)_4$, and $Pt(CO)_4$ were investigated by Osman, Ewig, and Van Wazer.[97] The method used was the model potential approach developed by these authors (see Section 7.2). The $Ni(CO)_4$ results were compared with all-electron SCF calculations and with experiment, and excellent agreement was found. Osman, Coffey, and Van Wazer used the same model potential approach for the investigation of the

P_2 and P_4 molecules.[98] The results were found to be comparable to all-electron calculations.

The pseudopotential method was used for the study of hypervalent compounds by Bartell, Rothman, Ewig, and Van Wazer.[99] The compounds investigated were XeF_2 and XeF_4. The method of approach was the model procedure developed by Ewig, Osman, and Van Wazer (see Section 7.2). The calculations carried out on XeF_2 and XeF_4 give valence orbital energies in fair agreement with those obtained from SCF-MO all-electron studies. Equilibrium structures of XeF_2 and XeF_4 provided by these modified potential calculations possess the correct symmetries. The authors concluded that the ability of the pseudopotential treatment to give a reasonable account of the structures of XeF_2 and XeF_4 justifies the application of this method to the higher fluorides.

Reviewing this small but representative sample of calculations we see that the pseudopotential method has been applied to a wide variety of cases. Specifically, we can emphasize the following points:

1. Ab initio methods have been applied in molecular calculations in the positive ion approximation (No. 2) as well as in the more complex neutral atom approximation (No. 5).

2. Model pseudopotentials of simple form (Nos. 1 and 3) as well as of more sophisticated structure (No. 4) have been applied successfully for molecules.

3. The applications range from molecules with very small number of binding electrons (Nos. 2 and 4) to molecules with a considerable number of binding electrons (Nos. 5 and 8).

4. The calculations involved molecules with only two centers (Nos. 2, 4, 5, 8) as well as structures with a considerable number of centers (Nos. 7 and 9).

In addition, we see that sophisticated computational methods have been applied to the approximate solutions of the molecular wave equation, Eq. (8.93). These methods included the LCAO-MO method, the generalized valence-bond (GVB) procedure, and configuration interaction (CI), as well as the use of density-functional formalism for the correlation energy. The numerical results obtained range from fairly crude approximations to values whose accuracy is on a par with the carefully executed all-electron results.

The goal of this chapter was to demonstrate that pseudopotential theory is a reliable tool for the calculation of molecular properties in the case of those molecules for which the basic assumptions of the theory can be applied. This demonstration has taken three steps. The first step was the derivation of the effective molecular Hamiltonian. The second step was the physical evaluation and interpretation of the Hamiltonian. The third step was the presentation of some representative calculations. Thus, it is clearly demonstrated that the theory is a reliable tool of molecular physics.

chapter 9

Excited States and
Rydberg States

9.1. INTRODUCTION

With the preceding chapter the conceptual presentation of pseudopotential theory is essentially completed. The application of the theory to the excited states of atoms and molecules can be carried out in a reasonably straightforward fashion on the basis of the preceding chapters. Most, if not all, parts of the theory are valid for excited states, although in the presentation we often considered primarily the ground state.

We shall discuss this fairly straightforward application in some detail because, as we put it in Chapter 1, the main goal of the pseudopotential theory is to provide a "valence only" formalism for atoms and molecules. Such a formalism can be considered as established only if it is demonstrated that it can handle the excited states as well as the ground state.

In addition, two important questions were left unanswered in the presentation of the formalism, questions which will have to be answered in this chapter. In Section 2.5 we saw the importance of correlation effects in the buildup of pseudopotential theory. However, we left unanswered the question: how large are the core-valence correlation effects? The answer to this question must be given in a comprehensive fashion, including the excited states. Then, in Section 8.3, in the pivotal derivation of the effective molecular Hamiltonian, we assumed that the one-electron, core-valence interaction operators, which were nonlocal in the exact theory, can be transformed exactly into local (or semilocal) potentials. The whole derivation of the effective molecular Hamiltonian rests on this assumption, which was expected to be valid for any (nl) state.

In the following presentation we shall demonstrate that the theory is applicable to the excited states. We shall also clear up the questions left unanswered in the preceding build-up of the formalism.

9.2. THE OPTICAL SPECTRUM OF ATOMS WITH ONE VALENCE ELECTRON

The exact, Phillips–Kleinman formulation of the theory was applied to the excited states of one-valence-electron atoms by McGinn.[84] The work of McGinn is the continuation of the work of Szasz and McGinn[11] in which the PK theory was applied to the ground states of atoms. McGinn investigated the optical spectrum of the atoms Li, Na, K, Rb, Be$^+$, Mg$^+$, Ca$^+$, Al^{++}, Cu, and Zn$^+$. We present his work here in detail.

In order to establish continuity with the preceding sections we quote the basic equations. The frozen core approximation, or, more accurately, the positive ion approximation, will be used. The core orbitals are the solutions of the HF equations [Eq. (2.2)]:

$$H_F \varphi_i = \epsilon_i \varphi_i, \quad (i = 1, 2, \ldots, N), \tag{9.1}$$

where

$$H_F = t + g + U, \tag{9.2}$$

with U being defined as the potential given by Eq. (2.40). The detailed form of U is given by Eq. (A.9). The HF equation for the valence electron is Eq. (2.15):

$$H_F \varphi_v = \epsilon_v \varphi_v, \tag{9.3}$$

where, in the frozen core (positive ion) approximation, H_F is the same operator as in Eq. (9.1). The equation above is valid for any state of the valence electron. We note that the solutions of Eq. (9.3) belonging to different eigenvalues are orthogonal:

$$\langle \varphi_{v_1} \mid \varphi_{v_2} \rangle = \delta_{v_1, v_2}, \tag{9.4}$$

where v_1 and v_2 are valence states.

Transforming Eq. (9.3) exactly into a pseudopotential equation, we obtain

$$(H_F + V_p)\psi_v = \epsilon_v \psi_v, \tag{9.5}$$

where V_p is the Phillips–Kleinman potential,

$$V_p = \sum_{i=1}^{N} \alpha_i \frac{(\epsilon_v - \epsilon_i)\varphi_i}{\psi_v}, \tag{9.6}$$

and

$$\alpha_i = \langle \varphi_i \mid \psi_v \rangle. \tag{9.7}$$

The detailed form of V_p is given by Eq. (A.11). The connection between the PO and the HF orbital is given by the relationship

$$\varphi_v = N_0 \left(\psi_v - \sum_{i=1}^{N} \alpha_i \varphi_i \right), \tag{9.8}$$

where

$$N_0 = \left(1 - \sum_{i=1}^{N} |\alpha_i|^2 \right)^{-1/2}. \tag{9.9}$$

It is evident from Eq. (9.6) that the exact pseudopotential depends on the quantum numbers n and l of the valence electron. Thus the total potential in Eq. (9.5) will be different for each valence state; the pseudo-orbitals belonging to different energy levels will not be orthogonal.

Taking a look at Eqs. (A.9) and (A.11), we see that both the exchange potential and the pseudopotential contain the term

$$\gamma(r) = \frac{\hat{P}_{n_i l_i}(r)}{P_{n_v l_v}(r)}. \tag{9.10}$$

In this expression, $\hat{P}_{n_i l_i}$ is the radial part of the core orbital φ_i, and $P_{n_v l_v}$ is the radial part of the pseudo-orbital ψ_v. These terms occur because we have changed the exchange and pseudopotential operators into local potentials. Since $P_{n_v l_v}$ will have nodes in the excited states, the presence of $\gamma(r)$ in the potentials presents a special problem.

McGinn solved Eq. (9.5) by numerical integration for a number of excited states of the atoms mentioned. The solution is an iteration, since the exchange potential depends on ψ_v and the pseudopotential depends on ϵ_v and ψ_v. The core orbitals were kept frozen throughout. The initial estimates were chosen as follows. The calculations have shown that the modified potential V_M, which is the sum of the HF potential and the pseudopotential, does not change much when one goes from n to $(n + 1)$ for the same l. Thus the initial estimate for the pseudo-orbital for (nl) was the self-consistent solution of the pseudopotential equation for $(n - 1, l)$. For the lowest excited states the ground state solutions of Appendix A were used. The empirical energy served as the initial input for ϵ_v. With this choice of input data, which enabled McGinn to proceed rapidly from the state (n, l) to the state $(n + 1, l)$, a fast convergence was achieved for all the states calculated.

The problem of discontinuities connected with Eq. (9.10) can be handled as follows. The lowest PO will have no nodes, since the pseudopotential replaces the orthogonality requirements with respect to the core orbitals. Similarly, the lowest HF valence orbital, with an azimuthal quantum number which does not occur in the core, will be nodeless. The lowest PO which will have a node will have it because the corresponding HF orbital is orthogonal to the HF orbital related to the lower PO. This means that the node of the lowest PO which has a node will be in the valence area where the bulk of the lower PO is located. In other words, the node will be well outside of the core. This is true for the lowest valence orbital with a node; it is true *a fortiori* for the higher valence levels.

Now let us look at Eq. (9.10). At the points where $P_{n_v l_v}$ is zero, the core orbitals will be extremely small. McGinn has shown that, since the interval length of the numerical integration increases with increasing r, the program will almost certainly avoid hitting the exact point in the r scale where γ goes to infinity; in fact, this has not happened in any of the calculations. Also, the area of singularity (the range of r in which γ is very large) appears to be extremely narrow due to the smallness of the core orbitals. By tabulating V_M in the area of one of the zero points of $P_{n_v l_v}$ (for the $5s$ state of the K atom) McGinn demonstrated that the presence of a zero point in the pseudo-orbital caused only a small "bump" in the modified potential.

In order to check whether the presence of the nodes in the pseudo-orbitals caused inaccuracies in the calculations, McGinn recalculated a number of states of the Na, K, Mg^+, and Zn^+ atoms with an alternative procedure in which the operators were not changed into potentials. We can rewrite Eq. (9.5) in the form

$$\left\{ -\frac{1}{2}\Delta - \frac{Z}{r} + V(r) \right\} \psi_v = \epsilon_v \psi_v - f(r), \qquad (9.11)$$

where $V(r)$ is the electrostatic potential of the core, and

$$f(1) = \sum_{i=1}^{N} \left(-\int \frac{\varphi_i(1)\varphi_i^*(2)\psi_v(2)}{r_{12}} \, dq_2 + \alpha_i(\epsilon_v - \epsilon_i)\varphi_i(1) \right). \qquad (9.12)$$

Eq. (9.11) is exactly the same as Eq. (9.5), the first term of f being the exchange operator and the second the PK operator. The solution of Eq. (9.11) can be carried out with an iterative procedure. Since this equation is inhomogeneous, the solution must be made self-consistent not only in the pseudo-orbital but also in the normalization of the pseudo-orbital. Thus Eq. (9.11) is more complicated to solve than Eq. (9.5). McGinn has shown that the eigenvalues obtained from Eq. (9.5) differed from the eigenvalues of Eq. (9.11) at most only in the fourth significant figure of the orbital energies.

The accuracy of the calculations was further checked by using Eq. (9.4). Having obtained the normalized, self-consistent pseudo-orbitals for a number of states, the corresponding HF orbitals were computed from Eq. (9.8). It was shown that the computed HF orbitals satisfied the orthogonality relationship, Eq. (9.4), very accurately.

Thus the calculations of McGinn can be viewed as highly accurate, exact, pseudopotential calculations, on a par with HF calculations, made in the frozen core (positive ion) approximation. We turn now to the evaluation of the results.

The energy levels obtained in the calculations are presented in Tables 9.1 through 9.5. In the first four tables we find the lowest six ^2S states, the lowest six ^2P states, and the lowest five or six ^2D states for the atoms mentioned above. In Table 9.5 we find the lowest four ^2F, ^2G, and ^2H levels for the atoms Al^{++}, Ca^+, and Rb. In the tables we have the absolute values of the energy levels in atomic units. On the left side we have the principal quantum number of the valence state. The calculated energies are compared with the empirical in such a way that the empirical value is the average of the doublet. This comparison is meaningful because in most, if not in all, cases, the doublet separation is small relative to the difference between the calculated and empirical values. The comparison is less meaningful, for example, in some of the ^2P states of Cu and Zn^+, where, in the higher states, the doublet separation is not negligibly small relative to the error of the calculations. Also, in some of these terms the spin-orbit splitting is inverted, meaning that the $j = 3/2$ level is lower than the $j = 1/2$. This is the reason for the anomalous results obtained in comparing these levels with the simple average of the empirical doublet. Apart from this, the comparison between the calculated and empirical levels is meaningful, keeping in mind that the calculations were made with the frozen core approximation. If the frozen core approximation were removed, some of the lower lying states would come out slightly lower.

In the tables the difference between the calculated and empirical values is given in percents, in the column headed by "Dif.". According to the discussions in Section 2.5.D, this difference is equal to the combined effects of the intra-core and core-valence correlations. As we see from the tables, the correlation effects are fairly large even in the excited states. Predictably, the correlation effects increase with increasing core size, showing the predominance of the core polarization. For the Rb atom the correlation effects amount to about 10% of the binding energy of the valence electron in the (5s) state and to 8% in the (4d) state. For the Cu atom correlation is about 18% of the (4s) state and 15% of the (4p) state. For the excited states the

TABLE 9.1. Excited States of the Atoms Li, Na, and K

Atom	n	S states			P states			D states		
		Calc.	Obs.	Dif.	Calc.	Obs.	Dif.	Calc.	Obs.	Dif.
Li	2	0.19657	0.19816	0.81	0.12863	0.13025	1.24			
	3	0.07384	0.07419	0.47	0.05677	0.05724	0.82	0.05556	0.05561	0.09
	4	0.03849	0.03862	0.37	0.03178	0.03198	0.63	0.03125	0.03128	0.10
	5	0.02358	0.02364	0.25	0.02028	0.02038	0.49	0.02000	0.02001	0.05
	6	0.01591	0.01595	0.25	0.01405	0.01411	0.43	0.01389	0.01390	0.07
	7	0.01146	0.01148	0.17	0.01031	0.01034	0.29	0.01020	0.01021	0.10
Na	3	0.18108	0.18886	4.12	0.10940	0.11156	1.94	0.05567	0.05594	0.48
	4	0.07002	0.07158	2.18	0.05031	0.05094	1.24	0.03132	0.03144	0.38
	5	0.03701	0.03759	1.54	0.02893	0.02920	0.92	0.02004	0.02011	0.35
	6	0.02286	0.02313	1.16	0.01878	0.01892	0.74	0.01391	0.01395	0.29
	7	0.01551	0.01566	0.96	0.01317	0.01325	0.60	0.01022	0.01025	0.29
	8	0.01121	0.01131	0.88	0.00975	0.00980	0.51	0.00782	0.00782	0.26
K	3							0.05812	0.06139	5.33
	4	0.14669	0.15952	8.04	0.09547	0.10022	4.74	0.03278	0.03468	5.22
	5	0.06090	0.06371	4.41	0.04554	0.04693	2.96	0.02096	0.02198	4.64
	6	0.03336	0.03444	3.14	0.02676	0.02737	2.23	0.01449	0.01510	4.04
	7	0.02105	0.02158	2.46	0.01762	0.01794	1.78	0.01060	0.01099	3.55
	8	0.01448	0.01478	2.03	0.01248	0.01267	1.50	0.00808	0.00834	3.12
	9	0.01057	0.01076	1.77	0.00931	0.00943	1.27			

TABLE 9.2. Excited States of the Atoms Rb, Cu, and Be⁺

Atom	n	S states Calc.	S states Obs.	S states Dif.	P states Calc.	P states Obs.	P states Dif.	D states Calc.	D states Obs.	D states Dif.
Rb	4							0.06007	0.06532	8.04
	5	0.13794	0.15351	10.14	0.09026	0.09565	5.64	0.03398	0.03640	6.65
	6	0.05832	0.06177	5.59	0.04369	0.04528	3.51	0.02157	0.02279	5.35
	7	0.03228	0.03362	3.99	0.02590	0.02660	2.63	0.01485	0.01554	4.44
	8	0.02050	0.02116	3.12	0.01716	0.01753	2.11	0.01083	0.01125	3.73
	9	0.01417	0.01454	2.54	0.01220	0.01242	1.77	0.00824	0.00852	3.29
	10	0.01038	0.01061	2.17	0.00912	0.00927	1.62			
Cu	4	0.23389	0.28394	17.63	0.12226	0.14424	15.24	0.05512	0.05640	2.27
	5	0.08007	0.08739	8.38	0.05419	0.05893	8.04	0.03100	0.03157	1.81
	6	0.04066	0.04314	5.75	0.03059	0.03377	9.42	0.01986	0.02015	1.44
	7	0.02458	0.02572	4.43	0.01965	0.02111	6.92	0.01380	0.01397	1.22
	8	0.01646	0.01708	3.63	0.01368	0.01375	0.51	0.01015	0.01026	1.07
	9	0.01179	0.01216	3.04	0.01007	0.01020	1.27	0.00777	0.00785	1.02
Be⁺	2	0.66484	0.66928	0.66	0.51941	0.52378	0.83			
	3	0.26647	0.26725	0.29	0.22841	0.22958	0.51	0.22229	0.22249	0.09
	4	0.14288	0.14315	0.19	0.12766	0.12814	0.37	0.12504	0.12512	0.06
	5	0.08894	0.08906	0.13	0.08137	0.08162	0.31	0.08002	0.08006	0.05
	6	0.06065	0.06071	0.10	0.05635	0.05648	0.23	0.05557	0.05559	0.04
	7	0.04399	0.04400	0.02	0.04132	0.04140	0.19	0.04083	0.04084	0.02

TABLE 9.3. Excited States of the Atoms Mg$^+$ and Ca$^+$

Atom	n	S states Calc.	S states Obs.	S states Dif.	P states Calc.	P states Obs.	P states Dif.	D states Calc.	D states Obs.	D states Dif.
Mg$^+$	3	0.54851	0.55255	0.73	0.38338	0.38981	1.65	0.22482	0.22680	0.87
	4	0.23334	0.23448	0.49	0.18322	0.18513	1.03	0.12648	0.12738	0.71
	5	0.12936	0.12977	0.32	0.10763	0.10847	0.77	0.08085	0.08132	0.58
	6	0.08215	0.08233	0.22	0.07082	0.07127	0.63	0.05607	0.05635	0.50
	7	0.05678	0.05688	0.18	0.05014			0.04115	0.04132	0.41
	8	0.04158	0.04164	0.14	0.03736					
Ca$^+$	3							0.33249	0.37393	11.08
	4	0.41386	0.43627	5.14	0.30994	0.32098	3.44	0.16908	0.17724	4.60
	5	0.19234	0.19857	3.14	0.15673	0.16027	2.21	0.10149	0.10490	3.25
	6	0.11159	0.11423	2.31	0.09519	0.09680	1.66	0.06759	0.06936	2.55
	7	0.07291	0.07426	1.82	0.06402			0.04823	0.04926	2.09
	8	0.05137	0.05215	1.50	0.04602			0.03613	0.03678	1.77
	9	0.03814			0.03468					

TABLE 9.4. Excited States of the Atoms Zn^+ and Al^{++}

Atom	n	S states			P states			D states		
		Calc.	Obs.	Dif.	Calc.	Obs.	Dif.	Calc.	Obs.	Dif.
Zn^+	4	0.60984	0.66018	7.63	0.40600	0.43729	7.16	0.21372	0.21851	2.19
	5	0.24628	0.25722	4.25	0.19015	0.19776	3.85	0.12063	0.12261	1.61
	6	0.13425	0.13848	3.05	0.11062	0.11376	2.76	0.07759	0.07862	1.31
	7	0.08454	0.08662	2.40	0.07239	0.07500	3.48	0.05411	0.05471	1.10
	8	0.05812	0.05930	1.99	0.05106	0.05051	-1.09	0.03989	0.04026	0.92
	9	0.04241	0.04314	1.69	0.03795			0.03062	0.03087	0.81
Al^{++}	3	1.04240	1.04549	0.30	0.79002	0.80035	1.29	0.51201	0.51714	0.99
	4	0.46907	0.47064	0.33	0.38760	0.39086	0.83	0.28785	0.29010	0.78
	5	0.26733	0.26800	0.25	0.23090	0.23240	0.65	0.18371	0.18489	0.64
	6	0.17267	0.17303	0.21	0.15329	0.15411	0.53	0.12725	0.12794	0.54
	7	0.12071	0.12097	0.21	0.10918	0.10978	0.55	0.09329	0.09374	0.48
	8	0.08912			0.08171					

TABLE 9.5. Excited States of the Atoms Al^{++}, Ca^{+}, and Rb

State	Al^{++}		Ca^{+}		Rb	
	Calc.	Obs.	Calc.	Obs.	Calc.	Obs.
4f	0.28132	0.28174	0.12518	0.12617	0.03126	0.03143
5f	0.18006	0.18034	0.08014	0.08074	0.02001	0.02011
6f	0.12504	0.12523	0.05566	0.05600	0.01389	0.01396
7f	0.09187	0.09201	0.04089	0.04110	0.01021	0.01025
5g	0.18000	0.18010	0.08000	0.08012	0.02000	0.02002
6g	0.12500	0.12509	0.05556	0.05562	0.01389	0.01392
7g	0.09184	0.09191	0.04082	0.04086	0.01020	
8g	0.07031	0.07036	0.03125	0.03128	0.00781	
6h	0.12500	0.12506	0.05556		0.01389	0.01389
7h	0.09184	0.09188	0.04082		0.01020	
8h	0.07031	0.07034	0.03125		0.00781	
9h	0.05556	0.05558	0.02469		0.00617	

correlation declines, but only slowly; for the Rb atom (10s) state it is still 2%. Thus we can conclude that the correlation effects between core and valence electron are generally not small, not even in the excited states.

In Table 9.5 we have some of the F, G, and H states. In these states the pseudopotential is zero, since none of the cores contain any f, g or h electrons. Thus these are conventional HF Values. As we see from the results, the correlation effects are negligibly small in these states.

We now have a clear-cut answer to the question of the size of the core-valence correlation effects. It is interesting to recall our discussion of the size of electron correlation in the binding energy of molecules. As we pointed out in Section 8.5, the molecular calculations show that accurate results for the binding energies of molecules can be obtained only if the correlation effects are taken into account. We see now that this is true even for atoms with one valence electron; accurate results for the binding energy (ionization potential) of a single valence electron in the field of an atomic core can be obtained only if the correlation effects are properly taken into account. Thus we can conclude that in any "valence only" effect, such as the binding of Class I molecules or the binding of one or more valence electrons in an atom, the correlation effects are important.

The calculations show that the modified potential is independent of the principal quantum number to a very high degree. Since the exchange potential is small relative to the electrostatic potential, this means that the pseudopotential is n-independent to a high degree. It was shown that using the modified potential containing the self-consistent pseudo-orbital for (n, l), one could start the calculation right away for the state $(n + 1, l)$. This clearly means that the modified potentials for (n, l) and $(n + 1, l)$ are very similar. The n-independence of the pseudopotential, which was assumed to be valid in most of pseudopotential theory, is demonstrated again in these calculations.

The calculations also show that, at least for the atoms investigated, the modified potential can be transformed exactly into local potentials for every n and l. This statement is shown to be true despite the fact that the valence electron orbitals have nodes in the excited states. We saw in Section 8.3 that the derivation of the effective molecular Hamiltonian rested on the assumption that the nonlocal interaction potentials between core and binding electrons can be transformed exactly into local potentials. The proof of this assumption—at least for those one-valence-electron atoms which were investigated by McGinn—constitutes a satisfactory piece of evidence for the correctness of the derivation presented in Section 8.3.

In addition to the energy values, McGinn calculated the integrals needed for the calculation of oscillator strengths. These integrals are of the form [Slater II, p. 222],

$$A(n, l, n', l + 1) = [(2l + 1)(2l + 3)]^{-1/2} \int_0^\infty r \hat{P}_{nl}(r) \hat{P}_{n'l+1}(r) \, dr, \qquad (9.13)$$

where $\hat{P}_{nl}$ is the HF orbital formed from the pseudo-orbital P_{nl}, using Eq. (9.8). The integrals were calculated for all allowed transitions between those S, P, and D states for which energy calculations were made.

Finally, it was noted by McGinn that the calculations were very simple; it took about one hour on a 1969-vintage small computer to obtain an excited state. Today such calculations could easily be done on a microcomputer.

9.3. CALCULATIONS FOR ATOMS WITH TWO VALENCE ELECTRONS

Pseudopotential theory was applied to atoms with two valence electrons by McGinn.[100] Since these calculations are completely analogous to the work presented in the preceding section, here we restrict ourselves to a short summary. Calculations were carried out for the $^1S(ns)^2$ ground states of the Li^-, Na^-, K^-, Rb^-, Cu^-, Ag^-, Be, Mg, Ca, Zn, and Al^+ atoms. Further, calculations were done for the $(nsml)$ $^{1,3}S$ and, $^{1,3}P$ states of Be and Ca and for the $^{1,3}D$ states of the Ca atom. Finally the doubly excited (npms) $^{1,3}P$ states of the atoms Be, Mg, Al^+, Si^{2+}, Ca, and Zn were investigated. In each case, McGinn reduced the problem to a one-electron equation in which the Pauli exclusion principle was represented by a PK-type potential. In the case of the (nsml) Rydberg states this reduction was accomplished by freezing not only the core orbitals but also the orbital which remains in the ground state.[101] In the doubly excited states one of the orbitals was frozen into the state of the (possibly excited) valence orbital of the $(N + 1)$-electron atom. The results were compared with the HF calculations as well as with empirical values. For the Rydberg states the comparison was made with the HF calculations of Froese.[102] In this case, the agreement was good in the lower Rydberg states and very good in the higher. Generally, the results were in line with the preceding section; that is, they were as good as can be expected in calculations in which correlation effects are neglected.

9.4. THE RYDBERG STATES OF ATOMS AND MOLECULES: THE METHOD OF HAZI AND RICE

A pseudopotential theory for the Rydberg states of atoms and molecules was developed by Hazi and Rice.[103] The starting point of the theory is the definition of a wave function for the atom or molecule under discussion. Let the number of electrons in the system be $2N$. From these $2N$ electrons $2N - 2$ are in the core, which has closed shells. The two remaining electrons are the valence electrons, which form a singlet or triplet. One of the valence electrons remains fixed in the ground state or in a low-lying state; the other valence electron, the Rydberg electron, moves into a higher excited state. If the Rydberg electron is removed, the remaining $(2N - 1)$-electron system forms a doublet.

Let $\varphi_1, \varphi_2, \ldots, \varphi_{N-1}$ be the spatial orbitals for the core, and let us form $(2N - 2)$ spin-orbitals by indicating the up spin by φ_1 and the down spin by $\bar{\varphi}_1$. Then the $(2N - 2)$ occupied core orbitals will be $\varphi_1, \bar{\varphi}_1, \varphi_2, \bar{\varphi}_2 \cdots \varphi_{N-1}, \bar{\varphi}_{N-1}$. We denote the spatial part of the valence orbital, which will remain fixed, by φ_N, and the Rydberg orbital by φ_R. The total wave function for the $2N$-electron system will be

$$\Psi_T = (2)^{-1/2}\{\tilde{A}[\varphi_1(1)\bar{\varphi}_1(2)\cdots\bar{\varphi}_{N-1}(2N-2)\varphi_N(2N-1)\bar{\varphi}_R(2N)]$$
$$\mp \tilde{A}[\varphi_1(1)\bar{\varphi}_1(2)\cdots\bar{\varphi}_{N-1}(2N-2)\bar{\varphi}_N(2N-1)\varphi_R(2N)]\}, \qquad (9.14)$$

where $\tilde{A}$ is an antisymmetrizer operator. The upper sign is for the singlet, the lower for the triplet state. We assume that all orbitals, including the Rydberg orbital φ_R, are orthonormal,

$$\langle \varphi_i \mid \varphi_j \rangle = \delta_{ij}, \quad (i, j = 1, 2, \ldots, N), \qquad (9.15)$$

and

$$\langle \varphi_i \mid \varphi_R \rangle = 0, \quad (i = 1, 2, \ldots, N). \qquad (9.16)$$

Next the expectation value of the Hamiltonian of the system is formed with the total wave function given by Eq. (9.14). The resulting energy expression is varied with respect to the Rydberg orbital φ_R^*. In the variation, the core orbitals $\varphi_1 \cdots \varphi_{N-1}$, as well as the frozen valence electron orbital φ_N, remain fixed. The subsidiary conditions are Eq. (9.16) and the normalization of the Rydberg orbital. The result is the HF equation of the Rydberg orbital,

$$H_R \varphi_R + \sum_{i=1}^{N} \lambda_i \varphi_i = \epsilon_R \varphi_R , \tag{9.17}$$

where the spin-less, HF Hamiltonian operator is given by

$$H_R = t + g + \sum_{i=1}^{N-1} (2J_i - K_i) + J_N \pm K_N , \tag{9.18}$$

where

$$J_i(1) = \int \frac{\varphi_i(2) \, dv_2}{r_{12}} , \tag{9.19}$$

and

$$K_i(1)f(1) = \int \frac{\varphi_i(1)\varphi_i^*(2)f(2) \, dv_2}{r_{12}} . \tag{9.20}$$

In the last term of Eq. (9.18) the upper sign is for the singlet, the lower for the triplet state.

In the following discussion we shall refer to the inner N electrons as the core electrons; that is, we shall call core electrons those which are actually in the core plus the frozen valence electron. We have not yet made any assumption about the nature of the core orbitals. Hazi and Rice mention two possibilities for these orbitals:

1. The core orbitals may be chosen to be the orbitals of the $(2N - 1)$-electron positive ion, or
2. The core orbitals may be chosen to be the orbitals of the ground state of the neutral, $2N$-electron atom or molecule.

Clearly, these two choices are what we call the positive ion and neutral atom approximations. The reader will observe that in the work of McGinn a slightly different form of the positive ion approximation is used.

We proceed now to the pseudopotential transformation of the HF equation [Eq. (9.17)]. First we want to incorporate the Lagrangian multipliers into the Hamiltonian, in order to get a homogeneous equation. Multiplying Eq. (9.17) from the left by φ_k^*, ($k = 1 \cdots N$), and integrating, we obtain

$$\lambda_i = -\langle \varphi_i | H_R | \varphi_R \rangle . \tag{9.21}$$

Using this expression, Eq. (9.17) becomes

$$\hat{H}_R \varphi_R = \epsilon_R \varphi_R, \tag{9.22}$$

where

$$\hat{H}_R = PH_R, \tag{9.23}$$

with

$$P = 1 - \Omega = 1 - \sum_{i=1}^{N} |\varphi_i\rangle\langle\varphi_i|. \tag{9.24}$$

It is easy to see that the eigenfunctions of Eq. (9.22) will be orthogonal to the core orbitals. Thus we can write φ_R in the form

$$\varphi_R = N_R \left(\psi_R - \sum_{i=1}^{N} \alpha_i \varphi_i \right), \tag{9.25}$$

where N_R is a normalization constant, ψ_R is the pseudo-orbital, and

$$\alpha_i = \langle \varphi_i | \psi_R \rangle. \tag{9.26}$$

On using Eq. (9.25) in Eq. (9.22) we obtain

$$\hat{H}_R \psi_R - \sum_{i=1}^{N} \alpha_i \hat{H}_R \varphi_i = \epsilon_R \psi_R - \epsilon_R \sum_{i=1}^{N} \alpha_i \varphi_i, \tag{9.27}$$

from which we obtain the equation for the pseudo-orbital

$$(\hat{H}_R + V_p)\psi_R = \epsilon_R \psi_R, \tag{9.28}$$

where

$$V_p \psi_R = \sum_{i=1}^{N} \alpha_i (\epsilon_R - \hat{H}_R) \varphi_i. \tag{9.29}$$

This can be further simplified if we introduce the operator H_c, which is defined in such a way that the core orbitals are eigenfunctions of this operator. Let

$$H_c \varphi_i = \epsilon_i \varphi_i, \quad (i = 1, 2, \ldots, N), \tag{9.30}$$

and let

$$\Delta = \hat{H}_R - H_c. \tag{9.31}$$

Then we obtain

$$V_p \psi_R = \sum_{i=1}^{N} \alpha_i (\epsilon_R - H_c - \Delta) \varphi_i = \sum_{i=1}^{N} \alpha_i (\epsilon_R - \epsilon_i - \Delta) \varphi_i . \tag{9.32}$$

Changing this operator into a local potential we obtain

$$V_p = \sum_{i=1}^{N} \frac{\alpha_i (\epsilon_R - \epsilon_i - \Delta) \varphi_i}{\psi_R} . \tag{9.33}$$

As we see from this formula, the first two terms are the Phillips–Kleinman potential. The term with Δ is the result of the core functions being the eigenfunctions of the operator H_c, which is different from the Hamiltonian of the Rydberg orbital, H_R. If the core orbitals were also the eigenfunctions of H_R, then we would get $\Delta = 0$, and V_p would become identical with the PK potential. We note that, in practical calculations, the Δ term can be simplified greatly because the operators H_R and H_c will differ only in a few terms.

Here we make a short digression to discuss briefly the pseudopotential given by Eq. (9.33). This is a potential for the case when the valence orbital (the Rydberg orbital) and the core orbitals are eigenfunctions of different operators. We have seen in Section 2.4 that the most general pseudopotential for uncorrelated core was given by the expressions in Eq. (2.115). In those expressions the core orbitals were completely arbitrary. The potential in Eq. (9.33) is somewhat less general, since in Eq. (9.33) it is assumed that the core orbitals are eigenfunctions of the operator H_c. In Eq. (2.115) it is not assumed that the core orbitals are eigenfunctions of any operator.

Hazi and Rice point out that Eq. (9.33) can be generalized, just as the Phillips–Kleinman potential was generalized in Eq. (2.79). Consider the operator

$$V_p = \sum_{i=1}^{N} \{ |\varphi_i\rangle\langle F_i| - \Delta |\varphi_i\rangle\langle\varphi_i| \} , \tag{9.34}$$

where the functions F_i are arbitrary. Hazi and Rice have shown that the eigenvalues of the wave equation

$$(\hat{H}_R + V_p)\psi = \epsilon\psi , \tag{9.35}$$

where V_p is given by Eq. (9.34), will be identical with the eigenvalues of Eq. (9.22), for any set of F_i functions. Thus Eq. (9.34) is the straightforward generalization of Eq. (2.79). This statement is valid if Δ has the form of Eq. (9.31) and the core orbitals satisfy Eq. (9.30). The potential operator in Eq. (9.32) is, of course, a special case of Eq. (9.34). In order to show this, we rewrite Eq. (9.32) as follows:

$$V_p \psi_R = \sum_{i=1}^{N} (\epsilon_R - \epsilon_i - \Delta)|\varphi_i\rangle\langle\varphi_i|\psi_R\rangle, \tag{9.36}$$

from which we see that

$$V_p = \sum_{i=1}^{N} (\epsilon_R - \epsilon_i - \Delta)|\varphi_i\rangle\langle\varphi_i|, \tag{9.37}$$

and comparing this with Eq. (9.34), we see that we obtain Eq. (9.37) from Eq. (9.34) by putting

$$\langle F_i| = (\epsilon_R - \epsilon_i)\langle\varphi_i|. \tag{9.38}$$

Hazi and Rice carried out test calculations, using the formalism above, for the triplet Rydberg states of the He and Be atoms. For the He atom, the core orbital used was the He^+ ground state wave function, which is known exactly. For the Be atom, the core orbitals were the analytic HF orbitals for the Be^+ atom in the $(1s)^2(2s)$ configuration. For the Rydberg orbitals, analytic expressions were constructed in terms of Slater functions, and the expansion coefficients determined in such a way that the orbitals were approximate solutions of Eq. (9.28). Here we quote some of the results for the absolute values of the energies of the Rydberg electron. The energies are in cm^{-1} units, with the empirical values given in parentheses. The results for the ^3S states of the He atom were: $E(1s, 2s) = 38$, 242 $(38,455)$; $E(1s, 3s) = 15,031$, $(15,074)$; $E(1s, 4s) = 7,997$ $(8,013)$; $E(1s, 5s) = 4,956$ $(4,964)$; $E(1s, 6s) = 3,370$ $(3,375)$. The results obtained for the Be atom's ^3S Rydberg states were as follows: $E(2s, 3s) = 22,010$ $(23,110)$; $E(2s, 4s) = 10,369$ $(10,685)$; $E(2s, 5s) = 6,046$ $(6,183)$; $E(2s, 6s) = 3,960$ $(4,031)$.

The reasonably good agreement between the computed and empirical values shows that the formalism is well suited for the treatment of Rydberg states. We have to point out, however, that just as in the case of McGinn's work, the correlation effects were neglected in the formalism of Hazi and Rice. We recognize the limitations of this method if we recall what we have said above about the importance of correlation effects.

Motivated by the desire to create a formalism which would be simpler than the method described above, Hazi and Rice have developed an alternative method in which model potentials were used.[38] We saw in Chapter 4 and in Section 7.2 how the model formulation of pseudopotential theory can be developed for atoms with one or more valence electrons. In Chapter 8 we saw how these atomic model potentials can be used in the effective Hamiltonian of Class I molecules. In the discussion which is to follow, we present the model formulation of Hazi and Rice developed for the Rydberg states of molecules in general, not just for the Rydberg states of Class I molecules.

Let us consider a diatomic molecule of *AB* type. The generalization for an arbitrary molecule will be straightforward. The molecule does not need to belong to the group that we have called Class I above. Thus there is no need to assume the presence of atomic cores in the molecule. Our goal is to develop a model formalism for the treatment of Rydberg states.

As we have seen, a pseudopotential formalism based on the HF method can be developed. The end result of that formalism is Eq. (9.28):

$$(\hat{H}_R + V_p)\psi = \epsilon\psi, \tag{9.39}$$

where ψ and ϵ were the pseudo-orbital and orbital parameter of the Rydberg electron. The potentials in $\hat{H}_R$ and V_p represented the interaction potential of the Rydberg electron with the core electrons. We note again that Eq. (9.39) is a one-electron equation which is valid for molecules as well as as for atoms.

We replace now Eq. (9.39) by the model equation

$$(t + V_m)\psi = \epsilon\psi, \tag{9.40}$$

where V_m is supposed to replace the complete potential of the core. The pseudopotential is supposed to be included in V_m. The next step is to set up V_m on the basis of physical plausibility and comparison with Eq. (9.39).

We have seen in Section 4.2 the guidelines for the construction of model potentials for atoms. The guidelines for the construction of the model potentials in Eq. (9.40) will be different to a certain extent, since here we are dealing with molecules which do not need to contain atomic cores. The first requirement for V_m is, of course, the assumption that the eigenvalues of Eq. (9.40) should match the eigenvalues of Eq. (9.39).

The term values of Rydberg states, in atomic units, are given by the formula[104]

$$I = \frac{1}{2}\frac{Z_c^2}{(n-\delta)^2}, \tag{9.41}$$

where Z_c is the charge of the core, n is the principal quantum number, and δ is the quantum defect. For neutral molecules, $Z_c = 1$. This formula suggests that the Rydberg states are essentially hydrogenic in character; that is, the Rydberg electron moves in the field of a monopole. Thus the potential V_m should have the form $(-Z_c/r)$ at large distances from the molecule (r being the distance from the center of the molecule).

It is clear from the form of the potentials in Eq. (9.39) that V_m must have the correct molecular spatial symmetry. Also, the presence of exchange operators in the potentials of H_R shows that Rydberg states with the same spatial symmetry but with different spin angular momentum will have different energies. Thus V_m will also depend on the spin of the Rydberg state. Finally, in Eq. (9.39), we have the pseudopotential V_p. This potential replaces the Pauli exclusion principle with respect to the core; and it is evident that V_m must have such a structure so as to prevent the Rydberg electron from falling into the core.

We can now formulate the requirements which must be satisfied by the model potential V_m. Besides the self-evident requirement of mathematical simplicity, the model should have the following properties:

1. The asymptotic behavior should be such that,

$$\lim_{r \to \infty} V_m = -\frac{Z_c}{r},$$ (9.42)

where r is the distance from the center of the molecule;

2. V_m should have the correct spatial and spin symmetry;
3. V_m should replace effectively the Pauli exclusion principle.

In addition to these requirements Hazi and Rice introduced the idea that, even though the molecule under discussion does not have atomic cores, the potential V_m is a sum of "atomic" contributions. Thus we put, for an AB type diatomic molecule,

$$V_m = V_A + V_B,$$ (9.43)

where V_A and V_B are model operators centered on nuclei A and B. The model potential [Eq. (9.43)] still has to satisfy the three requirements above. We assume that V_A and V_B will contain adjustable parameters which can be determined from the requirement that the model equation should match the empirical Rydberg spectrum of the molecule.

The general guidelines described here permit the use of many different types of model potentials. Hazi and Rice considered several types, from which we present the one that appears to be most consistent with the theory and has given good results in the calculations. Hazi and Rice adopted for V_A and V_B the model potential suggested by Abarenkov and Heine which is No. 1 in Table 4.3. Thus we put

$$\left.\begin{aligned} V_A &= -\frac{Z_A}{r_A}, & r_A \geq r_{A0}\,; \\ V_A &= C_A, & r_A < r_{A0}\,; \end{aligned}\right\}$$ (9.44)

where Z_A is the (fictitious) partial charge of center A, r_A is the distance of the electron from center A, and C_A and r_{A0} are constants. We have the analogous equation for center B. The partial charges satisfy the condition

$$Z_A + Z_B = Z_C.$$ (9.45)

The model Hamiltonian of the Rydberg electron will be

$$H = t + V_A + V_B, \tag{9.46}$$

and the equation for the Rydberg orbital

$$(t + V_A + V_B)\psi = \epsilon\psi. \tag{9.47}$$

This equation is a one-electron, molecular equation. As we see from Eq. (9.44), the model satisfies conditions 1 and 3. In order to satisfy condition 2, Hazi and Rice made the constants C_A, r_{A0}, C_B, r_{B0} dependent on the spatial symmetry and spin of the Rydberg state; that is, there are different sets for Rydberg states with different spatial symmetry and different spin. The constants for each state were determined by first choosing a physically reasonable r_{A0} and r_{B0} and then determining C_A and C_B from the requirement that the computed term values of higher Rydberg states should agree reasonably well with the empirical values.

Test calculations were carried out for the $^3\Pi_u$ states of H_2 at an internuclear distance $R = 2.0$ a.u. For this state, the constants of the model potential were

$$r_{A0} = r_{B0} = 1.5 \text{ a.u.},$$
$$C_A = C_B = -0.85 \text{ a.u.},$$
$$Z_A = Z_B = 1/2.$$

The approximate calculations for the solution of Eq. (9.47) were carried out using variation method. The trial function was expanded in terms of STO's centered on the molecular midpoint. The energy matrix was diagonalized to obtain the energy of the Rydberg electron plus the expansion coefficients of the trial function. The resulting energy terms along with the experimental values are given in Table 9.6.

Calculations were also carried out for the $^1\Sigma_u$ states of the N_2 molecule at the internuclear distance $R = 2.113$ a.u. This is the empirical internuclear distance of the $^2\Sigma_g$ state of N_2^+. It can be assumed that the Rydberg electron will have a negligible effect on the binding of the nuclei; therefore, the molecular constants of the Rydberg states are those found in the parent positive-ion state. The constants of the model potential were

$$r_{A0} = r_{B0} = 2.5 \text{ a.u.},$$
$$C_A = C_B = 0.125 \text{ a.u.},$$
$$Z_A = Z_b = \frac{1}{2}.$$

TABLE 9.6.   Energies of Some Rydberg States of H$_2$ and N$_2$ in eV Units

Rydberg state	H$_2$	
	Calc.	Obs.
$2p\pi$	3.566	3.689
$3p\pi$	1.571	1.588
$4p\pi$	0.877	0.885
$5p\pi$	0.558	0.565
$6p\pi$	0.387	0.388
	N$_2$	
$3p\sigma$	2.567	2.648
$4p\sigma$	1.223	1.253
$5p\sigma$	0.720	0.735
$6p\sigma$	0.475	0.481

The resulting energy terms, along with the experimental values, are also given in Table 9.6.

Comparing the calculated results with the empirical, Hazi and Rice concluded that the model equation reproduces the empirical Rydberg states reasonably well. Further, the conclusion was reached that the model potential does not depend on the principal quantum number of the Rydberg states. In this respect, the situation is very similar to atoms with one valuence electron. We have seen in Section 9.2 that the modified potential of the valence electron of atoms with one valence electron was n-independent in a very good approximation. We see now that a similar statement can be made for the Rydberg states of molecules.

The most important feature of the method developed by Hazi and Rice is its applicability to molecules which do not contain atomic cores. As we see from the results, the idea of building up the model potential from atomic contributions has proved to be useful, even in the case of molecules which do not exhibit atomic structure. Thus, at least for the Rydberg states, it can be generally assumed that the molecular model potential will be a sum of atomic contributions.

9.5.   FURTHER CONTRIBUTIONS TO THE THEORY OF EXCITED STATES

We have seen that the Rydberg states of atoms and molecules can be accurately treated by pseudopotential theory. In the treatment of these states the theory was applied in its one-electron formulation. For a general treatment of excited states (i.e., for the case when two or more valence electrons move out of their ground states into higher orbitals), the model Hamiltonian [Eq. (8.92)] can be used. In the derivation of the effective molecular Hamiltonian [Eq. (8.92)], there was nothing which would prohibit its use for excited states.

The procedure of using Eq. (8.92) for excited states was tested first by Melius, Goddard, and Kahn.[71] This work is the counterpart of the calculations of Kahn and Goddard[70] discussed in Section 8.2. In those calculations the molecular effective Hamiltonian was tested for the ground states of molecules. The work we shall discuss here is a test of the molecular effective Hamiltonian for excited states.

MGK carried out calculations for the first four $^1\Sigma^+$ and $^3\Sigma^+$ excited states and the first three $^1\Pi$ and $^3\Pi$ states of LiH. Two sets of calculations were performed. The first set consisted of ab initio all-electron calculations using the G1 method of Goddard (Section 2.5.C). The second set of calculations was based on the molecular effective Hamiltonian defined by Eqs. (8.7) and (8.8). In both sets of calculations, approximate solutions of the respective wave equations were obtained. The goal was to compare the results obtained with the effective Hamiltonian, Eq. (8.7), with the results of the full, all-electron calculations.

Comparing the two sets of results, Melius, Goddard, and Kahn concluded that the results of the pseudopotential calculations agreed well with the ab initio values. We quote here some of the total energies obtained in the pseudopotential calculations, along with the all-electron, ab initio results in parentheses (all energies in atomic units):

$$E(1\ ^1\Sigma^+) = -8.01377, (-8.01605);$$
$$E(2\ ^1\Sigma^+) = -7.89922, (-7.89982);$$
$$E(3\ ^1\Sigma^+) = -7.81868, (-7.81839);$$
$$E(4\ ^1\Sigma^+) = -7.80466, (-7.80486);$$
$$E(5\ ^1\Sigma^+) = -7.79875, (-7.79907);$$
$$E(1\ ^1\Pi) = -7.86076, (-7.86165);$$
$$E(2\ ^1\Pi) = -7.79549, (-7.79605);$$
$$E(3\ ^1\Pi) = -7.79314, (-7.79374);$$
$$E(1\ ^3\Pi) = -7.87483, (-7.87555);$$
$$E(2\ ^3\Pi) = -7.79814, (-7.79866);$$
$$E(3\ ^3\Pi) = -7.79594, (-7.79647).$$

The pseudopotential total energies were obtained by adding the energy of the Li atomic core to the energy of the two binding electrons.

The excellent agreement between the values obtained from the pseudopotential calculations and the values obtained from the ab initio calculations shows that the molecular effective Hamiltonian, in which the Pauli exclusion principle is replaced by the pseudopotential, can be used for the accurate calculation of molecular excited states.

Another approach for the treatement of atomic and molecular excited states was proposed by Tully.[105] Let H be the Hamiltonian of the excited states which we want to investigate. In the previous discussions we have replaced the orthogonality requirement with respect to the core by pseudo-potentials. Tully proposed that the orthogonality requirement between a highly excited state and the lower-lying states should also be replaced by pseudopotentials.

Let $\varphi_1 \cdots \varphi_{(k-1)}$ be the first few exact eigenfunctions of the Hamiltonian H. These may be atomic or molecular, n-electron, correlated wave functions. Let

$$H\varphi_j = E_j\varphi_j, \quad (j = 1, \ldots, (k-1)).\tag{9.48}$$

We want to determine φ_k, which is an excited state above the first few levels. Then we can form the projection operator

$$\Omega = \sum_{j=1}^{k-1} |\varphi_j\rangle\langle\varphi_j|,\tag{9.49}$$

which is, of course, a many-electron projection operator not separable into one-electron projection operators. We can define ψ_k, a many-electron pseudo-orbital, by the relationship:

$$\varphi_k = (1 - \Omega)\psi_k,\tag{9.50}$$

and putting this into the wave equation, we obtain

$$(H + V_p)\psi_k = E_k\psi_k,\tag{9.51}$$

where

$$V_p = (E_k - H)\Omega.\tag{9.52}$$

This argument is the analog of the argument leading from Eqs. (2.53) to (2.56). The difference is that, in Eq. (2.56) H_F and Ω are one-electron operators, while in Eq. (9.52) H and Ω are both many-electron operators. If the wave functions used in Eq. (9.49) are exact eigenfunctions of H, then we can write

$$V_p = \sum_{j=1}^{k-1} (E_k - E_j)|\varphi_j\rangle\langle\varphi_j|.\tag{9.53}$$

This expression is the generalization of the PK pseudopotential for many electron wave functions.

The further development of this idea was carried out by Tully for a scattering problem, and thus lies outside of the scope of this book. In the actual computations Tully suggested the replacement of V_p by a model potential, and concluded that the method leads to significant computational simplifications.

Finally we note that for the calculation of excited states, a pseudopotential method was developed by Simons.[106] The method of Simons rests on the model potential No. 5 in Table 4.3. We have seen that this model potential was used in molecular calculations with good results (Section 8.5). It can be shown that the valence orbitals can be obtained in analytic form for this potential. Calculations were carried out for excited states of atoms, and it was shown that the calculated oscillator strengths were in good agreement with HF results.

Exact Solutions of the Phillips–Kleinman Equation for Atoms

In this appendix we summarize the results of the calculations of Szasz and McGinn. First we derive the formulas for V_M. Let us start with Eq. (2.42):

$$\left(-\frac{1}{2}\Delta + V_M\right)\psi_v = \epsilon_v \psi_v, \tag{A.1}$$

where we have attached the v index to the valence electron quantities. The modified potential is given by Eq. (2.41):

$$V_M = -\frac{Z}{r} + U + V_p. \tag{A.2}$$

Using Eqs. (2.40), (2.4), and (2.5), we get for the total HF potential

$$U(1) = \sum_{i=1}^{N} \left\{ \iint \frac{|\varphi_i(2)|^2}{r_{12}} \, dq_2 - \frac{1}{\psi_v(1)} \int \frac{\varphi_i(1)\varphi_i^*(2)\psi_v(2) \, dq_2}{r_{12}} \right\}. \tag{A.3}$$

Using Eqs. (2.31) and (2.34), we get

$$V_p(1) = \sum_{i=1}^{N} \int \varphi_i^*(2)\psi_v(2)\, dq_2 (\epsilon_v - \epsilon_i) \frac{\varphi_i(1)}{\psi_v(1)}. \tag{A.4}$$

For the core orbitals we put, according to Eq. (2.18),

$$\varphi_i(q) = \hat{u}_i(r, \vartheta, \varphi)\eta_i(\sigma) = (\hat{P}_{n_i l_i}(r)/r)Y_{l_i m_{li}}(\vartheta, \varphi)\eta_{m_{si}}(\sigma), \tag{A.5}$$

where $\hat{u}_i$ and η_i are the spatial and spin parts, respectively, $\hat{P}_{n_i l_i}$ is the radial part of the HF orbital, $Y_{l_i m_{li}}$ is the normalized spherical harmonics, and $\eta_{m_{si}}$ is the spin function. As we have shown in Section 2.1, if the core orbitals are of the form (A.5), then the valence electron PO can also be written in the central field form:

$$\psi_v(q) = u_v(r, \vartheta, \varphi)\eta_v(\sigma) = (P_{n_v l_v}(r)/r)Y_{l_v m_{lv}}(\vartheta, \varphi)\eta_{m_{sv}}(\sigma), \tag{A.6}$$

where u_v is the spatial part of the pseudo-orbital and $P_{n_v l_v}$ is its radial part. Using Eqs. (A.5) and (A.6) in Eqs. (A.3) and (A.4), we can carry out the spin summations and obtain

$$U(1) = \sum_{i=1}^{N} \left\{ \int \frac{|\hat{u}_i(2)|^2}{r_{12}}\, dv_2 - \frac{1}{u_v(1)}\delta(m_{si}, m_{sv}) \int \frac{\hat{u}_i(1)\hat{u}_i^*(2)u_v(2)}{r_{12}}\, dv_2 \right\}, \tag{A.7}$$

and

$$V_p(1) = \sum_{i=1}^{N} \delta(m_{si}, m_{sv})(\epsilon_v - \epsilon_i) \int \hat{u}_i^*(2)u_v(2)\, dv_2 \frac{\hat{u}_i(1)}{u_v(1)}. \tag{A.8}$$

In these expressions $\delta(m_{si}, m_{sv})$ indicates that the summation is over those core orbitals which have parallel spins with the valence electron. Now we substitute the forms indicated by Eqs. (A.5) and (A.6) for the spatial parts $\hat{u}_i$ and u_v and carry out the spatial integrations. These operations yield the formulas for Eqs. (A.7) and (A.8). We obtain for Eq. (A.7) the formula [Slater I, p. 316; Slater II, p. 15]:

$$U(r_1) = \sum_{n_i} \sum_{l_i=0}^{n_i-1} \left\{ 2(2l_i + 1)\frac{1}{r_1} Y_0(n_i l_i, n_i l_i \mid r) \right.$$
$$\left. - \frac{\hat{P}_{n_i l_i}(r_1)}{P_{n_v l_v}(r_1)}\left(\frac{2l_i + 1}{2l_v + 1}\right)^{1/2} \sum_{k=0}^{l_i+l_v} c^k(l_v 0, l_i 0)\frac{1}{r_1} Y_k(n_i l_i, n_v l_v \mid r_1) \right\}. \tag{A.9}$$

The first term in curly bracket is the electrostatic potential, while the second, negative term is the exchange potential. The summation over n_i is for all occupied closed shells in the core. The function Y_k is defined as follows:

$$Y_k(nl, n'l' | r_1) = \frac{1}{r_1^k} \int\limits_0^{r_1} P_{nl}(r_2) P_{n'l'}(r_2) r_2^k \, dr_2 + r_1^{k+1} \int\limits_{r_1}^{\infty} P_{nl}(r_2) P_{n'l'}(r_2) r_2^{-(k+1)} \, dr_2 \,,$$

(A.10)

[Slater II, p. 17]. The constants c^k which are obtained from the integration over the angles are conveniently tabulated by Slater [Slater II, p. 281].

For the pseudopotential, Eq. (A.8), we obtain

$$V_p(r_1) = \sum_{n_i} (\epsilon_{n_v l_v} - \epsilon_{n_i l_v}) \int \hat{P}_{n_i l_v}(r_2) P_{n_v l_v}(r_2) \, dr_2 \, \frac{\hat{P}_{n_i l_v}(r_1)}{P_{n_v l_v}(r_1)} \,.$$

(A.11)

The summation is over those core states whose azimuthal quantum number is the same as the azimuthal quantum number of the valence electron. Thus, for example, in the case of the Na $3s$ electron the summation is over $1s$ and $2s$, for the $3p$ electron the summation is over the $2p$, and for the $3d$ electron the pseudopotential is zero.

Now let us return to Eq. (A.1). Since V_M is a function of r only, we see that our assumption of writing ψ_v in central field form is justified in a self-consistent fashion. Using Eq. (A.6) for Eq. (A.1), we obtain the radial equation for the PO:

$$\left\{ -\frac{1}{2} \frac{d^2}{dr^2} + \frac{l_v(l_v + 1)}{2r^2} + V_M(r) \right\} P_{n_v l_v}(r) = \epsilon_{n_v l_v} P_{n_v l_v}(r) \,.$$

(A.12)

If we want the equation in terms of the effective potential, we get

$$\left\{ -\frac{1}{2} \frac{d^2}{dr^2} + V_{\text{eff}}(r) \right\} P_{n_v l_v} = \epsilon_{n_v l_v} P_{n_v l_v} \,.$$

(A.13)

Here we note again that the effective potential, which fully determines the pseudo-orbital, is given by the formula

$$V_{\text{eff}} = V_M + \frac{l_v(l_v + 1)}{2r^2} = -\frac{Z}{r} + U(r) + V_p(r) + \frac{l_v(l_v + 1)}{2r^2} \,,$$

(A.14)

where the HF potential U and the pseudopotential V_p are given by Eqs. (A.9) and (A.11), respectively.

The results of the calculations, which were carried out for the atoms mentioned in the main text, are summarized in the tables at the end of this appendix. As we have seen, the pseudo-orbitals are obtained by numerical integration, the final results of which is a numerical table. In order to facilitate the easy applicability of the PO's, analytic fits were constructed which give them in terms of STO's. The analytic fits are of the form

$$(P_{nl}/r) = \sum_{n',p} C_{n'l}^{(p)} \chi_{n'l}^{(p)}, \tag{A.15}$$

where $\chi_{n'l}^{(p)}$ is a normalized STO of the form

$$\chi_{n'l}^{(p)} = A_{nl}^{(p)} r^{(n-1)} \exp[-\xi_{nl}^{(p)} r], \tag{A.16}$$

with

$$A_{nl}^{(p)} = \left[\frac{(2\xi_{nl}^{(p)})^{2n+1}}{(2n)!} \right]^{1/2}. \tag{A.17}$$

The $C_{nl}^{(p)}$ is an expansion coefficient determined in such a way as to give a good fit to the numerical PO. The superscript (p) is the serial number of the STO with the quantum numbers (nl). In Tables A.1, A.2, and A.3 we have all the data for the construction of the pseudo-orbitals. The layout of the tables is self-explanatory, in conjunction with the formulas defining the analytic fits.

As far as the modified or effective potentials are concerned, there are two ways of tabulating them. One can use Eq. (A.14), for which the HF potential is given by Eq. (A.9) and the pseudopotential by Eq. (A.11). In order to tabulate the effective potential from these formulas, the core orbitals as well as the PO's are needed. The PO's can be taken from our tables; the parameters of the core orbitals can be found in the work of Szasz and McGinn.[11] Another way of tabulating the potentials is to use the formula of Kahn, Baybutt, and Truhlar (see Section 7.1). Solving Eq. (A.12) for V_M, we obtain the formula which is valid for the exact pseudo-orbital

$$V_M = \frac{\left(\epsilon_{n_v l_v} + \frac{1}{2} \frac{d^2}{dr^2} - \frac{l_v(l_v+1)}{2r^2} \right) P_{n_v l_v}}{P_{n_v l_v}}. \tag{A.18}$$

Since the P_{nl} are given in analytic form, there is no difficulty in computing the derivatives for this formula. For this formula only the PO's but not the core orbitals are needed.

Next, in Tables A.4 and A.5 we present some of the important quantities which characterize the PO's. For all states which were computed, we present the coefficients $\alpha_{nln'l'}$ that are defined by Eq. (2.29). These coefficients show the degree of overlap between the PO and the core orbitals. We present also the expectation values of the projection operator Ω. This operator is defined by Eq. (2.50). Using that definition, we obtain the formula for the expectation value,

TABLE A.1. The Data for the Psuedo-Orbitals

STO	Li Exp.	Li Coeff.	Be$^+$ Exp.	Be$^+$ Coeff.	Na Exp.	Na Coeff.	Mg$^+$ Exp.	Mg$^+$ Coeff.	Al^{++} Exp.	Al^{++} Coeff.
1s	2.4830	0.0083	4.5000	0.0134	11.1330	0.0028	12.2200	0.0036	13.2500	0.0043
1s	4.7071	0.0024	6.9697	0.0013	18.0860	0.0002	19.3000	-0.00003	20.5000	-0.0006
2s	1.7350	-0.0871	1.2967	0.3229	4.0470	0.0509	4.3400	0.0808	4.8700	0.0985
2s	1.0000	0.0878	2.4750	-0.1388	10.0000	-0.0046	11.0350	-0.0073	11.9800	-0.0110
2s	0.6615	0.9659	1.1500	0.7727						
2s	0.3500	0.0032								
3s					1.1800	0.3287	1.4890	0.4707	1.7100	0.8060
3s					3.0600	0.0139	3.4920	0.0052	1.3900	0.1781
3s					0.7280	0.7024	1.1240	0.5258	3.9000	-0.0091
2p					6.7500	0.0043	4.7500	0.0139	6.2400	0.0102
2p					10.2100	-0.0003	11.1182	0.0012	12.0000	0.0018
3p					0.8900	0.2168	1.2000	0.3551	1.3200	0.6129
3p					0.5350	0.8169	0.9200	0.6554	1.4800	0.3854
3p					3.8145	0.0111	4.2546	0.0096	5.6600	0.0147

TABLE A.2. The Data for the Pseudo-Orbitals

STO	K Exp.	K Coeff.	Ca⁺ Exp.	Ca⁺ Coeff.	Rb Exp.	Rb Coeff.	Zn⁺ Exp.	Zn⁺ Coeff.	STO	Cu Exp.	Cu Coeff.
$1s$	32.0000	0.00004	20.0836	0.0016	39.6300	0.0002	29.7161	0.0011	$1s$	29.0000	0.0012
$1s$	19.0681	0.0009	31.9733	−0.0002			43.2400	0.0001	$3s$	35.9608	−0.0001
$2s$	6.5000	0.0159	17.3333	−0.0028	35.2236	−0.0003	24.0000	−0.0034	$3s$	24.9677	0.0021
$2s$	16.8706	−0.0013	8.4500	0.0162	18.4326	0.0018	13.2473	0.0179	$3s$	16.5078	0.0058
									$3s$	11.8924	−0.0021
									$3s$	6.4025	0.0362
$3s$	2.6123	0.0407	7.4282	−0.0146	17.4163	0.0011	11.4771	−0.0245	$3s$	4.1289	0.0749
$3s$	3.9530	0.0521	3.9378	0.1239	9.0582	−0.0230	6.8600	0.1532	$4s$	2.1400	0.3801
$3s$	6.5429	−0.0174	3.0000	−0.0319	7.1941	0.0640	5.6300	−0.0916	$4s$	1.1102	0.6888
$4s$	0.7500	0.5839	3.0739	0.0999	7.2812	−0.0371	4.9768	0.1303			
$4s$	1.1875	0.3765	1.4290	0.5294	4.1379	0.0861	2.1750	0.5585			
$4s$	0.5500	0.0673	0.9972	0.5105	2.6661	0.0291	1.3119	0.4823			
$4s$	2.5432	0.0452	0.8637	−0.0800			0.9218	−0.0497			
$5s$					3.0000	0.0605					
$5s$					2.0000	0.1104					
$5s$					1.0000	−16.1875					
$5s$					0.8500	1.7041					
$5s$					1.0150	15.4309					

TABLE A.3. The Data for the Pseudo-Orbitals

STO	K Exp.	K Coeff.	Ca⁺ Exp.	Ca⁺ Coeff.	Rb Exp.	Rb Coeff.	Zn⁺ Exp.	Zn⁺ Coeff.	STO	Cu Exp.	Cu Coeff.
2p	12.4039	0.00005	13.2000	−0.0006	24.7689	0.00005	13.2000	0.0004	2p	14.5000	0.0056
2p	20.7528	0.0001	21.5786	0.0004	16.9810	0.0001	21.5786	0.0001	2p	4.0000	−0.1239
3p	12.0405	0.0004	12.7400	0.0020	16.0646	0.00001	12.7400	−0.0019	4p	33.4766	0.0021
3p	7.7183	−0.0010	8.0305	−0.0051	10.4349	0.0007	8.0305	0.0064	4p	15.8162	0.0114
3p	4.8915	0.0075	5.2052	0.0216	6.7471	−0.0003	5.2052	0.0328	4p	10.7511	0.0337
4p	3.5184	0.0266	3.7648	0.0543	5.8819	0.0007	3.7648	0.0291	4p	6.8265	0.0712
4p	1.9693	0.0205	2.0779	0.0476	3.3859	0.0367	2.0560	0.2853	4p	4.0720	0.0685
4p	0.9500	0.2650	1.0950	0.6858	2.1006	0.0428	1.2400	0.7027	4p	1.4200	0.3779
5p	(4p) 0.57	0.7758	(4p) 0.80	0.2941	2.0000	0.0012	(4p) 0.77	0.0654	4p	0.7013	0.7246
5p					1.5000	0.0605					
5p					1.0000	0.2192					
5p					0.8000	0.3292					
5p					0.5620	0.4717					
3d					17.0514	0.0002	13.8723	0.00001	3d	9.6670	0.0102
3d					10.9469	−0.0012	7.7346	0.0006	3d	4.0000	−0.1578
3d					8.0479	0.0038	4.9344	−0.0028	5d	13.9350	0.0063
3d					5.4442	−0.0024	3.0000	0.0177	5d	7.8790	0.0756
4d					4.5000	0.0095	3.0000	0.0241	5d	4.2800	0.2073
4d					3.0000	0.0237	2.5000	0.0323	5d	3.6830	−0.1431
4d					2.0000	0.0808	2.0000	−0.0162	5d	2.2870	0.0486
4d					1.0000	0.1348	1.5000	0.0519	5d	0.8170	0.3068
4d					0.7500	0.1426	1.0000	0.3028	5d	0.4500	0.7749
4d					0.4292	0.7882	0.7150	0.6883			

TABLE A.4. Pseudo-Orbital Parameters for the s States.

n'	Li $\alpha_{n'0,20}$	Be$^+$ $\alpha_{n'0,20}$	Na $\alpha_{n'0,30}$	Mg$^+$ $\alpha_{n'0,30}$	Al^{++} $\alpha_{n'0,30}$	K $\alpha_{n'0,40}$	Ca$^+$ $\alpha_{n'0,40}$	Rb $\alpha_{n'0,50}$	Cu $\alpha_{n'0,40}$	Zn$^+$ $\alpha_{n'0,40}$
1	0.1658	0.2318	0.0214	0.0322	0.0389	0.0052	0.0076	0.0016	0.0063	0.0077
2			-0.2193	0.3326	0.3959	0.0544	0.0860	0.0143	0.0555	0.0764
3						0.3022	0.4369	0.0807	0.2697	0.3649
4								0.3613		
$\langle\Omega\rangle$	0.0275	0.0537	0.04855	0.1017	0.1578	0.0943	0.1983	0.1372	0.0758	0.1390
$-\epsilon(n_v l_v)$	0.1966	0.6648	0.1811	0.5481	1.0419	0.1467	0.4139	0.1379	0.2339	0.6097

TABLE A.5. Pseudo-Orbital Parameters for the p and d States.

n'	Na $\alpha_{n'1,31}$	Mg$^+$ $\alpha_{n'1,31}$	Al^{++} $\alpha_{n'1,31}$	K $\alpha_{n'1,41}$	Ca$^+$ $\alpha_{n'1,41}$	Rb $\alpha_{n'1,51}$	Cu $\alpha_{n'1,41}$	Zn$^+$ $\alpha_{n'1,41}$	Rb $\alpha_{n'2,42}$	Cu $\alpha_{n'2,42}$	Zn$^+$ $\alpha_{n'2,42}$
2	0.1110	0.2108	0.2744	0.0152	0.0345	0.0019	0.0104	0.0219			
3				0.1816	0.3288	0.0266	0.1181	0.2233	0.0372	0.0397	0.1100
4						0.2205					
$\langle\Omega\rangle$	0.0123	0.0444	0.0753	0.0332	0.1093	0.0493	0.0140	0.0503	0.0014	0.0016	0.0121
$-\epsilon(n_v l_v)$	0.1094	0.3834	0.7900	0.0955	0.3099	0.0902	0.1222	0.4059	0.0601	0.0551	0.2137

$$\langle \Omega \rangle = \langle \psi_{n_v l_v} | \Omega | \psi_{n_v l_v} \rangle = \sum_{\substack{i=1}}^{N} \langle \psi_{n_v l_v} | \varphi_i \rangle \langle \varphi_i | \psi_{n_v l_v} \rangle = \sum_{\substack{n_i \\ (\text{core})}} |\alpha_{n_i l_v, n_v l_v}|^2 , \qquad \text{(A.19)}$$

from which we see that $\langle \Omega \rangle$ can be computed simply from the α's. As the operator projects the PO onto the core states, the quantity $\langle \Omega \rangle$ is the measure of the overlap between the PO and the whole core. Finally, the tables contain the orbital parameters which are needed for the construction of the potentials from Eq. (A.18).

If we construct V_M either from Eq. (A.14) or from Eq. (A.18), we obtain the potential in a form from which it can be tabulated easily in any scale, but which is not convenient for using the potential in an atomic or molecular calculation. In order to put the most important modified potentials in a form which can be used readily in such calculations, convenient analytic fits were constructed for them. Specifically, for the potential functions of the lowest s states of the valence electron of the atoms Li, Na, and K, the following analytic fits were constructed. First, let us rewrite Eq. (A.2) as follows:

$$V_M = -\frac{(Z-N)}{r} + V' + V_p , \qquad \text{(A.20)}$$

where Z is the nuclear charge, N is the number of electrons in the core, and

$$V' = U - \frac{N}{r} . \qquad \text{(A.21)}$$

The exact V' and V_p, both of which decline exponentially, were represented by the expressions

$$V' = \sum_i \alpha_i \exp[-\beta_i r] , \qquad \text{(A.22)}$$

and

$$V_p = \sum_i \alpha_i' \exp[-\beta_i' r] . \qquad \text{(A.23)}$$

In addition to these expressions, analytic fits were also constructed for Callaway's polarization potentials (see Section 2.5.D). Similarly to the other expressions, the polarization potential V_C was put in the form:[18]

$$V_C = \sum_i \alpha_i'' \exp[-\beta_i'' r] . \qquad \text{(A.24)}$$

In Table A.6 we tabulated the parameters for the construction of V', V_p,

TABLE A.6. The Parameters of the Potentials for the Li Atom

$\alpha_1 =$	-2.77617	$\beta_1 =$	2.550
$\alpha_2 =$	-27.7510	$\beta_2 =$	4.625
$\alpha_3 =$	-26.5718	$\beta_3 =$	8.883
$\alpha_4 =$	68.5000	$\beta_4 =$	17.500
$\alpha_1' =$	5.27800	$\beta_1' =$	2.070
$\alpha_2' =$	35.8860	$\beta_2' =$	3.800
$\alpha_3' =$	64.7070	$\beta_3' =$	7.677
$\alpha_4' =$	-139.243	$\beta_4' =$	13.556
$\alpha_1'' =$	-0.023963	$\beta_1'' =$	1.020
$\alpha_2'' =$	-0.0787133	$\beta_2'' =$	2.730
$\alpha_3'' =$	0.557829	$\beta_3'' =$	3.990
$\alpha_4'' =$	0.395523	$\beta_4'' =$	7.380
$\alpha_5'' =$	-0.450000	$\beta_5'' =$	16.000

and V_C for the valence electron of Li; in Tables A.7 and A.8 we have the corresponding parameters for the Na and K. We emphasize that these functions are analytic fits accurately representing the exact potentials, and should not be confused with model potentials.

TABLE A.7. The Parameters of the Potentials for the Na Atom.

$\alpha_1 =$	-1.46778	$\beta_1 =$	1.865
$\alpha_2 =$	81.0830	$\beta_2 =$	4.170
$\alpha_3 =$	-109.114	$\beta_3 =$	3.620
$\alpha_4 =$	-206.752	$\beta_4 =$	12.750
$\alpha_1' =$	4.56769	$\beta_1' =$	1.810
$\alpha_2' =$	64.3109	$\beta_2' =$	3.100
$\alpha_3' =$	-301.732	$\beta_3' =$	6.300
$\alpha_4' =$	539.322	$\beta_4' =$	8.800
$\alpha_1'' =$	-0.087076	$\beta_1'' =$	0.960
$\alpha_2'' =$	2.45071	$\beta_2'' =$	2.47
$\alpha_3'' =$	5.76328	$\beta_3'' =$	3.60
$\alpha_4'' =$	-9.88059	$\beta_4'' =$	7.50
$\alpha_5'' =$	5.68942	$\beta_5'' =$	11.25

TABLE A.8.   The Parameters of the Potentials for the K Atom.

$\alpha_1 =$	-0.551184	$\beta_1 =$	1.19
$\alpha_2 =$	-53.7107	$\beta_2 =$	2.27
$\alpha_3 =$	97.0381	$\beta_3 =$	3.80
$\alpha_4 =$	-183.018	$\beta_4 =$	5.90
$\alpha_5 =$	9.0000	$\beta_5 =$	8.90
$\alpha_6 =$	-320.000	$\beta_6 =$	17.00
$\alpha_1' =$	1.62871	$\beta_1' =$	1.20
$\alpha_2' =$	52.9832	$\beta_2' =$	2.02
$\alpha_3' =$	-256.551	$\beta_3' =$	3.80
$\alpha_4' =$	407.9820	$\beta_4' =$	5.00
$\alpha_5' =$	-517.180	$\beta_5' =$	14.40
$\alpha_6' =$	1564.680	$\beta_6' =$	21.021
$\alpha_1'' =$	-0.0803875	$\beta_1'' =$	0.559
$\alpha_2'' =$	-2.52579	$\beta_2'' =$	1.44
$\alpha_3'' =$	12.1420	$\beta_3'' =$	3.18
$\alpha_4'' =$	-20.0540	$\beta_4'' =$	5.25
$\alpha_5'' =$	13.6500	$\beta_5'' =$	8.30
$\alpha_6'' =$	-7.0000	$\beta_6'' =$	22.00

Orthogonality Projection Operators for One Valence Electron

Let us consider the HF Hamiltonian operator which is defined in terms of Eqs. (2.2)–(2.5). For a fixed set of core orbitals, with which Eq. (2.2) is self-consistent, this operator is Hermitian and we assume that it has a complete set* of solutions $\varphi_1, \varphi_2, \ldots, \varphi_k$ which satisfy the equation

$$H_F \varphi_i = \epsilon_i \varphi_i, \quad (i = 1, \ldots, N, N+1, \ldots, \infty). \tag{B.1}$$

The lowest N orbitals are the self-consistent solutions of Eqs. (2.2)–(2.5) and represent the core orbitals for the ground state of the closed-shell atom. The higher orbitals $\varphi_{N+1} \ldots$ represent the valence electron states. The orbitals are orthonormal

---

* The energy spectrum will have discrete and continuous parts. Throughout the book the symbol $\Sigma_{i=1}^{\infty}$ means summation over the eigenfunctions of the discrete part and integration over the continuum.

$$\langle \varphi_i | \varphi_j \rangle = \delta_{ij}, \quad (i, j = 1, \ldots, \infty), \tag{B.2}$$

and they satisfy the closure relationship

$$\sum_{i=1}^{\infty} \varphi_i(1)\varphi_i^*(2) = \delta(\mathbf{r}_1 - \mathbf{r}_2), \tag{B.3}$$

which can also be written in the form

$$\sum_{i=1}^{\infty} |\varphi_i\rangle\langle\varphi_i| = 1, \tag{B.4}$$

where 1 stands for the identity operator.

From the core orbitals we form the operator given by Eq. (2.50). Let

$$\Omega = \sum_{i=1}^{N} |\varphi_i\rangle\langle\varphi_i|. \tag{B.5}$$

The operator Ω is Hermitian. It is easy to show that it is also idempotent. Let f be an arbitrary one-electron function, and let us operate on it by the operator Ω^2. Then we obtain

$$\Omega^2 f = \sum_{i=1}^{N} |\varphi_i\rangle\langle\varphi_i| \sum_{j=1}^{N} |\varphi_j\rangle\langle\varphi_j | f\rangle = \sum_{i,j} |\varphi_i\rangle\langle\varphi_i | \varphi_j\rangle\langle\varphi_j | f\rangle$$

$$= \sum_{i,j} |\varphi_i\rangle\delta_{ij}\langle\varphi_j | f\rangle = \sum_{i} |\varphi_i\rangle\langle\varphi_i | f\rangle = \Omega f, \tag{B.6}$$

where we have used Eq. (B.2). Thus we have

$$\Omega^2 = \Omega. \tag{B.7}$$

Now we define the complement of Ω, the operator P, according to Eq. (2.52):

$$P = 1 - \Omega, \tag{B.8}$$

and

$$\Omega = 1 - P. \tag{B.9}$$

It is evident that P is also Hermitian and idempotent. Thus we have

$$P^2 = P. \tag{B.10}$$

We note that Ω and P are projection operators; thus they have no inverse. Using the definitions, we obtain

$$P + \Omega = 1 , \tag{B.11}$$

which is equivalent to the closure relationship.

Now let us consider $\Omega\varphi_k$. From the definition we get

$$\Omega\varphi_k = \sum_{i=1}^{N} |\varphi_i\rangle\langle\varphi_i \mid \varphi_k\rangle = \sum_{i=1}^{N} \varphi_i\delta_{ik} . \tag{B.12}$$

Thus we obtain

$$\left.\begin{array}{l} \Omega\varphi_k = \varphi_k , \quad (k = 1, \ldots, N) ; \\[2mm] \Omega\varphi_k = 0 , \quad (k = N + 1, \ldots, \infty) . \end{array}\right\} \tag{B.13}$$

From this it follows that

$$\left.\begin{array}{l} P\varphi_k = (1 - \Omega)\varphi_k = \varphi_k - \varphi_k = 0 , \quad (k = 1, \ldots, N) ; \\[2mm] P\varphi_k = (1 - \Omega)\varphi_k = \varphi_k - 0 = \varphi_k , \quad (k = N + 1, \ldots, \infty) . \end{array}\right\} \tag{B.14}$$

The Ω annihilates the valence orbitals while P annihilates the core orbitals.

The physical meaning of Ω and P can be established as follows. Let f be again an arbitrary one-electron function, and let us expand f in terms of the complete set:

$$f = \sum_{i=1}^{\infty} c_i\varphi_i . \tag{B.15}$$

Operating on this equation with Ω we get, using Eq. (B.13),

$$\Omega f = \sum_{i=1}^{\infty} c_i \Omega\varphi_i = \sum_{i=1}^{N} c_i\varphi_i \tag{B.16}$$

From the complete expansion only the core orbitals are left. We see that Ω projects a function onto the core orbitals. Now operating on Eq. (B.15) with the operator P we get

$$Pf = (1 - \Omega)f = \sum_{i=N+1}^{\infty} c_i\varphi_i , \tag{B.17}$$

which means that P has removed the core orbitals from f. An arbitrary f is orthogonalized to the core orbitals by P. This can be seen from Eq. (B.17).

Using that equation we get

$$\int \varphi_k^*(1)[P(1)f(1)] \, dq_1 = 0, \quad (k = 1, \ldots, N),$$ (B.18)

regardless of the form of f.

Next we establish some relationships between the operators. First we have

$$P\Omega = (1 - \Omega)\Omega = \Omega - \Omega^2 = 0,$$ (B.19)

and thus

$$\Omega P = 0.$$ (B.20)

Consider the pseudopotential given by Eq. (2.17):

$$V_p = \sum_{i=1}^{N} (\epsilon - \epsilon_i)|\varphi_i\rangle\langle\varphi_i|.$$ (B.21)

Operating on it by Ω we get

$$\Omega V_p = \sum_{ij} (\epsilon - \epsilon_i)|\varphi_j\rangle\langle\varphi_j \mid \varphi_i\rangle\langle\varphi_i| = \sum_i (\epsilon - \epsilon_i)|\varphi_i\rangle\langle\varphi_i| = V_p,$$ (B.22)

and thus

$$PV_p = (1 - \Omega)V_p = 0.$$ (B.23)

The operator P annihilates the PK pseudopotential. Next consider $H_F\Omega$. Let f be arbitrary and let us expand it in the form of Eq. (B.15). Operate on that equation with Ω. Then we get, according to Eq. (B.16),

$$\Omega f = \sum_{i=1}^{N} c_i\varphi_i.$$ (B.24)

Operate on this from the left with H_F:

$$H_F\Omega f = \sum_{i=1}^{N} c_i H_F\varphi_i = \sum_{i=1}^{N} c_i\epsilon_i\varphi_i.$$ (B.25)

Now operate on Eq. (B.15) with ΩH_F:

$$\Omega H_F f = \Omega \sum_{i=1}^{\infty} c_i H_F\varphi_i = \Omega \sum_{i=1}^{\infty} c_i\epsilon_i\varphi_i = \sum_{i=1}^{\infty} c_i\epsilon_i \Omega\varphi_i = \sum_{i=1}^{N} c_i\epsilon_i\varphi_i.$$ (B.26)

Using the last two equations we get

$$(H_F\Omega - \Omega H_F)f = 0,\tag{B.27}$$

and since f is arbitrary, we conclude that Ω commutes with H_F:

$$H_F\Omega - \Omega H_F = 0.\tag{B.28}$$

Using this relationship we readily obtain that

$$H_FP - PH_F = 0,\tag{B.29}$$

that is, the operator P also commutes with H_F.

Now consider the general, PK-type potential [Eq. (2.79)]:

$$V_p = \sum_{i=1}^{N}|\varphi_i\rangle\langle F_i|.\tag{B.30}$$

It is easy to see from Eq. (B.14) that

$$PV_p = \sum_{i=1}^{N}P|\varphi_i\rangle\langle F_i| = 0.\tag{B.31}$$

The operator P annihilates the PK-type pseudopotential given by Eq. (B.30).

The Hermitian Character of the Operator PHP

In the derivation of the wave equation, Eq. (2.114), from the variation principle, Eq. (2.113), it must be assumed that the operator PHP is Hermitian [Slater I, p. 110]. Let us prove this. An operator P is Hermitian if

$$\langle f|F|g\rangle = \langle g|F|f\rangle^* , \qquad (C.1)$$

where f and g are arbitrary functions. Now using Eq. (2.110) we get

$$\langle f|PHP|g\rangle = \langle f|P(PH')P|g\rangle = \langle f|P^2 H' P|g\rangle = \langle f|PH'P|g\rangle . \qquad (C.2)$$

Using the Hermitian character of P and H' we obtain

$$\langle f|PH'P|g\rangle = \langle Pf|H'|Pg\rangle = \langle Pg|H'|Pf\rangle^* = \langle g|PH'P|f\rangle^* = \langle g|PHP|f\rangle^* . \qquad (C.3)$$

Putting together Eqs. (C.2) and (C.3) we get

$$\langle f|PHP|g\rangle = \langle g|PHP|f\rangle^* , \qquad (C.4)$$

which proves that *PHP* is Hermitian.

Orthogonality Projection Operators for the APM Model

We consider first the projection operators Ω^s, P^s, Ω_s, and P_s as defined by Eqs. (5.4), (5.5), (5.13), and (5.14). It is easy to see from the definitions that we have the following relationships:

$$\Omega^s \varphi_s = 0. \quad (s = 1, \ldots, N); \tag{D.1}$$

$$\Omega^s \varphi_k = \varphi_k, \quad (k \neq s, k = 1, \ldots, N); \tag{D.2}$$

$$P^s \varphi_s = (1 - \Omega^s)\varphi_s = \varphi_s, \quad (s = 1, \ldots, N); \tag{D.3}$$

$$P^s \varphi_k = (1 - \Omega^s)\varphi_k = 0, \quad (k \neq s, k = 1, \ldots, N). \tag{D.4}$$

That is, Ω^s annihilates φ_s but preserves any other occupied orbital; P^s preserves φ_s but annihilates all other occupied orbitals. Next we have

$$\Omega_s \varphi_s = 0, \quad (s = 1, \ldots, N); \tag{D.5}$$

$$\Omega_s \varphi_k = \varphi_k, \quad (k = 1, \ldots, (s-1)); \tag{D.6}$$

$$P_s \varphi_s = (1 - \Omega_s)\varphi_s = \varphi_s, \quad (s = 1, \ldots, N); \tag{D.7}$$

$$P_s \varphi_k = (1 - \Omega_s)\varphi_k = 0, \quad (k = 1, \ldots, (s-1)). \tag{D.8}$$

That is, Ω_s annihilates φ_s but preserves the orbitals below φ_s; P_s preserves φ_s but annihilates all orbitals below φ_s. These equations are analogous to Eqs. (B.13)–(B.14).

It is easy to see that since Ω_s is idempotent we have

$$P_s \Omega_s = (1 - \Omega_s)\Omega_s = \Omega_s - \Omega_s^2 = 0, \tag{D.9}$$

and similarly

$$\Omega_s P_s = 0. \tag{D.10}$$

These equations are analogous to Eqs. (B.19) and (B.20).

Next consider the relationships involving operators with upper and lower indices:

$$
\begin{aligned}
\Omega_s \Omega^s &= \sum_{t=1}^{s-1} \sum_{\substack{u=1 \\ (u \neq s)}}^{N} |\varphi_t\rangle\langle\varphi_t | \varphi_u\rangle\langle\varphi_u| \\
&= \sum_{t=1}^{s-1} \sum_{u=1}^{N} |\varphi_t\rangle\delta_{tu}\langle\varphi_u| \\
&= \sum_{t=1}^{s-1} |\varphi_t\rangle\langle\varphi_t| = \Omega_s.
\end{aligned}
\tag{D.11}
$$

Similarly we obtain

$$\Omega^s \Omega_s = \Omega_s. \tag{D.12}$$

Next consider

$$P^s P_s = (1 - \Omega^s)(1 - \Omega_s) = 1 - \Omega^s - \Omega_s + \Omega^s \Omega_s. \tag{D.13}$$

Using Eq. (D.12) we obtain

$$P^s P_s = 1 - \Omega^s - \Omega_s + \Omega_s = (1 - \Omega^s) = P^s, \tag{D.14}$$

and similarly

$$P_s P^s = P^s. \tag{D.15}$$

Orthogonality Projection Operators for Many Valence Electrons

In this appendix we summarize the properties of many-electron ortho-gonality projection operators. We start with the two-electron operators. Let us consider the one-electron orbitals $\varphi_1 \cdots \varphi_N$ which appear in the FVP trial function [Eq. (6.24)]. While in appendix B we have based the projection operators on the complete set of solutions of Eq. (B.1), here we have to proceed differently, since in Eq. (6.24) the one-electron orbitals are un-specified apart from the condition, Eq. (6.4). Let $\hat{H}$ be an auxiliary Her-mitian operator with a complete set of orthonormal eigenfunctions defined by the equations

$$\hat{H}\varphi_i = \epsilon_i\varphi_i, \quad (i = 1, \ldots, N, N + 1, \ldots, \infty). \tag{E.1}$$

Let us define $\hat{H}$ in such a way that the N orbitals in Eq. (6.24) should form the first N functions in the complete set of $\hat{H}$. We define the operator Ω_1, Eq. (6.16), and P_1, Eq. (6.17), in terms of the first N orbitals. Then the formulas of Eqs. (B.7), (B.10), (B.13), (B.14), (B.19), and (B.20) are valid without change, except that they are now defined in terms of the solutions of

Eq. (E.1) rather than those of Eq. (B.1). Equations (B.28) and (B.29) are also valid if we replace in them H_F by the operator $\hat{H}$.

Next we form the two-electron operators P and Ω by Eqs. (6.18) and (6.19). As we have seen in the main text, P can be used to orthogonalize any two-electron function to the core orbitals. Thus, if $f_0(1, 2)$ is not orthogonal to the core orbitals, then

$$f(1, 2) = Pf_0(1, 2) \tag{E.2}$$

will be strong-orthogonal to $\varphi_1 \cdots \varphi_N$.

From the definitions of Ω and P we obtain readily the formulas

$$\Omega^2 = \Omega, \tag{E.3}$$

$$P^2 = P, \tag{E.4}$$

and

$$P\Omega = \Omega P = 0. \tag{E.5}$$

From Eq. (E.4) it follows that

$$Pf = P^2f_0 = Pf_0 = f, \tag{E.6}$$

which means that the operator P does not change a function which is strong-orthogonalized to the core orbitals. On the other hand, we obtain from Eq. (E.5)

$$\Omega f = \Omega Pf_0 = 0, \tag{E.7}$$

meaning that Ω annihilates such functions.

Using the relevant formulas of Appendix B, we obtain the mixed expressions

$$P_iP = P, \quad (i = 1, 2); \tag{E.8}$$

$$\Omega_iP = 0, \quad (i = 1, 2); \tag{E.9}$$

$$\Omega_i\Omega = \Omega_i, \quad (i = 1, 2); \tag{E.10}$$

$$P_i\Omega = P_i - P, \quad (i = 1, 2). \tag{E.11}$$

If f is again the function defined by Eq. (E.2), then

$$P_if = P_iPf_0 = Pf_0 = f, \tag{E.12}$$

and

$$\Omega_i f = \Omega_i P f_0 = 0 . \tag{E.13}$$

The equations are analogous for the case of n valence electrons. The n-electron projection operators were defined by Eqs. (7.117) and (7.118). It is easy to see that Eqs. (E.3), (E.4), and (E.5) are again valid. Also, for any function which is orthogonalized to the core orbitals by P, Eq. (E.6) is valid.

The general rule is that all relationships involving n-electron operators can be built up using the one-electron and two-electron relationships. An example will elucidate the situation. In the main text we met the expression $P_k f$ where f is an n-electron function strong-orthogonalized to the core orbitals. Thus f will have the form

$$f(1, 2, \ldots, n) = P f_0(1, 2, \ldots, n), \tag{E.14}$$

where f_0 is arbitrary and P is given by Eq. (7.117). Using the idempotent character of P_k, we obtain

$$P_k f = P_k P f_0 = P_k (P_1 P_2 \cdots P_k \cdots P_n) f_0$$
$$= (P_1 P_2 \cdots P_k^2 \cdots P_n) f_0 = P f_0 = f , \tag{E.15}$$

which is the relationship quoted several times in the main text.

The Invariance of the Total Wave Function to the Orthogonalization of its Correlated Part

We want to prove the statement embodied in Eq. (6.23). According to that equation

$$\Psi_T = [(N+2)!]^{-1/2} \tilde{A}\{\Psi(1,2)\det[\varphi_1(3)\cdots\varphi_N(N+2)]\}$$
$$= [(N+2)!]^{-1/2} \tilde{A}\{\Phi(1,2)\det[\varphi_1(3)\cdots\varphi_N(N+2)]\}, \qquad (F.1)$$

where

$$\Phi = P\Psi. \qquad (F.2)$$

According to the definition of P we have

$$\Phi = P\Psi = (1-\Omega)\Psi = \Psi - \Omega\Psi. \qquad (F.3)$$

If we substitute this into the second line of Eq. (F.1), the first term of Eq.

(F.3) will yield the first line of Eq. (F.1). Thus Eq. (F.1) will be proved if the total wave function resulting from the substitution of $\Omega\Psi$ can be shown to be zero. In order to show this, we expand Ψ in terms of the complete set of the auxiliary operator $\hat{H}$, which we introduced in Appendix E, Eq. (E.1). We obtain

$$\Psi(1, 2) = \frac{1}{2} \sum_{s, t=1}^{\infty} a_{st} [\varphi_s(1)\varphi_t(2) - \varphi_s(2)\varphi_t(1)] . \tag{F.4}$$

We must operate on this with Ω. According to Eqs. (6.18) and (6.19) we have

$$\Omega = 1 - P = 1 - P_1 P_2 = 1 - (1 - \Omega_1)(1 - \Omega_2) = \Omega_1 + \Omega_2 - \Omega_1\Omega_2 . \tag{F.5}$$

Operating with this operator on Eq. (F.4) we obtain

$$\begin{aligned}
\Omega\Psi(1, 2) = \frac{1}{2} \sum_{s, t=1}^{\infty} a_{st} (&\{[\Omega_1\varphi_s(1)]\varphi_t(2) - [\Omega_2\varphi_s(2)]\varphi_t(1)\} \\
&+ \{\varphi_s(1)[\Omega_2\varphi_t(2)] - \varphi_s(2)[\Omega_1\varphi_t(1)]\} \\
&- \{[\Omega_1\varphi_s(1)][\Omega_2\varphi_t(2)] - [\Omega_2\varphi_s(2)][\Omega_1\varphi_t(1)]\}) .
\end{aligned} \tag{F.6}$$

In this expression each curly bracket contains one (2×2) determinant. Now if we substitute this into the second line of Eq. (F.1) in the place of Φ, then the antisymmetrizer $\tilde{A}$ will combine these (2×2) determinants with the $(N \times N)$ determinant, $\det[\varphi_1 \cdots \varphi_N]$, into $(N + 2) \times (N + 2)$ determinants. The first two rows of these determinants will contain the orbitals which are in Eq. (F.6), and the remaining N rows will contain the core orbitals coming from $\det[\varphi_1 \cdots \varphi_N]$. Looking at Eq. (F.6), we see that each product contains at least one of the Ω_i operators. According to Eq. (B.13) these operators preserve the core orbitals and annihilate the valence states. Thus in each of the (2×2) determinants of Eq. (F.6), at least one of the orbitals will be from the core. But then the $(N + 2) \times (N + 2)$ determinants will contain at least two identical rows, since they will also contain the core orbitals in the last N rows. Thus we obtain

$$\tilde{A}\{[\Omega\Psi(1, 2)] \det[\varphi_1(3) \cdots \varphi_N(N + 2)]\} = 0 , \tag{F.7}$$

which proves the statement contained in Eq. (F.1), Q.E.D.

The Matrix Components of the Hamiltonian Operator with Respect to Correlated Wave Functions

We give here the formulas for the matrix components of the standard many-electron atomic or molecular Hamiltonian with respect to correlated wave functions of the form of Eqs. (6.24), (7.121), and (8.18). Let

$$H = \sum_{i=1}^{N} (t_i + g_i) + \frac{1}{2} \sum_{i,j=1}^{N} \frac{1}{r_{ij}}, \qquad (G.1)$$

where t_i is the kinetic energy operator, g_i is the potential of the nucleus (nuclei), and the number of electrons is N. The total wave function of the system is

$$\Psi_T = (N!)^{-1/2} \tilde{A} \{ \Phi(1, 2, \ldots, n) \det[\varphi_1(N+1)\varphi_2(N+2) \cdots \varphi_K(N+K)] \}. \qquad (G.2)$$

where we have denoted the number of core electrons by K and the number of valence electrons by n. The one-electron orbitals are orthonormal

$$\langle \varphi_i \mid \varphi_k \rangle = \delta_{ik} \, , \tag{G.3}$$

and Φ is strong-orthogonal to the one-electron orbitals

$$\int \varphi_i^*(1)\Phi(1, 2, \ldots, n) \, dq_1 \equiv 0 \, , \quad (i = 1, 2, \ldots, K) \, , \tag{G.4}$$

and it is normalized. Taking into account Eqs. (G.3) and (G.4), we obtain[65]

$$\langle \Psi_T | \bar{H} | \Psi_T \rangle = \frac{1}{(n-1)!} \langle \Phi | t_1 + g_1 + \sum_{i=1}^{K} U_i(1) | \Phi \rangle + \frac{1}{2(n-2)!} \langle \Phi | \frac{1}{r_{12}} | \Phi \rangle$$

$$+ \frac{1}{n!} \left\{ \sum_{i=1}^{K} \langle \varphi_i | t + g | \varphi_i \rangle + \frac{1}{2} \sum_{i,j=1}^{K} \langle \varphi_i | U_j | \varphi_i \rangle \right\} , \tag{G.5}$$

and

$$\langle \Psi_T \mid \Psi_T \rangle = \frac{1}{n!} . \tag{G.6}$$

In Eq. (G.5), U_i is defined by Eq. (2.5).

In order to obtain Eq. (6.27), we put $n = 2$, and taking into account the antisymmetry of Φ we obtain immediately Eq. (6.27). Similarly, for a general n, we obtain, by using the antisymmetry of Φ, directly the expressions given by Eq. (7.124) or (8.24).

The Derivation of the Equation for the Correlated Pair Function of the Valence Electrons

Our goal here is to derive Eq. (6.31). According to the main text, we vary the total energy of the atom, as given by Eq. (6.27), with respect to the two-electron function Φ. The variation is carried out in such a way that the subsidiary conditions [Eqs. (6.25) and (6.26)] are taken into account and the one-electron orbitals are kept fixed. This task is straightforward except for the slightly unusual form of the subsidiary condition, Eq. (6.25), which requires special treatment.

According to the requirement of the strong orthogonality, Φ must satisfy the condition

$$\langle \varphi_i \mid \Phi \rangle = \int \varphi_i^*(1)\Phi(1, 2)\, dq_1 \equiv 0, \quad (i = 1, 2, \ldots, N). \qquad \text{(H.1)}$$

In this formula, a function must be identically zero for any value of q_2. In order to be able to apply the conventional Lagrangian multiplier method we transform Eq. (H.1) into integral conditions.[59] Let

$$\langle \varphi_i \mid \Phi \rangle = \sum_{s=1}^{\infty} c_{is} \varphi_s(2), \quad (i = 1, 2, \ldots, N), \tag{H.2}$$

where the expansion is in terms of the complete set of the auxiliary operator $\hat{H}$ we defined in Eq. (E.1). From Eq. (H.2) we get

$$c_{is} = \int \varphi_i^*(1) \varphi_s^*(2) \Phi(1, 2) \, dq_1 \, dq_2. \tag{H.3}$$

We replace the condition given by Eq. (H.1) with the equivalent integral conditions

$$c_{is} = 0, \quad \begin{Bmatrix} i = 1, 2, \ldots, N, \\ s = 1, 2, \ldots, \infty \end{Bmatrix}. \tag{H.4}$$

In addition to Eq. (H.4), we shall include the conjugate complex conditions which contain Φ^*. Doing this enables us to vary Φ and Φ^* independently. Thus we put

$$c_{is}^* = 0, \quad \begin{Bmatrix} i = 1, 2, \ldots, N, \\ s = 1, 2, \ldots, \infty \end{Bmatrix}. \tag{H.5}$$

Now let $(-E)$ be the Lagrangian multiplier for the subsidiary condition, Eq. (6.26), and let $(-2\lambda_{is})$ and $(-2\bar{\lambda}_{is})$ be the Lagrangian multipliers for Eqs. (H.4) and (H.5). Then our variational equation becomes

$$\delta \left[\langle \Phi | H_{12} | \Phi \rangle + E_c - E \langle \Phi \mid \Phi \rangle - \sum_{i=1}^{N} \sum_{s=1}^{\infty} 2\lambda_{is} c_{is} - \sum_{i=1}^{N} \sum_{s=1}^{\infty} 2\bar{\lambda}_{is} c_{is}^* \right] = 0. \tag{H.6}$$

In this equation, δ means a variation with respect to Φ^*; a variation with respect to Φ leads to the same result. In the variation it must be ensured that the antisymmetric character of Φ is preserved. This has to be taken into account explicitly for the following reason. The strong orthogonality requirement, to which $\Phi(1, 2)$ is subjected, refers, of course, to either coordinates. In Eq. (H.1) the strong orthogonality is expressed in terms of the coordinate q_1. This is what we have built into the variation principle [Eq. (H.6)]. Now the condition, Eq. (H.1), expresses the strong orthogonality in terms of q_2 also, if Φ is assumed to be antisymmetric beforehand. In building Eq. (H.1) into the variation principle, we have not ensured the strong orthogonality with respect to q_2, since the antisymmetry has not been built into the variation principle.

We take care of this problem by the simple device of writing the variation of Φ^* in the form

$$\delta\Phi^*(1,2) = \delta\varphi^*(1,2) - \delta\varphi^*(2,1), \qquad (H.7)$$

where $\delta\varphi^*$ is a completely arbitrary function. Carrying out the variation in Eq. (H.6) using Eq. (H.7) we get

$$\int \delta\varphi^*(1,2)\left\{\left(H_1' + H_2' + \frac{1}{r_{12}} - E\right)\Phi(1,2)\right.$$
$$\left. + \sum_{i=1}^{N}\sum_{s=1}^{\infty} \bar\lambda_{is}\varphi_i(1)\varphi_s(2) + \sum_{i=1}^{N}\sum_{s=1}^{\infty} \bar\lambda_{is}\varphi_i(2)\varphi_s(1)\right\} dq_1\, dq_2 = 0, \qquad (H.8)$$

where H_i' is given by Eq. (6.29). Since $\delta\varphi^*$ is completely arbitrary, from Eq. (H.8) it follows that

$$\left(H_1' + H_2' + \frac{1}{r_{12}} - E\right)\Phi(1,2) = \sum_{i=1}^{N}\sum_{s=1}^{\infty} \bar\lambda_{is}[\varphi_i(1)\varphi_s(2) - \varphi_i(2)\varphi_s(1)]. \quad (H.9)$$

On multiplying this equation from the left by the operator P we get

$$P\left(H_1' + H_2' + \frac{1}{r_{12}} - E\right)\Phi(1,2) = \sum_{i=1}^{N}\sum_{s=1}^{\infty} \bar\lambda_{is}P[\varphi_i(1)\varphi_s(2) - \varphi_i(2)\varphi_s(1)]. \qquad (H.10)$$

On the right side, the operator $P = P_1 P_2$ operates on the square bracket. From Eq. (B.14) we see that P annihilates the core orbitals. Since φ_i is always from the core set, we see that the right side will be zero. On the left side we have

$$PH_1'\Phi = P_1 P_2 H_1'\Phi = P_1 H_1' P_2 \Phi = P_1 H_1'\Phi, \qquad (H.11)$$

where we have used Eq. (E.12). Similarly we obtain

$$PH_2'\Phi = P_2 H_2'\Phi \qquad (H.12)$$

and

$$PE\Phi = E\Phi, \qquad (H.13)$$

where in the last equation we have used Eq. (E.6). Collecting Eqs. (H.11)–(H.13), we obtain for Eq. (H.10)

$$\left(H_1 + H_2 + P\frac{1}{r_{12}}\right)\Phi = E\Phi, \qquad (H.14)$$

where $H_i = PH_i'$. Our result is Eq. (6.31) quoted in the main text.

The Mathematical Properties of the Equation for the Correlated Pair Function

Here we discuss and prove the four statements made at the end of Section 6.1.B. The first statement said that the solutions of Eq. (6.31) must be strong-orthogonal to the core orbitals. This statement is sufficiently elucidated in the main text. Next we stated that the Hamiltonian operator in Eq. (6.31) is a "pH-operator." The abbreviation means "partially Hermitian"; that is, it means an operator that is Hermitian with respect to strong-orthogonal functions. Let us denote the operator in Eq. (6.31) by H:

$$H \equiv H_1 + H_2 + P\frac{1}{r_{12}}, \tag{I.1}$$

where

$$H_i = P_i H'_i, \quad (i = 1, 2), \tag{I.2}$$

and

$$P = P_1 P_2 . \qquad (I.3)$$

Let us prove that Eq. (I.1) is a pH-operator. Let f and g be arbitrary, strong-orthogonal, two-electron functions. By definition, they satisfy the equations

$$Pf = f, \qquad Pg = g . \qquad (I.4)$$

The operator H will be a pH-operator if

$$\langle f|H|g \rangle = \langle g|H|f \rangle^* . \qquad (I.5)$$

Now,

$$\langle f|H|g \rangle = \langle f|H_1 + H_2 + P\frac{1}{r_{12}}|g \rangle$$

$$= \langle f|P_1 H_1' + P_2 H_2' + P\frac{1}{r_{12}}|g \rangle$$

$$= \langle f|H_1' + H_2' + \frac{1}{r_{12}}|g \rangle \qquad (I.6)$$

where we have used the Hermitian property of P_i and P. The operator in Eq. (I.6) is Hermitian; thus we can write

$$\langle f|H_1' + H_2' + \frac{1}{r_{12}}|g \rangle = \langle g|H_1' + H_2' + \frac{1}{r_{12}}|f \rangle^* = \langle Pg|H_1' + H_2' + \frac{1}{r_{12}}|f \rangle^*$$

$$= \langle g|P_1 H_1' + P_2 H_2' + P\frac{1}{r_{12}}|f \rangle^* = \langle g|H|f \rangle^* , \qquad (I.7)$$

which proves that H is partially Hermitian, Q.E.D.

Next we prove that exact solutions of Eq. (6.31) exist. It was shown by the author[60] that this can be established by adopting the elegant proof of Merzbacher,[67] which can be applied to our equation with minor modifications. We start with Eq. (6.31):

$$H\Phi = E\Phi . \qquad (I.8)$$

Let us consider the real number

$$\lambda[\Phi] = \langle \Phi|H|\Phi \rangle / \langle \Phi | \Phi \rangle , \qquad (I.9)$$

where H is the pH-operator given by Eq. (I.1), and Φ is an arbitrary, strong-orthogonal, two-electron function. λ will be real, since H is partially Hermitian. Let us assume that λ is bounded and its greatest lower bound is λ_0, which is obtained from Eq. (I.9) for the strong-orthogonal function Φ_0:

$$\lambda_0 = \lambda[\Phi_0] = \frac{\langle\Phi_0|H|\Phi_0\rangle}{\langle\Phi_0|\Phi_0\rangle}, \tag{I.10}$$

where

$$\lambda_0 \leqq \lambda. \tag{I.11}$$

Now consider

$$\Phi = \Phi_0 + \epsilon\varphi, \tag{I.12}$$

where φ is an arbitrary, strong-orthogonal, two-electron function and ϵ is a small, positive number. Since λ_0 is the greatest lower bound, we have

$$\lambda_0 \leqq \lambda[\Phi_0 + \epsilon\varphi]. \tag{I.13}$$

Using Eqs. (I.9) and (I.13) we get

$$\pm\epsilon\{\langle\varphi|H - \lambda_0|\Phi_0\rangle + \langle\Phi_0|H - \lambda_0|\varphi\rangle \pm \epsilon\langle\varphi|H - \lambda_0|\varphi\rangle\} \geqq 0. \tag{I.14}$$

Since φ is strong-orthogonal, we obtain

$$S \equiv \langle\varphi|H - \lambda_0|\varphi\rangle \geqq 0, \tag{I.15}$$

and using the last equation, we obtain from Eq. (I.14),

$$-\epsilon S \leqq \langle\varphi|H - \lambda_0|\Phi_0\rangle + \langle\Phi_0|H - \lambda_0|\varphi\rangle \leqq +\epsilon S. \tag{I.16}$$

Now let $\epsilon \to 0$. Then

$$\langle\varphi|H - \lambda_0|\Phi_0\rangle + \langle\Phi_0|H - \lambda_0|\varphi\rangle = 0, \tag{I.17}$$

or

$$\langle\varphi|H - \lambda_0|\Phi_0\rangle + \langle\varphi|H - \lambda_0|\Phi_0\rangle^* = 0, \tag{I.18}$$

where, in the last transformation, we took into account that H is partially Hermitian.

Now let us choose

$$\varphi = (H - \lambda_0)\Phi_0. \tag{I.19}$$

According to our assumption, φ must be strong-orthogonal. Using the fact that Φ_0 is strong-orthogonal and taking into account the form of H, we obtain

$$P\varphi = P(H - \lambda_0)\Phi_0 = PH\Phi_0 - \lambda_0(P\Phi_0)$$

$$= P\left(H_1 + H_2 + P\frac{1}{r_{12}}\right)\Phi_0 - \lambda_0\Phi_0$$

$$= \left(H_1 + H_2 + P\frac{1}{r_{12}}\right)\Phi_0 - \lambda_0\Phi_0 = (H - \lambda_0)\Phi_0 = \varphi, \qquad (I.20)$$

which shows that φ is indeed strong-orthogonal. On putting Eq. (I.19) into Eq. (I.18), we get

$$\langle(H - \lambda_0)\Phi_0 \mid (H - \lambda_0)\Phi_0\rangle = 0, \qquad (I.21)$$

from which it follows that

$$(H - \lambda_0)\Phi_0 = 0. \qquad (I.22)$$

Our result is that the strong-orthogonal Φ_0, which has the property that in Eq. (I.9) it produces the greatest lower bound, is the exact solution of Eq. (I.22), which in turn is identical with Eq. (6.31) for $E = \lambda_0$. Clearly we obtained the exact solution of Eq. (6.31) for the ground state of the atom. Thus we see that the two-electron function, which is strong-orthogonal and makes the expression in Eq. (I.9) into an absolute minimum, is the exact solution of Eq. (6.31) for the ground state of the two valence electrons.

This existence proof is valid for the ground state. A similar argument can be constructed for the excited states.[67] The construction of solutions for the excited states hinges on the fact that the eigenfunctions of Eq. (6.31) belonging to different eigenvalues are orthogonal. It is easy to see that this statement is valid. Let Φ_i and Φ_k be eigenfunctions of Eq. (6.31) belonging to the different eigenvalues E_i and E_k. Then we can write

$$H\Phi_i = E_i\Phi_i, \qquad (I.23)$$

and

$$H^*\Phi_k^* = E_k\Phi_k^*. \qquad (I.24)$$

Multiply the first equation from the left by Φ_k^* and the second by Φ_i, integrate with respect to both particle coordinates, and subtract the second equation from the first. The result is

$$\langle\Phi_k|H|\Phi_i\rangle - \langle\Phi_i|H|\Phi_k\rangle^* = (E_i - E_k)\langle\Phi_k \mid \Phi_i\rangle. \qquad (I.25)$$

The left side is zero since H is partially Hermitian. Thus we obtain, since $E_i \neq E_k$,

$$\langle \Phi_k \mid \Phi_i \rangle = 0 \,, \quad (i \neq k) \,. \tag{I.26}$$

Finally we note that the arguments above can easily be generalized for the case when one or more of the eigenvalues of Eq. (6.31) are degenerate; that is, when there are more than one linearly independent eigenfunctions belonging to the eigenvalue. Such cases are discussed by Merzbacher.[67]

The reader will observe that, in general, the properties of partially Hermitian operators can be assumed to be the same, *mutatis mutandis*, as the properties of the ordinary Hermitian operators. The change which has to be made is to replace arbitrary wave functions by arbitrary, strong-orthogonal wave functions.

Returning to the four statements at the end of Section 6.1.B, we have now proved the first three. The last statement was that the conventional Rayleigh–Ritz variation procedure can be used for the construction of the solutions of Eq. (6.31). The proof of this is already given in the preceding derivation. Let us consider the ground state. We have shown that the strong-orthogonal Φ_0 which makes Eq. (I.9) to a minimum is the exact solution for the ground state. The actual calculation of Φ_0 then obviously can be carried out by the variation procedure. The task is to construct a function f, which contains a number of variational parameters, and to form, with f, the trial function

$$\Phi = Pf \,. \tag{I.27}$$

This function Φ must be substituted into Eq. (I.9) and the parameters determined from the energy minimum principle. The energy value obtained this way will be an upper limit to the exact eigenvalue. If the energy value obtained is the absolute minimum of Eq. (I.9), then the trial function Φ will be identical with the exact solution Φ_0. Thus, as we have stated in the main text, the calculation of approximate solutions, as well as the calculation of the exact solution of Eq. (6.31), can be carried out by the conventional variation procedure.

Some Mathematical Properties of the Exact Pseudopotential Equation for Two Valence Electrons

As we have seen in Section 6.2, the Hamiltonian of the exact, two-electron, pseudopotential equation, Eq. (6.115), is not Hermitian. In order to make sure that the basic principles of quantum mechancis are not violated, we must prove that the eigenvalues of Eq. (6.115) are real, regardless of the non-Hermitian nature of the Hamiltonian.

We divide the solutions of Eq. (6.115) into two groups. As we have seen in the main text, for the valence state solutions, $P\Psi \neq 0$. The eigenvalues belonging to these solutions are real since they are equal to the corresponding eigenvalues of Eq. (6.114). We have seen, however, that Eq. (6.115) also possesses solutions for which $P\Psi = 0$. We called these solutions core solutions, and it is easy to show that the eigenvalues belonging to such solutions are also real.

In order to prove this statement let us write down here Eq. (6.115):

$$\left(H_1 + V_1 + H_2 + V_2 + \frac{1}{r_{12}} P\right)\Psi = E\Psi. \tag{J.1}$$

For solutions which satisfy the condition $P\Psi = 0$, the last term on the left side is zero and we obtain

$$(H_1 + V_1 + H_2 + V_2)\Psi = E\Psi. \tag{J.2}$$

The eigenvalues of this equation are real because $H_1 + V_1$ is Hermitian. This is easy to show. Taking into account Eqs. (6.32) and (6.83), we obtain

$$H_1 + V_1 = P_1 H_1' - \Omega_1(P_1 H_1' - \epsilon) - P_1(P_1 H_1' - \epsilon)\Omega_1. \tag{J.3}$$

Taking into account that $\Omega_1 P_1 = P_1 \Omega_1 = 0$ and the idempotent character of P_1, we get

$$H_1 + V_1 = P_1 H_1'(1 - \Omega_1) + \epsilon\Omega_1 = P_1 H_1' P_1 + \epsilon\Omega_1. \tag{J.4}$$

We have shown in Appendix C that an operator of the form $P_1 H_1' P_1$ is Hermitian if H_1' is Hermitian. The ϵ parameter which comes from the pseudopotential V_1 is real, and Ω_1 is Hermitian. Thus the combination $H_1 + V_1$ is Hermitian. From this it follows that the eigenvalues of Eq. (J.2) will be real, and this, in turn, proves that the eigenvalues belonging to the core solutions of Eq. (J.1) will also be real.

Therefore it is proved that all eigenvalues of Eq. (J.1), those belonging to valence solutions as well as those belonging to core solutions, will be real. This means that, despite the fact that the Hamiltonian of Eq. (6.115) is not Hermitian, the introduction of Eq. (6.115) does not violate the basic principles of quantum mechanics. The condition of Hermiticity arises from the requirement that the eigenvalues of an operator which represents measurable quantities must be real numbers. The eigenvalues of Eq. (6.115) certainly represent measurable quantities; E represents the total energy of the two valence electrons. Thus it is imperative that these eigenvalues come out always as real numbers. We have now proved that this is indeed the case.

Next we want to take a look at the core solutions of Eq. (6.115). These solutions are defined by Eqs. (6.77) and (6.81). We have already discussed the meaning of these conditions in connection with the PK and WR-type pseudopotential equations, which likewise possess core solutions. We have seen that a two-electron function which satisfies Eq. (6.81) will have the form [Eq. (6.45)]:

$$\Psi = \Omega\Psi = \frac{1}{2}\left(\sum_{s=1}^{N}\sum_{t=1}^{N} + \sum_{s=1}^{N}\sum_{t=N+1}^{\infty} + \sum_{s=N+1}^{\infty}\sum_{t=1}^{N}\right)a_{st}\mu_{st}, \tag{J.5}$$

where $\mu_{st} = \det[\varphi_s(1)\varphi_t(2)]$. Thus we see from Eq. (J.5)—and it is also evident from Eq. (J.2)—that the core solutions will be built from (2×2) determinants in which either both orbitals, or at least one of the orbitals, will be selected from the core functions. It is easy to see, for example, that if φ_α and φ_β are any two core orbitals, then

$$\Psi = \det[\varphi_\alpha(1)\varphi_\beta(2)] \tag{J.6}$$

will be a core solution with the eigenvalue

$$E = 2\epsilon, \tag{J.7}$$

where ϵ is the energy parameter appearing in the WR potential [Eq. (6.83)].

We have seen in Section 6.1 and in Appendix I that the solutions of Eq. (6.34) can be obtained using the conventional variation procedure. We want to outline here a method for obtaining the solutions of Eq. (J.1). Since we do not have a Hermitian or partially Hermitian operator in this case, the variation method cannot be used. It was shown by Merzbacher[67] that, in the case of non-Hermitian operators, a computational procedure which is formally similar to the variation procedure can be used. We shall demonstrate here that Merzbacher's suggestions are well suited for the handling of Eq. (J.1).

Let us denote the Hamiltonian of Eq. (J.1) by H. Our equation is then

$$H\Psi = E\Psi. \tag{J.8}$$

Let us expand Ψ in terms of a complete set of two-electron functions:

$$\Psi = \sum_{i=1}^{\infty} c_i \psi_i. \tag{J.9}$$

The set ψ_i is chosen by physical plausibility; that is, we are basically proceeding in the same way that we chose the trial functions for the variation procedure in Section 6.1.C. The basic difference is that, unlike those trial functions, our set does not have to be subjected to any orthogonality conditions; that is, the step given by Eq. (6.48) is not necessary. On the other hand, we shall have to exclude the core functions from the calculations. Thus we subject the set ψ_i to the condition

$$P\psi_i \neq 0, \quad (i = 1, 2, \ldots, \infty). \tag{J.10}$$

Substituting Eq. (J.9) into (J.8), the differential equation is transformed into the algebraic equation

$$\sum_{i=1}^{\infty} (H_{ji} - ES_{ji})c_i = 0, \quad (j = 1, 2, \ldots, \infty), \tag{J.11}$$

where

$$H_{ji} = \langle \psi_j | H | \psi_i \rangle, \qquad (\text{J.12})$$

and

$$S_{ji} = \langle \psi_j | \psi_i \rangle. \qquad (\text{J.13})$$

The system given by Eq. (J.11) will have nontrial solutions if

$$\det[H_{ji} - ES_{ji}] = 0, \quad (j, i = 1, 2, \ldots, \infty). \qquad (\text{J.14})$$

This equation is formally the same as the secular equation which is obtained in the variation method. In practical calculations the expansion of Eq. (J.9) must be truncated after a finite number of terms. Then the energy value can be computed from Eq. (J.14) and the corresponding wave function from Eq. (J.11). These will be approximations to the exact solutions. If the set ψ_i is properly chosen, the computed data will converge toward the exact as the number of terms in the expansion increases. The lowest root of the secular equation will approximate the eigenvalues of the ground state, while the higher roots will approximate the eigenvalues of the excited states. Thus the calculation of approximate solutions of the pseudopotential equation, Eq. (J.1), can be carried out with a method very similar to the variation procedure.

There are two points in which this method differs from the variation procedure. First, since the Hamiltonian is now non-Hermitian, the wave functions belonging to different eigenvalues will not be orthogonal. Indeed, we have seen that the pseudo-orbitals Ψ_i and Ψ_k satisfy the relationship [Eq. (I.26)]:

$$\langle P\Psi_i | P\Psi_k \rangle = \langle \Psi_i | P | \Psi_k \rangle = 0, \quad (i \neq k). \qquad (\text{J.15})$$

Second, the method presented here does not possess the elegant property of the variation procedure that the approximate energy values are upper limits to the exact eigenvalue. When we carry out the calculations as we have described them here, the calculated values may very well fluctuate about the exact eigenvalue instead of converging toward it from above. In this respect, the method is similar to perturbation theory.

Maximum-Smoothness Pseudo-Orbitals (*MSPO's*)

In this appendix we summarize the results of the MSPO calculations of Kahn, Baybutt, and Truhlar.[48] Calculations were carried out for the atoms C, N, O, F, Cl, Fe, Br, and I. First we list the coefficients of the MSPO's, which are defined by Eq. (7.33). In Tables K.1 and K.2 we have these expansion coefficients yielding normalized MSPO's for the *s*, *p*, and *d* states. Also given in the tables are the values of $\langle \Omega \rangle$ and λ. Assuming that the HF orbitals are available, the MSPO's can be constructed from these tables.

The modified potentials are given in analytic form. These were obtained by constructing analytic fits to the numerically given potentials. Let us rewrite Eq. (7.19) in the following form:

$$V^l = -\frac{Z-N}{r} + \hat{V}^l, \tag{K.1}$$

where Z is the nuclear charge, N is the number of core electrons, and

$$\hat{V}^l = \frac{(U + V_p)\psi_v}{\psi_v} - \frac{N}{r}. \tag{K.2}$$

TABLE K.1. The Data for the Construction of the MSPO's for the s States. [In the cases indicated by "a", Eq. (7.34) completely determines the coefficients so it is not necessary to use Eq. (7.35).]

n'	C $a_{n'0,20}$	N $a_{n'0,20}$	O $a_{n'0,20}$	F $a_{n'0,20}$	Cl $a_{n'0,30}$	Fe $a_{n'0,40}$	Br $a_{n'0,40}$	I $a_{n'0,50}$
1	−0.2084	−0.2152	−0.2212	−0.2255	0.0076	−0.00072	−0.0013	0.0003
2	0.9780	0.9766	0.9752	0.9742	−0.3145	0.0202	0.0347	−0.0062
3					0.9492	−0.2549	−0.3620	0.0603
4						0.9668	0.9315	−0.4159
5								0.9074
$\langle \Omega \rangle$	0.04343	0.0463	0.0489	0.0508	0.0990	0.0654	0.1323	0.1766
λ	0.0[a]	0.0[a]	0.0[a]	0.0[a]	0.052	∞	∞	∞

TABLE K.2. **The Data for the Construction of the MSPO's for the *p* and *d* States. [In the cases indicated by "a", Eq. (7.34) completely determines the coefficients so it is not necessary to use Eq. (7.35).]**

	Cl	Fe	Br		I	
n'	$a_{n'1,31}$	$a_{n'1,31}$	$a_{n'1,41}$	$a_{n'2,42}$	$a_{n'1,51}$	$a_{n'2,52}$
2	-0.25995	0.0073	0.0136		-0.0008	
3	0.9656	-0.2101	-0.2820	-0.1184	0.0288	0.0077
4		0.9776	0.9593	0.9930	0.3389	-0.1926
5					0.9404	0.9812
$\langle \Omega \rangle$	0.0676	0.0442	0.0797	0.0140	0.1157	0.0371
λ	0.0^a	∞	0.0737	0.0^a	∞	∞

The $\hat{V}^l$ functions, which decline exponentially for large r, are written in the following analytic form:

$$\hat{V}^l = \frac{1}{r^2} \sum_k d_{kl} (r^{n_k} e^{-\xi_k r^2}). \tag{K.3}$$

Using this expression we obtain for V^l:

$$V^l = -\frac{Z-N}{r} + \frac{1}{r^2} \sum_k d_{kl} (r^{n_k} e^{-\xi_k r^2}). \tag{K.4}$$

The potentials $\tilde{V}_l$, which do not contain a pseudopotential, can be put in the same form,

$$\tilde{V}_l = -\frac{Z-N}{r} + \frac{1}{r^2} \sum_k d_{kl} (r^{n_k} e^{-\xi_k r^2}). \tag{K.5}$$

In Tables K.3 through K.6 we have the coefficients of the potential functions for those states of the atoms mentioned above which are needed for the construction of Eq. (7.32). With the aid of these tables the modified potentials can be tabulated easily, or the analytic expressions can be used directly in any calculation.

TABLE K.3. The Parameters of the Modified Potentials for the C and N Atoms.

		C					N		
k	n_k	ξ_k	d_{k0}	d_{k1}	k	n_k	ξ_k	d_{k0}	d_{k1}
1	2	0.0575	0.0046	-0.0013	1	2	0.0779	0.0084	-0.0015
2	2	0.3710	-0.0118	-0.0146	2	2	0.4720	-0.0254	-0.0149
3	2	1.4750	1.0069	-0.0356	3	2	1.7450	1.0036	-0.0393
4	2	5.2360	10.6260	-0.5379	4	2	6.3370	12.2796	-0.4646
5	2	19.7510	55.6162	-5.0104	5	2	23.9640	67.2719	-5.0366
6	2	76.2670	252.7352	-13.9637	6	2	92.9680	308.1171	-15.6271
7	1	342.1470	-2.4686	-2.0000	7	1	424.8090	-2.7896	-2.0000
8	0	163.7310	3.0000	0.0	8	0	200.6100	3.0000	0.0

TABLE K.4. The Parameters of the Modified Potentials for the O and F Atoms

		O					F		
k	n_k	ξ_k	d_{k0}	d_{k1}	k	n_k	ξ_k	d_{k0}	d_{k1}
1	2	0.1440	0.0174	-0.0027	1	2	0.1320	-0.0001	-0.0016
2	2	0.6960	-0.0810	-0.0159	2	2	0.5570	0.0104	-0.0102
3	2	1.7550	0.7986	-0.0314	3	2	1.8460	0.5126	-0.0423
4	2	6.8470	12.1107	-0.3483	4	2	7.8060	13.4495	-0.3010
5	2	26.1920	72.8487	-4.5026	5	2	30.0570	82.1202	-4.3737
6	2	103.1350	342.2458	-16.8052	6	2	119.0860	397.0347	-18.0616
7	1	485.5420	-3.1612	-2.0000	7	1	565.2560	-3.5744	-2.0000
8	0	225.9710	3.0	0.0	8	0	262.6890	3.0000	0.0

TABLE K.5. The Parameters of the Modified Potential for the Cl and Br Atoms.

			Cl		
k	n_k	ξ_k	d_{k0}	d_{k1}	d_{k2}
1	2	0.0638	−0.0053	−0.0046	−0.0039
2	2	0.3156	0.0104	−0.0102	−0.0837
3	2	1.4064	1.9815	1.6954	−0.2269
4	2	3.1230	7.2967	4.6101	−0.1268
5	2	6.7232	23.6614	14.2881	−6.3862
6	1	14.6830	24.8275	9.7987	−4.8972
7	1	52.4284	22.8360	−17.4396	−1.1641
8	1	196.9380	54.3932	−6.7302	−3.9387
9	0	32.6075	3.0000	5.0000	0.0

			Br		
k	n_k	ξ_k	d_{k0}	d_{k1}	d_{k2}
1	2	1.4200	4.2025	3.6120	3.4843
2	2	3.5600	13.1946	10.7203	−0.2112
3	2	6.3500	10.7959	5.8280	9.1999
4	2	10.4000	26.6483	34.3910	−3.5422
5	2	20.7000	33.1302	61.0182	43.7134
6	2	32.4000	45.3591	115.3490	77.0768
7	2	76.5000	35.3468	257.0062	245.6628
8	2	142.1000	−31.7670	161.6316	509.8986
9	2	341.8710	−633.1200	−144.6760	−330.3994
10	1	1000.2050	−47.2853	−9.4137	−21.5362
11	0	220.0000	3.0000	5.0000	7.0000

TABLE K.6.   The Parameters of the Modified Potentials for the I and Fe Atoms

I

k	n_k	ξ_k	d_{k0}	d_{k1}	d_{k2}
1	2	0.8700	2.6441	2.5733	3.8661
2	2	1.9800	9.6042	8.4557	2.8703
3	2	5.8090	36.1086	38.6401	29.5380
4	2	14.4110	47.4864	125.4153	63.3212
5	2	25.8310	-92.7379	-22.3777	123.5685
6	2	100.0000	395.7460	-284.0830	567.6384
7	2	175.9890	-173.6977	768.0786	-45.3179
8	1	227.4220	39.8806	-10.5854	-0.6913
9	1	842.5430	-41.8655	-16.4732	-300.3966
10	1	4521.8760	-2.5592	-22.6215	-555.5922
11	0	429.4000	3.0000	5.0000	7.0000

Fe

k	n_k	ξ_k	d_{k0}	d_{k1}	d_{k2}
1	2	0.4116	0.4596	0.7201	-0.3325
2	2	0.9548	3.1064	2.5941	-0.3738
3	2	1.9623	5.9262	6.0027	-3.5613
4	1	3.2257	8.0034	10.0076	-2.7867
5	1	8.8846	5.4686	16.6092	-5.5956
6	1	96.7586	-23.8090	-14.9267	-9.6177
7	0	118.6423	3.0000	5.0000	0.0

References

1. C. Moore, *Atomic Energy Levels*, National Bureau of Standards Circular No. 467, U.S. Government Printing Office, Washington, D.C., 1949; reissued 1971.

2. H. Hellmann, *J. Chem. Phys.*, **3**, 61 (1935); *Acta Fizicochem. USSR*, **1**, 913 (1935); **4**, 225 (1936); H. Hellmann and W. Kassatotschkin, *Acta Fizicochem. USSR*, **5**, 23 (1936).

3. C. Froese-Fischer, *Atomic Data*, **4**, 301 (1972).

4. P. Szepfalusy, *Acta Phys. Hung.*, **5**, 325 (1955).

5. L. Szasz and G. McGinn, J. Chem. Phys., **45**, 2898 (1966).

6. V. Fock, *Z. Physik*, **81**, 195 (1933).

7. C. Herring, *Phys. Rev.*, **57**, 1169 (1940).

8. J.C. Phillips and L. Kleinman, *Phys. Rev.*, **116**, 287 (1959); **118**, 1153 (1960).

9. J.C. Slater, *Phys. Rev.*, **81**, 385 (1951); **82**, 538 (1951); **91**, 528 (1953).

10. J.C. Slater, *Adv. Quant. Chem.*, **6**, 1 (1972).

11. L. Szasz and G. McGinn, *J. Chem. Phys.*, **47**, 3495 (1967).

12. M.H. Cohen and V. Heine, *Phys. Rev.*, **122**, 1821 (1961).

13. B.J. Austin, V. Heine, and L.J. Sham, *Phys. Rev.*, **127**, 276 (1962).

14. J.D. Weeks and S.A. Rice, *J. Chem. Phys.*, **49**, 2741 (1968).

15. Y. Öhrn and R. McWeeny, *Arkiv f. Fysik*, **31**, 461 (1966).

16. W.A. Goddard, *Phys. Rev.*, **157**, 81 (1967); **169**, 120 (1968); **174**, 659 (1968).

17. J. Callaway, *Phys. Rev.*, **106**, 868 (1957); J. Callaway, R.W. LaBahn, R.T. Pu, and W.M. Duxler, *Phys. Rev.*, **168**, 12 (1968).

18. L. Szasz and G. McGinn, *J. Chem. Phys.*, **48**, 2997 (1968).

19. H. Hellmann, *Einfuerung in die Quantenchemie*, Franz Deutike, Leipzig und Wien, 1937, p. 36.

20. P. Gombas, *Z. Physik*, **118**, 164 (1941).

21. S. Topiol, A. Zunger, and M.A. Ratner, *Chem. Phys. Letters*, **49**, 367 (1977).

22. W. Kohn and L.J. Sham, *Phys. Rev.*, **140**, A1133 (1964); B.Y. Tong and L.J. Sham, *Phys. Rev.*, **144**, 1 (1966).

23. L. Szasz, *Z. Naturforschung*, **35a**, 628 (1980).

24. P. Gombas, *Acta Phys. Hung.*, **3**, 127 (1953).

25. L. Szasz, *J. Chem. Phys.*, **73**, 5212 (1980).

26. P. Gombas, *Acta Phys. Hung.*, **1**, 285 (1952).

27. P. Gombas, *Z. Physik*, **172**, 293 (1963).

28. E. Fermi, *Z. Physik*, **48**, 73 (1928).

29. P. Gombas, *Fortschritte der Physik*, **13**, 137 (1965).

30. P. Gombas and D. Kisdi, *Theor. Chim. Acta* (*Berl.*), **5**, 127 (1966).

31. T. Szondi, *Acta Phys. Hung.*, **15**, 193 (1962).

32. V. Fock, *Z. Physik*, **63**, 855 (1930); *Phys. Z. USSR*, **1**, 747 (1932).

33. P. Gombas, *Z. Physik*, **119**, 318 (1942); B. Kozma and A. Konya, *Z. Physik*, **118**, 153 (1941); G. Peter, *Z. Physik*, **119**, 713 (1942).

34. W.H.E. Schwarz, *Acta Phys. Hung.*, **27**, 391 (1969).

35. P. Gombas, *Theor. Chim. Acta* (*Berl.*), **11**, 210 (1968).

36. P. Csavinszky and R. Hucek, *Int. J. Quant. Chem. Symp.*, **8**, 37 (1974); *Int. J. Quantum Chem. Symp.*, **9**, 215 (1975).

37. I.V. Abarenkov and V. Heine, *Phil. Mag.*, **12**, 529 (1965).

38. A.U. Hazi and S.A. Rice, *J. Chem. Phys.*, **48**, 495 (1968).

39. L. Szasz and G. McGinn, *J. Chem. Phys.* **42**, 2363 (1965).

40. W.H.E. Schwarz, *Theor. Chim. Acta* (*Berl.*), **11**, 307 (1968).

41. G. Simons, *J. Chem. Phys.*, **55**, 756 (1971).

42. K. Freed, *Semiempirical Methods of Electronic Structure Calculations*, Plenum, New York, 1977, p. 201.

43. J.N. Bardsley, *Chem. Phys. Letters*, **7**, 517 (1970).

44. G. Iafrate, *J. Chem. Phys.*, **46**, 728 (1967).

45. K. Ladanyi, *Acta Phys. Hung.*, **5**, 361 (1956).

46. J. Callaway and P.S. Laghos, *Phys. Rev.*, **187**, 192 (1969).

47. G. McGinn, *J. Chem. Phys.* **58**, 772 (1973).

48. L.R. Kahn, P. Baybutt, and D.G. Truhlar, *J. Chem. Phys.*, **65**, 3826 (1976).

49. L. Szasz, *J. Chem. Phys.*, **64**, 492 (1976); **73**, 5212 (1980); *Z. Naturforschung*, **35a**, 628 (1980).

50. P. Gombas and R. Gaspar, *Acta Phys. Hung.*, **1**, 317 (1952).

51. P. Gombas and K. Ladanyi, *Acta Phys. Hung.*, **5**, 313 (1955). See also Gombas II, p. 90.

52. P. Gombas and T. Szondi, *Solutions of the Simplified Self-Consistent Field*, Adam Hilger, London, 1970.

53. P. Szepfalusy, *Acta Phys. Hung.*, **6**, 273 (1956).

54. C.C.J. Roothaan, *Rev. Mod. Phys.*, **23**, 69 (1951); **32**, 179 (1960). See also Slater III, p. 97.

55. J.C. Slater and J.H. Wood, *Int. J. Quantum Chem. Symp.*, **4**, 3 (1971).

56. M. Ross, *Phys. Rev.*, **179**, 612 (1969); W.E. Rudge, *Phys. Rev.*, **181**, 1033 (1969); L.J. Sham, *Phys. Rev.*, **A1**, 169 (1970).

57. V. Fock, *Z. Physik*, **61**, 126 (1930).

58. V. Fock, M. Veselov, and M. Petrashen, *J. Exptl. Theor. Phys.* (*USSR*), **10**, 723 (1940).

59. L. Szasz, *Z. Naturforschung*, **14a**, 1014 (1959).

60. L. Szasz, *Z. Naturforschung*, **32a**, 829 (1977).

61. E.A. Hylleraas, *Z. Physik*, **48**, 469 (1928); **54**, 347 (1929); **60**, 624 (1930); **63**, 291 (1930); **65**, 209 (1930).

62. T. Kinoshita, *Phys. Rev.*, **105**, 1490 (1957); **115**, 366 (1959).

63. L.C. Pekeris, *Phys. Rev.*, **112**, 1649 (1958).

64. L. Szasz, *J. Math. Phys.*, **3**, 1147 (1962).

65. L. Szasz, *Z. Naturforschung*, **15a**, 909 (1960); *Phys. Rev.*, **126**, 169 (1962).

66. L. Szasz and L. Brown, *J. Chem. Phys.*, **63**, 4560 (1975); **65**, 1393 (1976).

67. E. Merzbacher, *Quantum Mechanics*, 2nd ed., Wiley, New York, 1970. See Chapter 14.

68. W.H.E. Schwarz, *Theor. Chim. Acta (Berl.)*, **11**, 377 (1968).

69. J.N. Bardsley, *Case Studies in Atomic Physics*, vol. 4, no. 5, North-Holland, New York, 1974, pp. 299–368.

70. L.R. Kahn and W.A. Goddard, *Chem. Phys. Letters*, **2**, 667 (1968); *J. Chem. Phys.*, **56**, 2685 (1972).

71. C.F. Melius, W.A. Goddard, and L.R. Kahn, *J. Chem. Phys.*, **56**, 3342 (1972).

72. C.F. Melius and W.A. Goddard, *Phys. Rev.*, **A10**, 1528 (1974).

73. P.J. Hay, W.R. Wadt, and L.R. Kahn, *J. Chem. Phys.*, **68**, 3059 (1978).

74. L. Szasz, *Z. Naturforschung*, **32a**, 252 (1977).

75. S. Topiol, J.W. Moskowitz, and C.F. Melius, *J. Chem. Phys.*, **68**, 2364 (1978); **70**, 3008 (1979); **73**, 5191 (1980).

76. P. Durand and J.C. Berthelat, *Chem. Phys. Letters*, **27**, 191 (1974); *Theor. Chim. Acta*, **38**, 283 (1975).

77. P. Coffey, C.S. Ewig, and J.R. Van Wazer, *J. Am. Chem. Soc.*, **97**, 1656 (1975).

78. C.S. Ewig, R. Osman, and J.R. Van Wazer, *J. Chem. Phys.*, **66**, 3557 (1977).

79. P.A. Christiansen, Y.S. Lee, and K.S. Pitzer, *J. Chem. Phys.* **71**, 4445 (1979).

80. A.P. Jucys, *J. Exptl. Theor. Phys. (USSR)*, **23**, 371 (1952).

81. M. Kleiner and R. McWeeny, *Chem. Phys. Letters*, **19**, 476 (1973).

82. L. Szasz, *J. Chem. Phys.*, **49**, 679 (1968).

83. G. Herzberg, *Molecular Spectra and Molecular Structure*, Van Nostrand, New York, 1951.

84. G. McGinn, *J. Chem. Phys.*, **50**, 1404 (1969); **51**, 1681 (1969).

85. H. Preuss, *Z. Naturforschung*, **10a**, 365 (1955).

86. G. Simons, *Chem. Phys. Letters*, **18**, 315 (1973).

87. J.N. Bardsley, B.R. Junker, and D.W. Norcross, *Chem. Phys. Letters*, **37**, 502 (1976).

88. W.J. Stevens, A.M. Karo, and J.R. Hiskes, *J. Chem. Phys.*, **74**, 3989 (1981).

89. W.H.E. Schwarz, T.C. Chang, and P. Habitz, *Theor. Chim. Acta*, **44**, 61 (1977).

90. T.C. Chang, P. Habitz, B. Pittel, and W.H.E. Schwarz, *Theor. Chim. Acta*, **34**, 263 (1974).

91. J. Flad, H. Stoll, and H. Preuss, *J. Chem. Phys.*, **71**, 3042 (1979).

92. H. Stoll, M.E. Pavlidou, and H. Preuss, *Theor. Chim. Acta*, **49**, 143 (1978); H. Stoll, E. Golka, and H. Preuss, *Theor. Chim. Acta*, **55**, 29 (1980).

93. H. Preuss, H. Stoll, V. Wedig, and T. Krueger, *Int. J. Quantum Chem.*, **19**, 113 (1981).

94. B.C. Laskowski, S.R. Langhoff, and J.R. Stallcop, *J. Chem. Phys.*, **75**, 815 (1981).

95. M. Pelissier, *J. Chem. Phys.*, **75**, 775 (1981).

96. P. A. Christiansen, K.S. Pitzer, Y.S. Lee, H.J. Yates, W.C. Emler, and N.W. Winter, *J. Chem. Phys.*, **75**, 5410 (1981).

97. R. Osman, C.S. Ewig, and J.R. Van Wazer, *Chem. Phys. Letters*, **39**, 27 (1976).

98. R. Osman, P. Coffey, and J.R. Van Wazer, *Inorganic Chemistry*, **15**, 287 (1976).

99. L.S. Bartell, M.J. Rothman, C.S. Ewig, and J.R. Van Wazer, *J. Chem. Phys.*, **73**, 367 (1980).

100. G. McGinn, *J. Chem. Phys.*, **51**, 5090 (1969); **52**, 3358 (1970).

101. M. Cohen and P. Kelly, *Canad. J. Phys.*, 43, 1867 (1965); **44**, 3227 (1966); **45**, 1661, 2079 (1967).

102. C. Froese-Fischer, *J. Chem. Phys.*, **47**, 4010 (1967).

103. A.U. Hazi and S.A. Rice, *J. Chem. Phys.*, **45**, 3004 (1966); **47**, 1125 (1967).

104. R.S. Mulliken, *J. Am. Chem. Soc.*, **86**, 3183 (1964).

105. J.C. Tully, *Phys. Rev.*, **181**, 7 (1969).

106. G. Simons, *J. Chem. Phys.*, **55**, 756 (1971); **60**, 645 (1974); G. Simons and I. Martin, *J. Chem. Phys.*, **62**, 4799 (1975).

107. W.A. Bingel, R.J. Koch, and W. Kutzelnigg, *Acta Phys. Hung.*, **27**, 323 (1969).

108. J. D. Switalski and M.E. Schwartz, *J. Chem. Phys.*, **62**, 1521 (1975).

109. A. Dolgarno, C. Bottcher, and G.A. Victor, *Chem. Phys. Letters*, **7**, 265 (1970); C. Bottcher, *J. Phys.*, **B4**, 1140 (1971).

110. P. Schwerdtfeger, H. Stoll, and H. Preuss, *J. Phys.*, **B15**, 1061 (1982).

111. P. Fuentealba, H. Preuss, H. Stoll, and L.V. Szentpaly, *Chem. Phys. Letters*, **89**, 418 (1982).

112. L. Szasz, to be published.

Index